PHOTONIC SWITCHING AND INTERCONNECTS

OPTICAL ENGINEERING

Series Editor

Brian J. Thompson
Provost
University of Rochester
Rochester, New York

1. Electron and Ion Microscopy and Microanalysis: Principles and Applications, *Lawrence E. Murr*
2. Acousto-Optic Signal Processing: Theory and Implementation, *edited by Norman J. Berg and John N. Lee*
3. Electro-Optic and Acousto-Optic Scanning and Deflection, *Milton Gottlieb, Clive L. M. Ireland, and John Martin Ley*
4. Single-Mode Fiber Optics: Principles and Applications, *Luc B. Jeunhomme*
5. Pulse Code Formats for Fiber Optical Data Communication: Basic Principles and Applications, *David J. Morris*
6. Optical Materials: An Introduction to Selection and Application, *Solomon Musikant*
7. Infrared Methods for Gaseous Measurements: Theory and Practice, *edited by Joda Wormhoudt*
8. Laser Beam Scanning: Opto-Mechanical Devices, Systems, and Data Storage Optics, *edited by Gerald F. Marshall*
9. Opto-Mechanical Systems Design, *Paul R. Yoder, Jr.*
10. Optical Fiber Splices and Connectors: Theory and Methods, *Calvin M. Miller with Stephen C. Mettler and Ian A. White*
11. Laser Spectroscopy and Its Applications, *edited by Leon J. Radziemski, Richard W. Solarz and Jeffrey A. Paisner*
12. Infrared Optoelectronics: Devices and Applications, *William Nunley and J. Scott Bechtel*
13. Integrated Optical Circuits and Components: Design and Applications, *edited by Lynn D. Hutcheson*
14. Handbook of Molecular Lasers, *edited by Peter K. Cheo*
15. Handbook of Optical Fibers and Cables, *Hiroshi Murata*
16. Acousto-Optics, *Adrian Korpel*
17. Procedures in Applied Optics, *John Strong*
18. Handbook of Solid-State Lasers, *edited by Peter K. Cheo*
19. Optical Computing: Digital and Symbolic, *edited by Raymond Arrathoon*
20. Laser Applications in Physical Chemistry, *edited by D. K. Evans*

21. Laser-Induced Plasmas and Applications, *edited by Leon J. Radziemski and David A. Cremers*
22. Infrared Technology Fundamentals, *Irving J. Spiro and Monroe Schlessinger*
23. Single-Mode Fiber Optics: Principles and Applications, Second Edition, Revised and Expanded, *Luc B. Jeunhomme*
24. Image Analysis Applications, *edited by Rangachar Kasturi and Mohan M. Trivedi*
25. Photoconductivity: Art, Science, and Technology, *N. V. Joshi*
26. Principles of Optical Circuit Engineering, *Mark A. Mentzer*
27. Lens Design, *Milton Laikin*
28. Optical Components, Systems, and Measurement Techniques, *Rajpal S. Sirohi and M. P. Kothiyal*
29. Electron and Ion Microscopy and Microanalysis: Principles and Applications, Second Edition, Revised and Expanded, *Lawrence E. Murr*
30. Handbook of Infrared Optical Materials, *edited by Paul Klocek*
31. Optical Scanning, *edited by Gerald F. Marshall*
32. Polymers for Lightwave and Integrated Optics: Technology and Applications, *edited by Lawrence A. Hornak*
33. Electro-Optical Displays, *edited by Mohammad A. Karim*
34. Mathematical Morphology in Image Processing, *edited by Edward R. Dougherty*
35. Opto-Mechanical Systems Design: Second Edition, Revised and Expanded, *Paul R. Yoder, Jr.*
36. Polarized Light: Fundamentals and Applications, *Edward Collett*
37. Rare Earth Doped Fiber Lasers and Amplifiers, *edited by Michel J. F. Digonnet*
38. Speckle Metrology, *edited by Rajpal S. Sirohi*
39. Organic Photoreceptors for Imaging Systems, *Paul M. Borsenberger and David S. Weiss*
40. Photonic Switching and Interconnects, *edited by Abdellatif Marrakchi*

Additional Volumes in Preparation

Design and Fabrication of Acousto-Optic Devices, *edited by Akis Goutzoulis, Dennis Pape, and Sergei Kulakov*

Digital Image Processing Methods, *edited by Edward R. Dougherty*

Spatial Ligh Modulator Technology, *edited by Uzi Efron*

Visual Science and Technology: Models and Applications, *edited by Donald Kelly*

PHOTONIC SWITCHING AND INTERCONNECTS

EDITED BY
ABDELLATIF MARRAKCHI
*Siemens Corporate Research, Inc.
Princeton, New Jersey*

Marcel Dekker, Inc.   New York•Basel•Hong Kong

Library of Congress Cataloging-in-Publication Data

Photonic switching and interconnects / edited by Abdellatif Marrakchi.
    p. cm. -- (Optical engineering)
  Includes bibliographical references and index.
  ISBN 0-8247-8931-8 (acid-free paper)
  1. Telecommunication--Switching systems. 2. Optical communications. 3. Switching circuits. 4. Optical fibers--Joints. I. Marrakchi, Abdellatif. II. Series: Optical engineers (Marcel Dekker, Inc.)
TK5103.8.P52 1994
621.382'7--dc20   93-35643
                                                                                                                                              CIP

The publisher offers discounts on this book when ordered in bulk quantities. For more information, write to Special Sales/Professional Marketing at the address below.

This book is printed on acid-free paper.

Copyright © 1994 by Marcel Dekker, Inc. All Rights Reserved.

Neither this book nor any part may be reproduced or transmitted in any form or by any means, electronic or mechanical, including photocopying, microfilming, and recording, or by any information storage and retrieval system, without permission in writing from the publisher.

Marcel Dekker, Inc.
270 Madison Avenue, New York, New York 10016

Current printing (last digit):
10 9 8 7 6 5 4 3 2 1

PRINTED IN THE UNITED STATES OF AMERICA

About the Series

The series came of age with the publication of our twenty-first volume in 1989. The twenty-first volume was entitled *Laser-Induced Plasmas and Applications* and was a multi-authored work involving some twenty contributors and two editors: as such it represents one end of the spectrum of books that range from single-authored texts to multi-authored volumes. However, the philosophy of the series has remained the same: to discuss topics in optical engineering at the level that will be useful to those working in the field or attempting to design subsystems that are based on optical techniques or that have significant optical subsystems. The concept is not to provide detailed monographs on narrow subject areas but to deal with the material at a level that makes it immediately useful to the practicing scientist and engineer. These are not research monographs, although we expect that workers in optical research will find them extremely valuable.

There is no doubt that optical engineering is now established as an important discipline in its own right. The range of topics that can and should be included continues to grow. In the "About the Series" that I wrote for earlier volumes, I noted that the series covers "the topics that have been part of the rapid expansion of optical engineering." I then followed this with a list of such topics which we have already outgrown. I will not repeat that mistake this time! Since the series now exists, the topics that are appropriate are best exemplified by the titles of the volumes listed in the front of this book. More topics and volumes are forthcoming.

Brian J. Thompson
University of Rochester
Rochester, New York

Preface

The introduction of the optical fiber in the telecommunications network has opened up opportunities for the implementation of many new and improved services. These developments have occurred even though the optical technology was limited to transmission systems. By permeating other areas, such as switching and crossconnects, the research community hopes to achieve a more efficient use of the large bandwidth afforded with the optical fiber. Following a similar path, the field of computer science has also realized that optics can alleviate some of its bottlenecks, especially in the interconnection of high-speed systems. Consequently, as the telecommunications and computer networks merge into one (as the trend seems to indicate), many applications will rely on the advantages of photonic switching and interconnects.

The integrated optics device technology has emerged as the most promising owing to its rapid and numerous successes. Many components are now commercially available, ranging from optical modulators to polarization controllers, to crossbar switches. For the switching application, arrays as large as 16×16 have been fabricated and field tested, and problems such as polarization and wavelength sensitivity have been successfully resolved. Research is now moving from the $LiNbO_3$ substrate material to GaAs and InP for the obvious reason of facilitating the integration of guided wave optic components with electronics. This trend has resulted in research activities whose objective is the study and development of devices known as PICs (photonic integrated circuits) and OEICs (optoelectronic integrated circuits) with higher integration density and increased functionality.

In the systems area, novel architectures are being devised that rely on the high performance of photonic devices. A variety of optical switched networks can now be realized with commercially available components. Switching in these systems is achieved using time-, frequency-, or wavelength-division multiplexing schemes. Although some important issues such as optimal switch control and contention resolution are still under investigation, switching applications have been demonstrated in LANs (local area networks) and in μANs (microarea networks). Many laboratories now have systems capable of carrying voice, video, and data at high bit rates and with large channel capacity.

For data rates in the 50 Gb/s range and beyond, it is difficult to imagine that an optoelectronic device would be suitable for switching. All-optical nonlinearities, on the other hand, can be extremely fast, albeit not very efficient. Nevertheless, taking advantage of the long interaction length of nonlinear optical fibers, complete switching has been achieved in a variety of experiments. These switching elements can be classified as either routing devices (e.g., Kerr gates, nonlinear directional couplers) or logic devices (e.g., soliton gates). Applications that would use such a high data rate include picosecond circuit synchronization, multiplexing front-ends and demultiplexing back-ends of high-speed local area networks, add–drop multiplexers in recirculating optical fiber databases, and ultrafast all-optical processing of headers in a packet-switched network.

The availability of large arrays of optoelectronic devices and microoptic components make it possible to fully exploit the parallelism of optics in high-capacity interconnect applications. Adding the third spatial dimension helps alleviate the bottlenecks generally attributed to the limited number of electrical pin-outs and buses in printed circuit boards and in VLSI chips. This is accomplished with free-space switching fabrics that are now implemented using various schemes. Some rely on physically distinct optical gates, such as SEEDs (self-electrooptic effect devices) and surface-emitting laser arrays, others on the spatial multiplexing of diffractive elements such as phase gratings in photorefractive media. The reconfigurability of these photonic interconnects has also facilitated the implementation of many computing architectures based on neural networks algorithms. In some instances, it is desirable to "fold" free space to achieve more compact and rugged systems. In substrate-mode interconnects this is accomplished by means of multiple reflections, which leads to a device with reduced sensitivity to alignment fluctuations.

PREFACE

With the "information age" around the corner, the advances in photonic switching and interconnects will undoubtedly enhance the computing and communication capabilities of future high-performance systems. Parallel distributed computing and broadband communication networks based on fiber-to-the-home concept are stimulating factors and driving forces behind some of the research described in this book.

Finally, the editor wishes to thank all the contributors for sharing their knowledge and expertise in the field of photonic switching and interconnects. This book is meant to be a representative sampling of the vast amount of research that is now carried out all around the world. It is a compilation by some of the leading scientists who have the most comprehensive and extensive experience in this area.

Abdellatif Marrakchi

Contents

About the Series	*iii*
Preface	*v*
Contributors	*xi*

1. LiNbO$_3$ and Semiconductor Guided Wave Optics in Switching and Interconnects — 1
   Lars Thylén

2. Wavelength-Division Multiplexing Technology in Photonic Switching — 77
   Masahiko Fujiwara and Shuji Suzuki

3. Time-Division Optical Microarea Networks — 115
   Paul R. Prucnal, Raymond K. Boncek, Steven T. Johns, Mark F. Krol, and John L. Stacy

4. Ultrafast All-Optical Switching Devices — 163
   Mohammed N. Islam

5 Digital Switching Systems Based on the SEED Technology
   and Free-Space Optical Interconnects 213
   *H. Scott Hinton, Thomas J. Cloonan, Frederick B.
   McCormick, Jr., Anthony L. Lentine, Rick L. Morrison, R. A.
   Nordin, and Gaylord W. Richards*

6 Free-Space Holographic Grating Interconnects 249
   Abdellatif Marrakchi and Kasra Rastani

7 Substrate-Mode Diffractive Optical Elements
   and Interconnects 323
   Raymond K. Kostuk, Yang-Tung Huang, and Masayuki Kato

8 Optical Interconnects in Japan 363
   Takakiyo Nakagami and Satoshi Ishihara

9 Photonic Interconnect Applications in
   High-Performance Systems 401
   Sarry Fouad Habiby and Gail R. Lalk

Index 433

Contributors

Raymond K. Boncek Photonics Center, Rome Laboratories, Griffiss Air Force Base, New York

Thomas J. Cloonan AT&T Bell Laboratories, Naperville, Illinois

Masahiko Fujiwara Opto-Electronics Research Laboratories, NEC Corporation, Kawasaki, Japan

Sarry Fouad Habiby Optical Information Technology Research, Bellcore, Red Bank, New Jersey

H. Scott Hinton Department of Electrical Engineering, McGill University, Montreal, Quebec, Canada

Yang-Tung Huang Department of Electronics Engineering and Institute of Electronics, National Chiao Tung University, Taiwan, Republic of China

Satoshi Ishihara Optoelectronics Division, Electrotechnical Laboratory, Tsukuba Science City, Japan

Mohammed N. Islam University of Michigan, Ann Arbor, Michigan

xi

CONTRIBUTORS

Steven T. Johns Photonics Center, Rome Laboratories, Griffiss Air Force Base, New York

Masayuki Kato Holography and Color-Imaging Lab., Fujitsu Laboratories Ltd., Atsugi, Japan

Raymond K. Kostuk Departments of Electrical and Computer Engineering and Optical Sciences Center, The University of Arizona, Tucson, Arizona

Mark F. Krol Photonics Center, Rome Laboratories, Griffiss Air Force Base, New York

Gail R. Lalk Interconnection and Access Technology Research, Bellcore, Red Bank, New Jersey

Anthony L. Lentine AT&T Bell Laboratories, Naperville, Illinois

Abdellatif Marrakchi Siemens Corporate Research, Inc., Princeton, New Jersey

Frederick B. McCormick, Jr. AT&T Bell Laboratories, Naperville, Illinois

Rick L. Morrison AT&T Bell Laboratories, Naperville, Illinois

Takakiyo Nakagami Optical Interconnection Division, Fujitsu Laboratories Ltd., Kawasaki, Japan

R. A. Nordin AT&T Bell Laboratories, Naperville, Illinois

Paul R. Prucnal Department of Electrical Engineering, Center for Photonics and Optoelectronic Materials, Princeton University, Princeton, New Jersey

Kasra Rastani Lightwave Networks Systems Research, Bellcore, Red Bank, New Jersey

Gaylord W. Richards AT&T Bell Laboratories, Naperville, Illinois

CONTRIBUTORS

John L. Stacy Photonics Center, Rome Laboratories, Griffiss Air Force Base, New York

Shuji Suzuki C&C Systems Research Laboratories, NEC Corporation, Kawasaki, Japan

Lars Thylén* Fiber Optics Research Center, Ericsson Telecom AB, Stockholm, Sweden

---

**Present affiliation:* Department of Photonics and Microwave Engineering, Royal Institute of Technology, Stockholm, Sweden

LiNbO₃ and Sem
Optics in Sw

I. INTRODUCTIO

The intention of this
device principles an
space-division pho
reported and envi

II. GUIDED V

A. Introduc

The origin of
though the fi
tures was re
the early 19
this work c
the optical
semicond
waveleng
[5]. It w

*Presen

2

successful i
processing
optics was e
that in some
off-chip elect
nect medium

However,
1980s, one wa
optics devices.
as well as in co
circuits or its a
vant. The term
ena in the term
ena in the devi
this is the design
of state of the ar

LiNbO₃
- 16 × 16 pola
  arrays
- 8 × 8 polar
  arrays
- 8 × 8 polariz
  space switch,
- A hybrid inte
  epitaxial lift-of

Compound semicond
- 4 × 4 polarizat
- Integration of s
  sers, modulators
- Integration of pa
- Switch elements
- Integration of las
  (e.g., 4 × 4 switc
  grated amplifiers)

It should be noted tha
conditioned on that of t
cient use of PICs require

Why guided wave optic
- Flexible interconnect w
  ing due to propagation.
- The small transverse i

LiNbO₃ AND SEMICONDUCTOR GUIDED WAVE OPTICS

lengths decrease the energy required to control the devices by orders of magnitude [1], and ensures compatibility with electron confinement volumes.

But there are also drawbacks:

- Guided wave optics is (at least today) a planar technology; i.e., the inherent 3-D parallelism of optics is not utilized. Systems and technology implementing such parallelism are treated in other chapters in this book.
- The polarization sensitivity is in general aggravated in guided wave optics, since modal confinement factors and propagation constants are mode dependent.

This brings us to a further important subject in PICs for photonic switching and interconnects: the problem of *polarization*. The ordinary standard single-mode fiber does not preserve a defined state of polarization, as does the so-called polarization-holding fiber. The latter fiber is a nonstandard fiber, which is currently not used for telecom networks, and it is not likely to find widespread use there in the future. In some applications of photonic switching and interconnects, *functional polarization independence* is required. By functional polarization independence, we mean that the optical circuit can be interfaced to standard single-mode fibers. The need for functional polarization independence, however, is highly application sensitive: For transport network applications, compatibility with the ordinary single-mode fiber is a requirement. For applications in the switching layer of the telecom network (see Section III.A), this might not be the case. Further, in optical interconnects in a local environment, polarization independence might not be needed. As a general statement, however, standardization issues as well as simplicity of logistics of systems will make functional polarization independence a highly desirable feature.

There are in principle five possibilities of solving the polarization problem:

- Strictly polarization-independent devices: Devices independent of the state and degree of polarization; i.e., rapid polarization fluctuations are allowed.
- Polarization control, i.e., feedback stabilization, which requires an assessment of the state of polarization.
- Polarization diversity, i.e., separating the polarizations and processing them separately.
- Polarization-holding (PH) fibers.
- Polarization scrambling; i.e., the polarization is randomized before entering the device in question.

Each method has its merits and demerits, and specific application segments.

Strictly polarization-independent devices are appropriate in multiport devices and in general where the control signal derivation is difficult. The obvious advantage is straightforward interfacing to standard single-mode systems, which facilitates systems experiments (Section III). In LiNbO$_3$, this is, however, bought at the expense of a *voltage × length* product that is 3–10 times higher than for polarization-dependent devices, and usually, but not in all cases, requires tighter fabrication tolerances as well as more intricate electrode structures. The frequency transparency of a LiNbO$_3$ switch could also be compromised; see Section II.B.

Polarization control is preferable when the signals required to control the polarization are readily derivable and the number of ports to be controlled is low. An example is a coherent system. The method might be difficult to implement in a multiwavelength system, however.

Polarization diversity requires low-loss polarization splitters and in general increases complexity by a factor of 2.

Linear polarization-holding fibers are immediately compatible with PICs, and they have consequently been used in many of the reported device experiments. Their in general elliptic near fields improve the fiber to chip coupling efficiency for LiNbO$_3$ chips. As mentioned, however, these fibers are and will in all likelihood remain nonstandard fibers, leaving their possible applications with PICs to specialized systems, such as systems of limited physical extent, where neither the deployment of new fiber nor the comparatively high price of these fibers are issues. Due to the high drive voltage penalty in LiNbO$_3$ polarization-independent devices, the polarization-holding fibers are more or less indispensable in high-speed applications, such as when connecting a laser diode to an optical multiplexer; see, e.g., [6].

Polarization scrambling involves changing the state of polarization at a rate faster than the bit rate, an obvious disadvantage for high-speed networks. A polarizer ensures that a defined state of polarization and power level reaches the device. A 3-dB loss is inherent in this scheme.

Of course, one could envisage a large scale PIC, e.g., for interconnecting electronic ICs, where the entire optical fabric is single polarization mode; this is an example of a "stand alone" system of limited physical extent.

The assertion of an analogy between PICs and electronic ICs can essentially only be made from a processing and fabrication point of view. Functionally, PICs and ICs differ markedly: For the foreseeable future, functions like storage and logic are difficult to implement in a viable way with the wave propagation based PICs, compared with the charge transport based ICs; see, e.g., [7]. Hence, a *combination* of electronics and optics should combine the best of the two technologies, and *optoelectronic integrated circuits* (OEICs,

LiNbO₃ AND SEMICONDUCTOR GUIDED WAVE OPTICS

i.e., the integration of electronic and optics devices on one monolithic chip) is the subject of intense research.

Interestingly enough, and very relevant when assessing the role of photonics as a complement to electronics, there is a growing effort in applying guided wave concepts to *electronic devices* relying on the wave nature of the electron, so-called quantum interference devices, QIDs, also named electron wave devices [8]. Utilizing such *mesoscopic* devices (by which is meant that the device dimensions are smaller than or comparable to the mean free path of the electrons), electronic analogues to optical devices such as interferometers and directional couplers are envisaged, with major (predicted) performance improvements with regard to switch energy and speed in relation to current electronics devices [9].

B. LiNbO₃ Devices

1. General Comments

Since the first fabrication of indiffused waveguides in LiNbO₃ in 1974 [10], this material has been extensively used for integrated photonics research. It became possible to simply fabricate low-loss waveguides in the ferroelectric material LiNbO₃, which has excellent optical and electrooptical properties. Consequently, high-performance devices (modulators, switches, etc.) were envisaged. The growing importance of the single-mode fiber in the early 1980s further increased the interest in LiNbO₃ integrated optics devices, since they, like most other waveguide devices, basically require single-mode operation to be efficient. The LiNbO₃ research has encompassed material development of the LiNbO₃ crystal (impurity contents, doping, etc.) and improved understanding of the basic titanium indiffusion process as well as development of new fabrication processes (e.g., proton exchange [11], ion implantation [12]). Parallel to this development, progress was made in understanding complex devices and in device development as well [13,14]. As a result, a number of complex high-performance devices have been developed over the past years. This has made LiNbO₃ more mature than the corresponding semiconductor devices (in GaAs/GaAlAs or InP/InGaAsP) with their complicated fabrication technology. Also, pigtailed and packaged LiNbO₃ devices are available today, either commercially from a number of sources, or developed for in-house use by a number of laboratories. This had made LiNbO₃ devices the only realistic choice for systems experiments involving space switches; examples of this are given in Section III.

In this section, we first present the general features and principles of operation of some LiNbO₃ integrated photonics devices; then a short review of device fabrication technique is given. The rest of the section is devoted to the description of devices developed or usable for photonic switching, concentrating on space switches. In addition, reliability and stability issues relating to the use of LiNbO$_3$ devices in systems are discussed. The treatment is largely based on [15], with appropriate additions; it is interesting to note that the development in the LiNbO$_3$ area has been rapid since the writing of this paper late in 1987, in spite of the competition from semiconductors.

2. Device Principles

LiNbO$_3$ integrated photonics devices incorporate light guiding channels in the LiNbO$_3$ crystal, having higher refractive index than the surrounding medium. The manipulation of light is based on the linear electrooptic effect [16]:

$$\Delta B_{ij} = r_{ijk} E_k \tag{1}$$

where r_{ijk} is the electrooptic tensor and E_k is the electric field. Further,

$$B_{ij} K_{jk} = \delta_{ik} \tag{2}$$

where B and K are the relative dielectric impermeability and permeability tensors, respectively, and δ_{ik} is the Kronecker δ; the permeability tensor is generally referred to as the dielectric tensor. The change in refractive index is derived from the change in the dielectric constant [16]. Thus, an applied voltage perturbs the indicatrix, and hence the refractive index, seen by the optical fields [16]. In a common orientation in LiNbO$_3$ (electric field along the z-axis), the index change for polarization along the z-axis is

$$\Delta n = -\frac{1}{2} n_{33}^3 r_{33} E_3 \tag{3}$$

Using numerical values at $\lambda = 1.3 \mu$m ($r_{33} \approx 30 \times 10^{-12}$ m/V, $n_{33} = 2.15$) and with an electric field of 1 V/μm, it is seen that the resulting index change is only 1.5×10^{-4}. This is one of the important boundary conditions for LiNbO$_3$ photonic integrated devices, which should be contrasted to semiconductors (Section II.C) and which sets a limit to the level of integration obtainable, since device operation relies on index change and requires (order of magnitude)

$$k_0 L \Delta n = \pi \tag{4}$$

where $k_0 = 2\pi/\lambda_0$ and L is the device length. Since $\Delta n \propto E$, we can formulate an important figure of merit for switches: The *voltage* × *length* product, which is required to be as low as possible; an example is 5 V cm [17].

The electrically controlled refractive index change can be used for phase modulation, which can, in turn, be converted to amplitude modulation in several ways, e.g., by employing

- Interferometric operation (Mach–Zehnder modulator, balanced bridge switch)
- Phase match control (directional coupler)

If the index perturbations are large enough in relation to the light confining index variations, however, they will cause changes in the optical modal fields, and this can be employed to make cutoff modulators [18].

The perturbation of the indicatrix is also useful for polarization conversion. Here, one employs, e.g., the r_{51} or r_{42} coefficients, which are equal and of the same order of magnitude as r_{33} and hence give the same order of magnitude for ΔB_{ij} as r_{33} for the same applied electric field.

The difference between a modulator and a switch (in integrated photonics terminology) should be pointed out: a modulator features on/off modulation of light in a one output port device, whereas a switch effects light switching between input port(s) and several output ports. A switch is thus a more versatile device.

The "historically" dominating integrated photonics switch structure is the coupled waveguide structure of Figure 1, the *electrooptic directional*

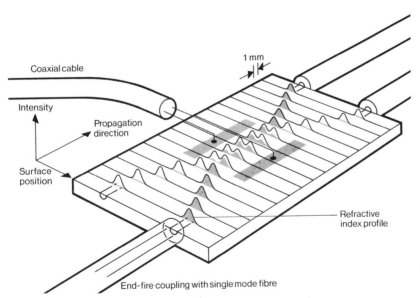

Figure 1 Electrooptic directional coupler in LiNbO$_3$. The index profiles of two waveguides in the crystal surface are shown together with the light power (shaded areas) in the guides with the coupler in the cross state. The incident light from a fiber (lower right) is switched between the output ports by applying a voltage to the electrodes (depicted as shaded stripes).

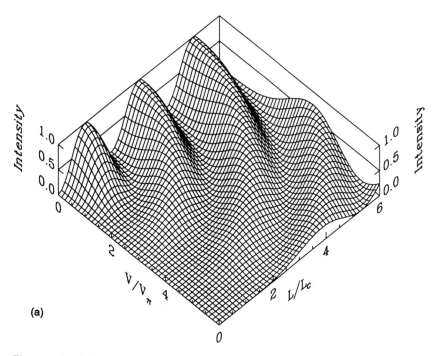

Figure 2 (a) Directional coupler switching diagram, showing the crossover power $P_\otimes$ versus normalized length L/L_c (where L_c is the coupling length, Eq. 8), and the voltage V_π needed to induce π phase shift between the waveguides. Here, $(V/V_\pi)\pi = \Delta\beta L$, where $\Delta\beta$ is the difference in propagation constants in the two guides. $P_\otimes \sim \sin^2[(\pi/2)(L/L_c)]$ with no applied voltage. Zeros in $P_\otimes$ are obtained for $[(\Delta\beta/2)^2 + \kappa^2]L = m\pi$, m an integer. (b) Same as (a), but utilizing the stepped $\Delta\beta$ structure; note that there now is a locus in the L, V-plane where the cross state can be obtained.

coupler [1]. Here, two phase-matched optical waveguides are arranged at such a small separation that light, by evanescent coupling, is periodically coupled back and forth between the waveguides in the direction of light propagation. By arranging electrodes at the waveguides, the refractive index can be changed via an applied electric field, employing the electro-optic effect. Hence, the original phase matching between the waveguides is destroyed, and full switching of light between the output ports can no longer be achieved. Figure 2 shows the switching diagrams.

In order to ease fabrication tolerances (the coupler otherwise has to be exactly an odd multiple of coupling or crossover lengths in physical length to implement the cross state), the stepped $\Delta\beta$ configuration [19] was de-

LiNbO₃ AND SEMICONDUCTOR GUIDED WAVE OPTICS

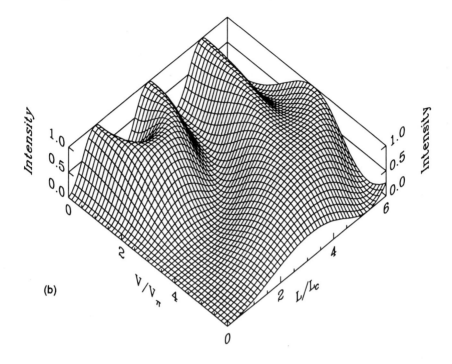

(b)

vised; and it has been extensively used since it was a decisive step toward making directional couplers practical.

There are other four-port switching devices, however, and Figure 3 shows, in addition to the directional coupler an X-switch [20,21], and a "BOA"-switch [22]. The electrooptically induced index change has different effects in the latter two (a BOA is essentially an elongated X-switch), and this can be analyzed as follows [23].

The applied voltage will, depending on the electrode configuration, give rise to an even (Δn_e) or odd (Δn_o) index perturbation, referring to the switch transverse coordinate. The even perturbation will change the βs of the modes of the composite two-channel structure (which are sometimes called the even and odd "supermodes" [1]):

$$\Delta \beta_e \propto (\psi_e, \Delta n_e \psi_e) \tag{5}$$
$$\Delta \beta_o \propto (\psi_o, \Delta n_e \psi_o) \tag{6}$$

where the brackets denote scalar products involving suitably normalized even ψ_e and odd ψ_o supermodes. We assume that the unperturbed coupler is

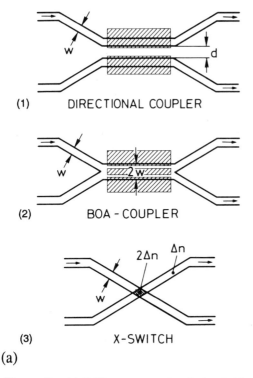

Figure 3 (a) Different types of optical switching devices, based on induced (e.g., by an electric field) refractive index changes. Note the double refractive index change of the intersection region in the X-switch, which ensures two-mode operation. (b) Field evolution in an X-switch, as analyzed with the beam propagation method (BPM) [13,23], showing the performance with no applied voltage as well as with applied voltage. Note the change in beat pattern in the waveguide crossing region, effecting the electrooptic switching. Also note the residual power in the lower arm when applying a voltage; the BPM is an excellent tool to analyze crosstalk, loss, etc., in optical switches.

symmetric. Similarly, the coupling between the even and odd supermodes, $\kappa_{e,o}$, is given by

$$\kappa_{e,o} \propto (\psi_o, \Delta n_e \, \psi_e) = 0 \tag{7}$$

for symmetry reasons. An odd perturbation will, complementarily, leave β_e and β_o unchanged (to first order) but couple ψ_e, ψ_o.

Hence, the BOA and X-switches with their strong coupling and consequently large difference between β_e and β_o are preferably operated by applying an even index perturbation, changing β_e and β_o by different

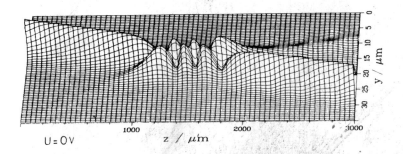

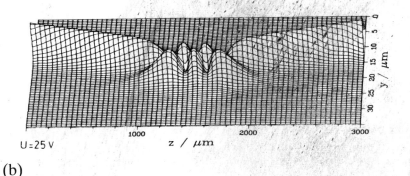

(b)

amounts, without coupling of the supermodes. This also gives a periodic transfer function (light power versus voltage) since the output power is proportional to $\cos^2\{(\beta_e - \beta_o)L/2\}$ [21], and this makes electric tuning possible without relying on the somewhat more complex stepped $\Delta\beta$ structure that has to be employed for a directional coupler. The BOA could in principle be operated with an odd perturbation; since strong coupling of the supermodes is then required, large voltages are needed; however, these only reflect the fact that the device length, taken as the coupling length

$$L_c = \frac{\pi}{\beta_e - \beta_o} \tag{8}$$

is short and hence the switch voltage obeys the usual length scaling law, Eq. (4). In this case, the nonperiodic directional coupler type response of Figure 2 results. The directional coupler structure, on the other hand, has a larger coupling length (smaller $\beta_e - \beta_o$) and should be operated with an odd perturbation Δn_o, to couple the supermodes to one another. Here, an even perturbation is less efficient for switching.

Equation (8) above, describing the coupling length, can also be written, as is customary, in terms of κ, the coupling coefficient [24]:

$$\kappa L_c = \frac{\pi}{2} \tag{9}$$

Figure 3 shows the required electrode structure on z-cut $LiNbO_3$ for achieving an odd perturbation for the directional coupler (top), and an even perturbation for the BOA (middle).

It should be obvious from the preceding that it is the electrode arrangement that determines whether a device is operated as a directional coupler, with a response as in Figure 2, or a BOA, with a periodic light output power versus applied voltage.

The switch types that have been described are all basically two-mode interference type switches: When exciting one input port, one excites a superposition of the even and odd supermodes, see, e.g. [25], and the subsequent interference and coupling of these supermodes, controlled by the electrooptic effect, brings about the switching between the output ports as previously described. The obvious disadvantage of such an interferometric switch is *limited optical bandwidth* and *periodic or quasiperiodic light output versus control voltage transfer characteristics,* characteristic of interferometric phenomena (Figure 2). An alternative to such interferometric switches is the *digital switch* [26–28], where a field incident on one port excites *one* mode of the composite structure and this mode is ideally adiabatically propagated though the structure; i.e., no mode conversion, causing crosstalk, takes place; this is further discussed below. The consequence of the one mode evolution is that

- Very large total bandwidths can be achieved, e.g., > 100 nm, and these do not scale inversely as device length (cf. [15]).
- The transfer characteristic is step-like or "digital," hence the name of the switch

These are two significant advantages: The first due to the interest in broadband wavelength-division multiplex (WDM) systems where space and wavelength switching are simultaneously utilized. Here, the digital switch can simultaneously switch extremely broadband information. The second feature, the digital response, also means that polarization independence can be achieved by simply increasing the voltage, such that both polarizations are switched, as is further treated below. Also, significantly simpler control is achieved since one common control voltage can be used for an entire switch array rather than having to tailor the voltages to the individual switches. The large voltage tolerance is of special importance in $LiNbO_3$ with the drift problems of this material, as discussed later.

LiNbO₃ AND SEMICONDUCTOR GUIDED WAVE OPTICS

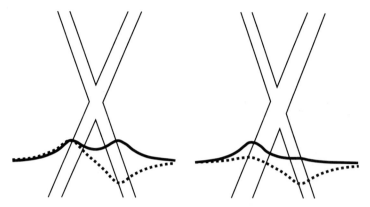

Figure 4 Digital switch operation: The parity of the incident mode is retained by the adiabatic propagation in this noninterferometric switch. Left: Supermodes in a symmetric switch, such as a directional coupler (even mode, solid line; odd mode, dashed line); Right: Supermodes in an asymmetric switch. Excitation in one input port for an asymmetric structure excites only *one* supermode, rather than *two* as in the directional coupler. A symmetric structure can be turned into an asymmetric one by e.g., applying an electric field, utilizing the electrooptic effect.

The principle of operation of the digital switch can be explained with reference to Figure 4: In a symmetrical two-waveguide structure, the even and odd modes look as in the left of Figure 4. In an *asymmetric* structure, however, the fields change and the "quasi-even" and "quasi-odd" modes get more and more confined to their respective waveguides, reflecting the increasing phase mismatch between the guides (Fig. 4, right). This is the key to understanding the operation of the digital switch: Instead of exciting *two* supermodes when exciting one input port, one (ideally) excites only *one* mode. Further, one employs a very shallow angle between the crossing waveguides (typically a few milliradians in LiNbO₃) such that there is ideally no coupling between the supermodes [29]. Obviously, any residual power in the "wrong" supermode at the input or excited along the structure will show up in the "wrong" output guide, giving rise to crosstalk. Switching is effected by electronically altering the refractive index at the output waveguides, such that the physical positions of the "even" and "odd" modes are interchanged. Instead of using a 2×2 switch one could employ a 1×2 switch, where there is only one mode at the input. With no applied voltage, the device acts as a power splitter; when applying a voltage, the index changes asymmetrically (increases in one waveguide, decreases in the other), such that the exciting even lowest order mode propagates into the waveguide supporting the quasi-even mode. Figure 5 shows beam propa-

Applied voltage: 0 V

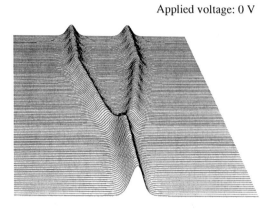

Applied voltage: 30 V

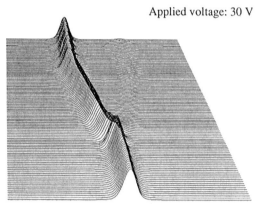

Figure 5 Beam propagation method simulation of a 1 × 2 digital optical switch, with no applied voltage and with 30 V applied. Note the crosstalk in the case of 30 V applied voltage; this crosstalk can be reduced by further increasing the voltage. The simulations are carried out on the effective index profile [13,28].

gation method simulations of such a switch. Switch arrays can be constructed from this basic building block, as described below.

The devices outlined here form the basic building blocks for time, space, and in some cases wavelength switches and are used to generate switch arrays. Before dealing with these, a short discussion on fabrication issues as well as the problem of polarization with reference to $LiNbO_3$ is given.

3. Fabrication of $LiNbO_3$ Devices

The fabrication of $LiNbO_3$ guided wave devices is dominated by the titanium indiffusion process with, e.g., the lift-off technique used to delineate waveguides as well as electrodes. Representative data to produce single-mode waveguides: 600 Å thick and 5 μm wide Ti stripes are evaporated on to the LiNbO$_3$ crystal and indiffused at 1050°C for 8 h (the Curie temperature of LiNbO$_3$ is ≈ 1142°C, although lowered somewhat by the Ti indiffusion process), creating indiffused waveguides with a Gaussian index distribution in depth (typically several micrometers) and with a width determined by the Ti stripe width. The maximum index increase at the surface of the LiNbO$_3$ waveguide is typically a few hundredths. This makes it possible to match the LiNbO$_3$ waveguides fairly well to conventional single-mode fibers, and insertion losses below 1 dB fiber-to-fiber have been demonstrated for straight channels [30]. The thicker electrodes (several micrometers) required for high frequency operation (coplanar microwave transmission lines) are usually deposited by electroplating. The Ti:LiNbO$_3$ technology has made possible the simple fabrication of low loss ($\leq$ 0.3 dB/cm at λ = 1.3 μm) waveguides of excellent stability (cf. the diffusion temperatures given above) without compromising the electrooptic effect. The available index increases are, however, as noted, low, restraining waveguide bend radii, available index mismatch for wavelength filtering, and light confinement. Proton exchange, on the other hand, offers higher index steps ($\Delta n \approx 0.1$) and simple fabrication. Only the extraordinary refractive index is increased in this case, however, making it all but impossible to fabricate polarization-independent devices. Other fabrication techniques have been of minor importance so far.

The Ti indiffusion process was investigated early [31,32] as to Ti concentration profile and index increase versus Ti concentration. Additional data on this as well as on the dispersion of the refractive index increase has been reported [33,34]. This has made possible an accurate modeling of integrated photonics devices in LiNbO$_3$ [13,14].

4. The Polarization Problem

As mentioned earlier, integrated photonics devices have to operate in a single transverse mode in order to be efficient, making them compatible with the dominant transmission fiber type. Ordinary single mode fibers do not preserve a defined state of polarization, however. As is obvious from Eq. (1), the induced index changes in LiNbO$_3$ are in general anisotropic: different polarizations will "see" different elements of the electrooptic tensor. The LiNbO$_3$ crystal has low enough symmetry (equivalent point group 3m) that the tensor is fairly well filled with elements that are in general different in magnitude. (In the semiconductors of cubic symmetry, such as GaAs and InP, with equivalent point group 43m, it is easy to find propagation directions where the index changes are equal for the two polarizations; see below). This means that a simple LiNbO$_3$ integrated

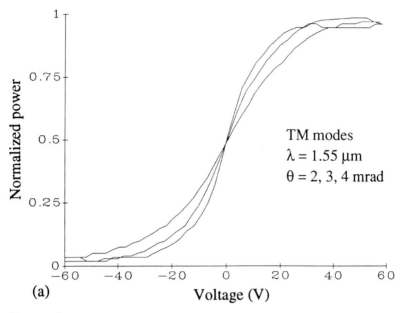

Figure 6 Measured transfer functions of a digital switch, fabricated in X-cut LiNbO$_3$, showing the response curves for TM (a) as well as TE (b) modes.

photonics circuit like the one in Figure 1 cannot simultaneously switch both polarizations, and this feature of LiNbO$_3$ has been a distinct disadvantage in practical applications. Below we concentrate on the strictly polarization independent devices (Section II.A) due to their practicality for systems experiments.

In general, polarization-independent operation can be achieved in a number of ways:

1. By employing polarization-independent physical switching effects *and* structures. The free carrier effect, Section II.C, is an example of the former, but if the structure employed is polarization dependent, such as a directional coupler, where the coupling coefficients are in general different for the two polarization modes, the entire device will still be polarization sensitive. A better approach is the following:

2. Cancel the effect of a polarization-sensitive physical switching mechanism by a suitable structure, and the digital switch is a good example of this: even though the transfer functions are different for the two polarizations (Fig. 6), polarization-insensitive operation is achieved by simply increasing the voltage, which is the normal price one has to pay for polarization independence.

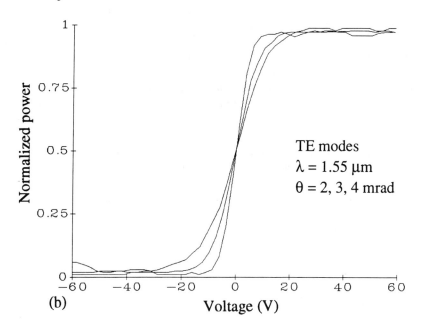
(b)

Two examples of polarization-independent switches will be given below:

- Polarization-independent directional couplers
- The digital switch

Several approaches to polarization-independent directional couplers have been taken, mostly relying on narrow fabrication tolerances [35–38]. In [38], the coupling lengths are required to be equal for the two polarizations and further equal to device length to implement the cross state with zero applied voltage. In addition, low bar state crosstalk for both polarizations requires control of the relations between the induced phase mismatch for the two polarizations ($\Delta\beta_{TM}$, $\Delta\beta_{TE}$), giving added fabrication constraints. In another case [36], a directional coupler (Fig. 1) with tapered coupling and equal coupling lengths (Eq. 9) for the TE and TM modes is fabricated on z-cut $LiNbO_3$ such that the cross state again is obtained in the absence of an applied voltage (i.e., the length is matched to the coupling length). The TE mode is controlled by the r_{13} coefficient, which is a factor of 3 weaker than the TM controlling r_{33} value, and hence (ideally) a factor of 3 larger voltage is required to reach the straight through or bar state. The tapered structure has the effect of attenuating the "sine"-like peaks of Figure 2, relaxing the tolerances as compared with the previous switch. Worst-case reported crosstalk is -10 dB.

In the switch of [35], by utilizing a stepped $\Delta\beta$ switch, Alferness was able to employ the features of the transfer characteristics of this device for further relaxing the tolerances for achieving cross as well as bar states in a polarization-independent manner. The coupling lengths are only required to be approximately equal and equal to device length.

In contrast to the preceding approaches, a switch has been reported [39] with fabrication tolerances relaxed down to those of the ordinary stepped $\Delta\beta$ switch. Record low crosstalk was reported for this switch (worst case -29 dB) and a *voltage $\times$ length* product of 16 V cm, to be compared to 5 V cm [17] for a polarization dependent switch. This performance was achieved at the expense of larger complexity: six electrodes/switch and two or three independent voltages/switch element as well as careful control of the individual voltages.

The digital switch [26–28] offers polarization independence in a much more brute force way. Figure 6 shows experimental transfer characteristics of a digital switch, demonstrating the way the digital response brings about the polarization-independent operation simply by increasing the voltage by (ideally) a factor of 3, i.e., the quotient between r_{33} and r_{13}. However, the disadvantage is that there is a trade-off between crosstalk and voltage: The digital switch with straight arms is, from a supermode point of view, formally analogous to a directional coupler with a tapered gap [29]; there is also a switch structure that is analogous to the constant gap directional coupler. For this structure, an analytic expression for the *voltage $\times$ length* product versus crosstalk can be found [40]; see Figure 7. It is seen that for a 15 dB crosstalk, roughly a factor of 2 higher voltage, compared with a stepped $\Delta\beta$ switch, is required, and each additional 10 dB costs a factor of 3 in voltage, in this idealized case. An important difference between the two-mode interference switches and the digital switch is that in the former case infinitely low crosstalk is *theoretically* possible, but for the digital switch this would require infinite voltages, which would in turn interfere with the adiabatic mode of operation. In practice, we have found, both from experiments and theoretical simulations, that the digital switch is not very efficient to use if very low crosstalk (< -25 dB) is required. Low crosstalk levels in switch *arrays*, however, can be achieved by means of special architectures, as described below.

In summary, the most important characteristics of a switch element are

- *Excess loss*: Should be below fractions of 1 dB.
- *Crosstalk*: Depends on application and switch architecture, in general, below -15 dB, will be required, in some cases below -25 dB.
- *Drive voltage*, characterized by a voltage length product: 5–10 V cm for

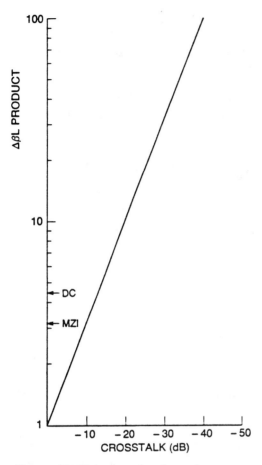

Figure 7 *Voltage × length* product versus crosstalk of the "shaped" digital switch, aimed at optimizing this figure of merit [40].

polarization-dependent devices, and roughly 3 times as much for polarization insensitive devices.
- *Switching speed*: Depends on the application, everything from many seconds to nanoseconds can be required.
- *Transparency* (i.e., the optical bandwidth): Matching of the Er fiber amplifier window ($\approx$ 30 nm) is desired.
- *Polarization sensitivity*: Differential insertion loss and crosstalk with respect to different polarizations. Lower than a few decibels desired.
- *Ease of control*: One common voltage desired.
- *Stability*: In many cases required for several years.
- *Environmental characteristics*.

5. Switch Arrays

One of the most intriguing properties of PIC space switches is their capability to route optical information practically irrespective of its bandwidth and coding format, generally termed *frequency and code transparency*. This is in marked contrast to electronic switches. The space-switching fabric can be viewed as a continuation of the "optical ether" of the fiber in this respect, and the space switches offer nearly the same extreme broadband capabilities as the optical fiber. This feature has been the subject of considerable research in photonic switching, and it has maybe not attracted due attention from systems researchers, partly because it is at odds with prevalent thinking in the time-division multiplexed (TDM) type networks of today. While it can be argued that frequency and code transparency is also a property of *mechanical* switches, which are in addition polarization independent without resorting to elaborate structures, these do not offer the decisive advantages of *integration* and *high-speed switching* (i.e., high-speed rearrangement of the switching fabric). These two features are in conflict, at least in $LiNbO_3$, however, since large-scale integration means short switch elements, giving high drive voltages (Eqs. 3 and 4) and hence impracticality for high-speed operation. Even so, very interesting systems applications exist for large (≥ 10 ports) low-speed switch ($> 1 \mu s$) as well as smaller (< 10 ports) high-speed ($< 1 \mu s$) switch arrays, as will be elaborated on in Section III. The design of a switch array involves a number of trade-offs (Fig. 8), and the optimum structure is highly application depen-

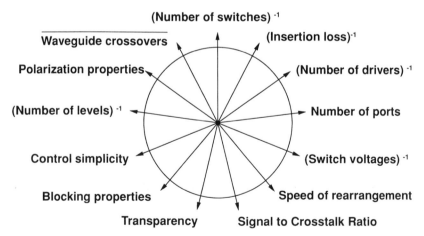

Figure 8 Trade-offs involved in photonic switch matrices. Larger distance from the center means higher performance of the parameter; bar means logical negation. Improvement in all parameters simultaneously is in general impossible.

LiNbO₃ AND SEMICONDUCTOR GUIDED WAVE OPTICS 21

dent. As an example, the requirements on blocking (strictly, rearrangeably nonblocking, blocking, etc.) will strongly influence its size (physical and in terms of ports) and characteristics (e.g., in terms of crosstalk). By strictly nonblocking, we mean that any unoccupied input/output port pair can be connected without rearranging existing signal paths; in a rearrangeably nonblocking structure, existing paths may have to be rerouted to give the nonblocking feature, temporarily interrupting connections, which can be of particular severity in high-speed systems ($\gg 1$ Gb/s). In a synchronous time switch array, this is obviously of less concern. Blocking arrays can of course be made with a larger number of ports for a given chip area. The number of switches per se, however, is not the most important parameter in LiNbO₃ integrated photonics, but rather the number of switches in cascade ("depth" of the array), since the size of the substrate is limited to 3–4 in. diameter (the lateral dimension is obviously of less concern).

The crossbar type is strictly nonblocking, but requires a large depth. This was the structure chosen for the first large switch array 8×8 [17]. The Benes structure saves switches but is not strictly nonblocking, whereas the tree structure [41,42] is interesting in that it trades length for width, which in view of the above discussion is advantageous. Thus, the depth grows only logarithmically with the number of ports. It does require more switches than the crossbar, but it is convenient for broadcasting and has excellent crosstalk properties (dilated structure): since the signal has to go wrong twice in order to contribute to crosstalk, the array's crosstalk (in decibels) is multiplied by two in relation to the element crosstalk. Figure 9 summarizes the properties of these three representative switch arrays [43]. Crosstalk should not be confused with noise, since it has a non-Gaussian distribution with a defined upper bound, which means that significantly lower *signal-to-crosstalk* ratios than signal-to-Gaussian-noise ratios can be accepted (≈ 10 dB [44]).

After the initial breakthrough in the switch array area in the shape of the 8×8 switch [17], development in the area has been rapid, and Fig. 10 summarizes the state of the art. The switches to date clearly illustrate the trade-offs in Fig. 8. The largest switch arrays (16×16) [45,46] are polarization sensitive and rearrangeably nonblocking, and the largest polarization-independent strictly nonblocking array [47] involves a nondilated structure, based on strict fabrication tolerances of the switch elements [38]. A different approach to the trade-offs involved is exhibited in the first reported switch array based on the digital switch [48]. In 1993, an 8×8 tree structure digital switch array was reported [48a]. Here, port number was traded for dilation and the high yield, broadband optical operation, and simply obtainable polarization insensitivity of the digital switch element. Figure 11(a) shows the layout of the switch elements in a tree structure, and Fig. 11(b) shows a packaged device,

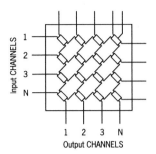

Blocking   : Strictly nonblocking
Interconnectivity : Point-to-point
Switch elements   : N^2
Electrical drivers: N^2
Insertion loss   : $(2N-1) \cdot L + 2W$
Δ Insertion loss : $(2N-2) \cdot L$
SNR   : $[X] - 10\log_{10}(N-1)$
Crossovers   : None

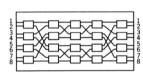

Blocking   : Rearrangeable nonblocking
Interconnectivity : Point-to-point
Switch elements   : $(N/2) \cdot (2\log_2(N) - 1)$
Electrical drivers: $(N/2) \cdot (2\log_2(N) - 1)$
Insertion loss   : $(2\log_2 N - 1) \cdot L + W$
Δ Insertion loss : 0
SNR   : $[X] - 10\log_{10}(2\log_2(N) - 1)$
Crossovers   : Yes

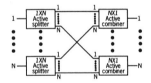

Blocking   : Strictly nonblocking
Interconnectivity : Point-to-point
Switch elements   : $2 \cdot N \cdot (N-1)$
Electrical drivers: $2 \cdot N \cdot \log_2 N$
Insertion loss   : $(2\log_2 N) \cdot L + 4W$
Δ Insertion loss : 0
SNR   : $[2 \cdot X] - 10 \cdot \log_{10}(\log_2 N)$
Crossovers   : Yes

Figure 9 Properties of three representative switch array architectures: Crossbar, Beues, and tree structure [43].

mounted on a standard circuit board and provided with standard single-mode fiber connectors. This is the device used in the systems demonstrators described in Section III. The conditions for making large switch arrays in $LiNbO_3$ can further be described with reference to Figure 12, which shows the number of (one sided) ports versus switch element length with the interconnect radius of curvature as a parameter for a crossbar structure. Normal values for the radius are 20–30 mm, but values below 5 mm can be reached by special methods [49]. It is seen that with reasonable element length (2 mm), sizes on the order of 20×20 can be reached for the 3-in. wafer assumed. These conditions are different in semiconductors, as treated below.

LiNbO₃ AND SEMICONDUCTOR GUIDED WAVE OPTICS

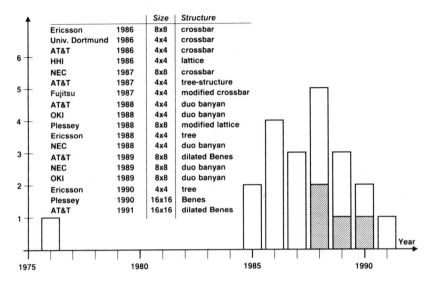

Figure 10 Number of published switch arrays ($\geq 4 \times 4$) versus year of publication for polarization-dependent as well as polarization-independent (shaded bars) LiNbO₃ switch arrays.

It should be noted that the term *frequency transparent* has to be used with some caution since LiNbO₃ is strongly birefringent [1] and dispersive, which gives restrictions on the *instantaneous* bandwidth, or bandwidth per channel (but not on the total bandwidth). The instantaneous bandwidth length performance can be estimated as follows:

For unipolarization operation or propagation along the z-axis:

$$B\sqrt{L} = 1 \text{ THz } \sqrt{\text{cm}}$$

However, for random polarization in a principal plane, we instead have

$$BL = 10 \text{ GHz dm}$$

The reason or this difference is the huge birefringence of LiNbO₃: $n_e = 2.203$, $n_o = 2.295$ at $\lambda = 0.6328$ μm.

In view of the success of the Er-fiber amplifier and the possibilities to dope LiNbO₃ with Er [50], the *loss* effects are of no fundamental significance, nor does crosstalk present an insurmountable obstacle to applying LiNbO₃ to large switching fabrics [51]. In fact, LiNbO₃ technology appears to meet a broad range of applications requirements (Section III). It is

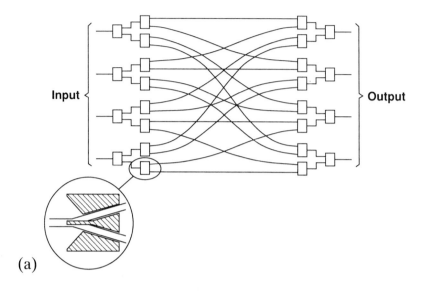

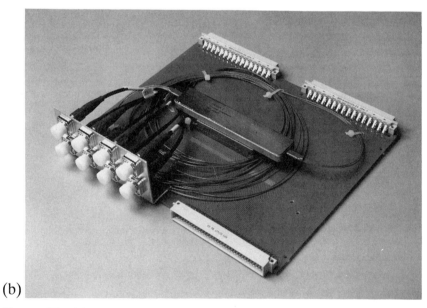

Figure 11 (a) Layout of the switch elements on the $LiNbO_3$ chip in the tree structure array of digital switches. (b) Packaged switch array with standard single-mode fiber pigtails mounted on a standard circuit board.

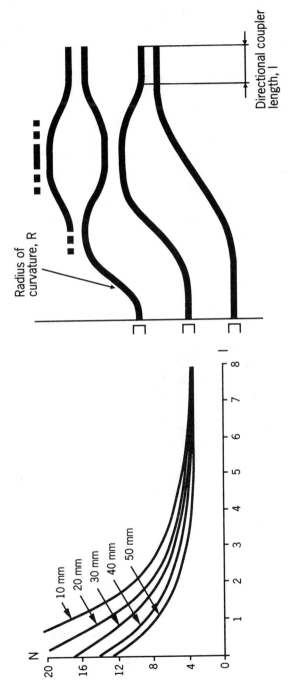

Figure 12 Design considerations for a 60-mm $N \times N$ crossbar switch matrix: maximum number of ports N versus crosspoint length with radius of curvature R of the waveguides as a parameter [15].

therefore appropriate to highlight some of the lingering problems of LiNbO$_3$: *optical damage*; *thermal* and *DC instabilities*.

The optical damage effect (light-induced index changes), which plagues short-wavelength devices (< 1 μm) seems to be largely but not fully eliminated at the longer communications wavelengths ($\lambda \geq 1.3$ μm) [52,53]. The thermal instabilities [54,55] are caused by the strong pyroelectric effect in LiNbO$_3$. X-cut devices are more tolerant to pyroelectric charging effects, due to the longer distances between the c-faces, than c-cut devices. In general, it is not the absolute temperature ranges that are of concern here but rather the *temporal rate of change* that matters. These thermal effects, however, can be checked to a large extent by deposition of conductive buffer layers [56], by temperature stabilization of the LiNbO$_3$ chip in its package, or by mounting it in a sufficiently thermally inert environment. The *DC drift*, however, fundamental to an insulator, poses a worse problem. Here, charge transport in the entire device tends to screen the applied external voltage, changing the effective applied voltage governing the response. These drift phenomena occur with widely different time ranges, from *milliseconds* to *months*, and reflect the dielectric relaxation times involved. The electrical properties of the host crystal and buffer layer (if used) as well as the electrode structure are of importance. Various equivalent circuits have been deviced to model DC drift [57,58]; according to these models, DC drift would be avoided if the instantaneous (capacitive) and asymptotic (resistive) partitioning of the applied external voltage would be equal. Apart from the fact that this would be difficult to achieve in practical fabrication due to the tolerances involved, the model is probably not sufficiently accurate. Utilizing devices such as the digital switch, sensitivity to *thermal* as well as *DC instabilities* are markedly decreased, however, and such LiNbO$_3$ devices are entirely practical for systems experiments. Figure 13 shows the results of temperature cycling of a digital switch element, fabricated in our laboratory, exhibiting excellent stability. Such switches have been the subject of long term (years) drift tests, and they have exhibited good stability. Systems experiments [59] have been conducted involving switch arrays based on the digital switch, as described in Section III.

Packaging of LiNbO$_3$ devices is instrumental to maintain the performance of the device by means of a well-controlled environment as well as to make the devices useful in practice. Substantial progress has been made in this area, reflected in the commercially available high- and low-frequency devices as well as in the emerging systems demonstrators to be described in Section III. For electrical connections, conventional bonding is applicable on LiNbO$_3$ devices. Several approaches for optical fiber pigtailing have been demonstrated: The straightforward "UV gluing" method,

LiNbO₃ AND SEMICONDUCTOR GUIDED WAVE OPTICS

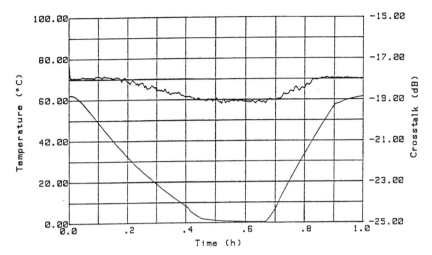

Figure 13 Temperature cycling of a digital switch element with 30 V applied to the switch. The crosstalk penalty is 1 dB over the 60 degree temperature range. The temperature slope is ≈ 2 deg/min.

which is fast and convenient but limits the available temperature range due to the low transition temperatures of the glues; and soldering, which requires preparation of the crystal as well as the fiber and is more critical due to the high thermal sensitivity of the crystal. The latter method does, however, result in devices meeting environmental specifications comparable with those of electronics.

C. Semiconductor Devices

1. General Comments

As noted above, LiNbO₃ devices suffer from a number of shortcomings, and attention was focused early on semiconductors for device performance improvements. LiNbO₃ is a passive material where only the refractive index (or strictly speaking the relative impermeability tensor) can be manipulated by changing an applied electric field (Pockels and Kerr effects). Semiconductors offer, due to the resonant interband *photon–electron interaction* in these materials, a multitude of physical effects in addition to the above mentioned (which were discovered late in the nineteenth century), allowing also controlled light amplification, absorption, and detection. In addition, free carrier effects can be used for index changes. Table 1 summarizes the physical mechanisms available in LiNbO₃ and in semiconductors to achieve modulation of the real and imaginary parts of the refractive index. It is obvious that

Table 1 Physical Mechanisms Available in LiNbO$_3$ and in Semiconductors (SC) to Achieve Modulation of the Real and Imaginary Parts of the Refractive Index

Physical mechanism	Operation mode	Characteristics	Materials
Index change through Pocket's effect (linear EO effect)	Voltage applied over dielectric or reverse-biased *pn* junction	Speed usually limited by walk-off or RC constants. Small nonresonant effect, $\Delta n/n \approx$ 0.0001; typically, no associated absorption	All materials lacking inversion symmetry: LN, III–V semiconductors
Index change through nonresonant Kerr effect (quadratic EO effect)	See above	See above. Usually very small	All materials
Free carrier, intraband transitions ("plasma" effect): Index and absorption change	Carrier injection or depletion through injection and reverse biasing, respectively	Usually smaller than other carrier-induced effects. Nonresonant effect, increases with the square of the wavelength	All semiconductors and other materials in which free carriers can be induced
Absorption/gain change through bandfilling	Unipolar through reverse biasing of *pin* junction (depletion); bipolar through current injection	Unipolar operation is fast (RC-limited) but cannot give gain; bipolar operation limited by carrier recombination. Strong λ dependence	III–V (and other SC)
Index change through bandfilling	See above	See above. Large index changes achievable, of the order $\Delta n/n \approx$ 0.001–0.01. Strong λ dependence	See above

Table 1 Continued

Physical mechanism	Operation mode	Characteristics	Materials
Absorption change by quantum confined Stark effect (QCSE)	Reverse biased *pin* junctions	RC- or transport-time limited response (fast, depends on number of wells and barrier heights, etc.). Strong λ dependence	Quantum well SC
Index change through QCSE	See above	RC-limited response. $\Delta/n \approx$ 0.001–0.01 Strong λ dependence	See above
Absorption change by Franz–Keldysh (FK) effect	See above	RC- or transport-time limited response. Strong λ dependence	All SC
Temperature induced absorption and index changes for λs close the bandgap	Change of external temperature, *and* heat dissipation associated with carrier recombination	Large effect that partially counteracts the index and absorption changes induced by electronic effects. Depends on heat sinking and the ratio between radiative and nonradiative recombination channels	All SC

semiconductors offer a far greater flexibility, and they allow (in principle) the integration of elements covering a wide spectrum of functionality, hence offering the possibility of truly integrated photonics.

This versatility makes semiconductor devices the prime candidate for employment in future photonic switching networks. Furthermore, the semiconductor system offers a considerable flexibility in *waveguide parameters* in that significantly larger refractive index steps than in LiNbO$_3$ can be achieved (order of magnitude 0.5 instead of 0.02), offering better optical field confinement, and as a consequence, the potential packing density is higher due to the fact that much smaller bend radii are possible [60]. As an example, typical bend radii for LiNbO$_3$ are on the order of tens of millimeters (Section II.B), whereas the submillimeter range is attainable in a semiconductor. The drawback with this tight confinement is of course the low coupling efficiency to an ordinary single-mode fiber, a problem that has plagued semiconductor integrated optics from the start and received much attention.

Whereas the potential of semiconductors in this area was recognized early, progress with regard to PICs has been remarkably slow, and one could compare this with the extremely rapid development in the field of lasers, detectors, and other *discrete* devices for optical communications. The reason for this state of affairs can be attributed partly to the fact that the integration of several discrete devices onto one chip implies a considerably increased complexity of the processing involved, and partly to the less immediate need for these types of devices in applications. As an example, even the superficially simple problem of integrating a "passive" waveguide and an amplifying waveguide segment has required much work.

The architectural options for semiconductor switches are the same as for LiNbO$_3$ (Section II.B), but the different boundary conditions for semiconductor PICs make the choice of architectures different:

- The "depth" dimension (length of the chip) is not so critical, since shorter (< 1 mm) switch element lengths are feasible (by virtue of the larger index changes or by employing semiconductor laser amplifier (SCLA) gates) and due to smaller bend radii or the employment of integrated mirrors [61,62].
- There is not the same trade-off in switching speed versus size (in number of parts) as in LiNbO$_3$, since the SCLAs and e.g., switches based on bandfilling (see below) do not require the same excessive powers for high (nanosecond) speeds.
- The routing bandwidth (i.e., the allowable bandwidth of the controlled optical signal) is in general limited in comparison with LiNbO$_3$.
- Switches based on SCLA gates will be susceptible to crosstalk in multiwavelength networks if they are not operated far below saturation output.
- Losses can be compensated with integrated amplifiers.

Reported switch arrays sizes are confined to 4 × 4 [63–66], implemented using the effects listed in Table 1. None of these arrays are polarization independent and generally suffer from high ($\approx$ 10 dB) fiber-to-fiber loss as

LiNbO₃ AND SEMICONDUCTOR GUIDED WAVE OPTICS

well as poor crosstalk. The last feature is not an inherent feature of semiconductor switches but rather a consequence of the way the switches are fabricated and characterized. Semiconductor switch arrays are currently less developed than LiNbO$_3$ arrays but are the subject of a large research effort. Figure 14 depicts the number of reported switches $\geq 1 \times 2$, showing the growing effort in the area.

Below, several types of semiconductor PICs are covered, concentrating on those with relevance for space-division photonic switching. First, the fabrication aspects are briefly treated (there are several excellent reviews in this area [1,67]); then, LiNbO$_3$-like PICs, based (more or less, since free carrier and bandgap effect will always play a role) on the Pockels electro-optic effect, are dealt with. The full potential of the semiconductor material can only be realized when using the effects based on the resonant bandgap effect, however, and devices based on these and free carrier effects are treated next.

2. Fabrication of Semiconductor PICs

Fabrication of semiconductor PICs is considerably more involved than that of LiNbO$_3$, both in the processes concerned and with regard to the multitude of different possibilities that exist. The basic waveguides were initially

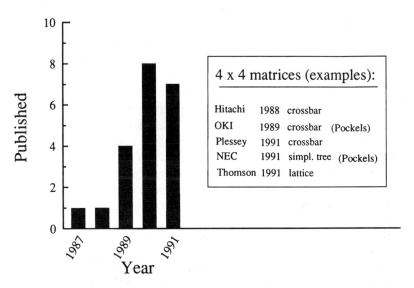

Figure 14 Number of reported semiconductor switches (excluding switches mainly based on the Pockels effect) $\geq 1 \times 2$ versus year of publication. The insert shows reported 4×4 switches, including Pockels effect-based devices.

produced by employing higher doping in the cladding layers, such that the free carrier index decrease effected the waveguiding. The obvious disadvantage of this, however, was the high attenuation due to absorption, as well as the limited index steps available [1]. With the advent of the currently prevalent heterostructures, the waveguiding was instead achieved by employing the different refractive indices of binary, ternary, or quarternary materials of different compositions [1,67]. This results in a considerable flexibility in adjusting the refractive index step to the application in mind. In this case, absorption losses enter *ideally* only in so far as one has to select the operating wavelength at an appropriate distance from the bandedge. (Of course, waveguide losses due to scattering are always present). In many complex circuits, however, doping of passive waveguides is required to permit charge transport, giving additional free carrier losses. Hence, figures on propagation losses in semiconductor losses have to be interpreted against the application in question; e.g., in the cases where optical amplifiers are integrated, losses are obviously of less concern.

The waveguide structure layers are grown by different epitaxial techniques: LPE—liquid phase epitaxy; MOCVD—metal–organic chemical vapor deposition; MBE—molecular beam epitaxy; to name the most representative, and the general device structure is delineated using a variety of processing, including wet and dry etching and various doping techniques. In general, to fully exploit the possibilities offered by the semiconductor system, one requires:

1. *Advanced epitaxial techniques*, permitting uniform dimensions, composition as well as doping over large, e.g., several square centimeter, areas. The employment of MOCVD and MBE is of special significance, since they allow fabrication of structures with sufficient control of thickness (layers down to the size of one atomic layer can be fabricated) to make possible the utilization of quantum effects (quantum well devices), modifying the density of states and exhibiting phenomena such as clearly distinguishable excitonic absorption peaks at room temperature and the quantum confined Stark effect (QCSE) [68]. Of further importance is epitaxial regrowth, such as semiinsulating layer regrowth, so that different elements in the same PIC can be operated independently of one another, e.g., in forward- and reversed-bias modes on the same chip.

2. *Processing* to accurately delineate the structures, such as wet and dry etching (RIE—reactive ion etching, RIBE—reactive ion beam etching); especially dry etching has gained widespread use due the possibilities of fabrication of narrow structures with extreme precision. Furthermore, we have different types of doping for contacting, e.g., ion implantation for contacting through a semiinsulating layer

3. *Advanced lithography*, e.g., electron beam lithography, allowing the required precision and flexibility in mask fabrication as well as the possibility of writing in situ patterns on the chips.

The progress of semiconductor technology in this area is shown by the achievement of single-mode waveguides with propagation losses below 0.5 dB/cm [69].

3. "LiNbO$_3$-like" PICs

These devices are mainly based on the electrooptic effect, as stated above. Early efforts in this area [70–74] focused on developing directional couplers, interferometers, etc., familiar from LiNbO$_3$ with little consideration for integration issues. In general, performance could at best match LiNbO$_3$, with generally high insertion loss. A further problem is due to the smaller electrooptic coefficient (1.4×10^{-12} m/V in comparison with 30×10^{-12} m/V for LiNbO$_3$), which is only partially offset by a higher refractive index (3.6 versus 2.2 for LiNbO$_3$) since $\Delta n = 0.5 n^3 rE$, as given by Eq. (3). For equal fields, the ratio of the Δns is 0.2. The possibilities of tighter confinement in semiconductors with higher electric fields and better overlap, however, can partially compensate for this disadvantage with respect to LiNbO$_3$. For high-speed applications, where a large dielectric constant ϵ is a disadvantage, either by increasing the capacitance or by slowing the microwave down in relation to the modulated light wave in traveling wave devices, semiconductors have a decided advantage in that for, e.g., GaAs, $\epsilon = 11$, whereas this number is 35 in standard orientation at microwave frequencies in LiNbO$_3$.

As an example [72], a directional coupler has been reported with a switch voltage of 12 V for an 8 mm long device at 1.6 μm, and 16-dB crosstalk, making the *voltage × length* product a factor of 2 higher than for the reference LiNbO$_3$ switch in Section II.B. It should be noted that for these devices essentially the same problems as in LiNbO$_3$ regarding polarization sensitivity exist. There are, however, specific crystal orientations where polarization-insensitive operation can be obtained ($\bar{2}11$) [75]. It should be borne in mind that the polarization independence of the basic refractive index change mechanism is not a sufficient requirement for polarization-independent operation.

4. PICs Based on Interband Resonant Effects and on Free Carrier Effects

Clearly, in order to achieve improvements over LiNbO$_3$ (except for sheer integration issues, such as integration of switches with laser, etc.), one will have to improve over LiNbO$_3$ in the physical effects employed in, e.g., switches. In this area, one has basically two possibilities (Table 1): free carrier effects and effects due to the interband transitions. The free carrier effects, which are polarization independent, can be used in two modes: carrier injection (in the forward-biased mode of operation) and carrier depletion (reverse biased mode of operation) in heterostructure *pn* junctions. In the former case, speed is limited to the spontaneous recombination time: nanoseconds. The forward-biased mode of operation tends to increase the power consumption of the devices. In the reverse-biased case, large index changes require high doping, giving high absorption in the absence of carrier outsweep, but high speeds can be achieved. For some switches, such as the digital switch, the high absorption plays a minor role, since the light is switched to the arm with higher index (where the carriers are swept out).

Of greater significance in this area is the utilization, in bulk as well as in quantum well devices, of interband effects (Table 1). Bandfilling in the forward-biased mode of operation is fairly slow, since it is limited by spontaneous recombination times to the order of gigahertz. It is further associated with heat dissipation, which is disadvantageous, since the temperature increase causes an index change opposing that due to the bandfilling contribution. In the interesting *barrier reservoir and quantum well electron transfer* structure [76], the large index changes obtained by bandfilling are *combined* with high speed, since the transported carriers are blocked from recombination, these devices being unipolar.

The Franz–Keldysh (FK) effect [77] and the QCSE [68] work in reverse bias and are in practice limited by RC constants, thus giving the potential for very fast operation. The FK effect, which is a bulk phenomenon, is polarization independent per se, but all effects relying on propagation parallel to the layers of a quantum well structure are generally speaking polarization dependent, due to the intrinsic polarization dependence of the matrix element. For the resonant effects, there is a relation between the real and imaginary parts of the susceptibility or the refractive index: the Kramers–Krönig relation, which in a general way expresses the connection between the real and imaginary parts of the Fourier transform (transfer function in filter theory language) of the impulse response of a causal system. This relation can be written

$$\chi'(\omega) = \frac{1}{\pi} \mathrm{PV} \int_{-\infty}^{+\infty} \frac{\chi''(\omega')}{\omega' - \omega} d\omega' \tag{10}$$

and

$$\chi''(\omega) = -\frac{1}{\pi} \mathrm{PV} \int_{-\infty}^{+\infty} \frac{\chi'(\omega')}{\omega' - \omega} d\omega' \tag{11}$$

LiNbO₃ AND SEMICONDUCTOR GUIDED WAVE OPTICS

where PV denotes principal value of the integrals. χ is the complex dielectric susceptibility. When employing resonant bandgap effects, the absorption spectrum is changed by changing the external potential (FK, QCSE) or by "removing" oscillators (bleaching), as in bandfilling. The latter is effected by virtue of the Pauli exclusion principle, which blocks transitions to already occupied states. Hence, a *change in the absorption spectrum* will cause a *change in the refractive index spectrum*, and this change can be calculated as follows:

$$\Delta n = \frac{1}{2\pi n} \int_{-\infty}^{+\infty} \frac{\chi''(\omega') - \chi''(\omega)}{\omega' - \omega} d\omega' \qquad (12)$$

where the singularity is removed since we are calculating the *change* in index. Figure 15 illustrates the above equations for an absorption spectrum in the shape of a Lorentzian function and for different absorption *change* spectra. It is seen that there is a trade-off between the achievable index changes and absorption, and in general larger index changes are paid for by higher absorption. The application at hand will determine the exact "operating point," depending on how tolerable large losses are. Figure 15 also illustrates why the bandfilling effect is more efficient: When changing the external potential, each individual oscillator, the sum of which constitute the absorption spectrum, will move, giving a change in the refractive index, e.g., in the low energy tail of Figure 15I. A larger change in refractive index results if the oscillator is removed altogether (Figure 15I), such as in bandfilling. A discussion of Stark effect versus bandfilling in the context of photonic switching devices is given in [78].

While the various quantum well structure devices have been the subject of intense research, it has to be borne in mind that the active regions of these devices are smaller than the total waveguide "core," lowering the effective confinement factor, whereas in bulk devices, the whole core region is active in changing the modal index.

In this area, a number of switching devices have been reported [63,79–89], and as an example Figure 16 shows a switch element based on carrier injection [63], which has been used to make a crossbar switch matrix. It should be noted that the operation of these switches relies on a combination of free carrier and bandgap effects; in this case, with operation far from the bandgap ($\lambda = 1.31$ μm and $\lambda_g = 1.15$ μm), these switch elements require comparatively large currents (200 mA) for switching; lower currents require different structures and/or operation closer to the bandgap (with higher losses). As an example, Figure 17 shows the transfer characteristic of the first digital optical switch (DOS) in InP/InGaAsP, requiring only 15–20 mA for good switching in the TM polarization [86]. Similar low-

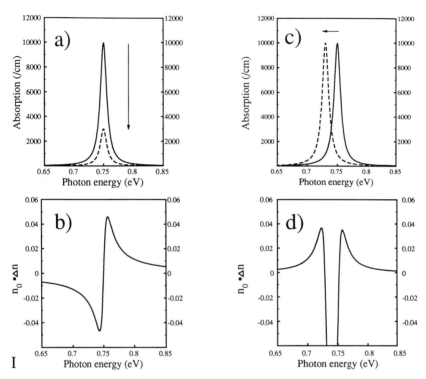

Figure 15 (I) Different principles of resonantly induced large refractive index changes Δn: Bleaching of an absorption line (a and b) and energy shift of an absorption line (c and d). In both cases, a single, isolated homogeneously broadened absorption line is assumed. In (a) the absorption is decreased by, for instance, carrier injection, and (b) shows the corresponding index changes (n_0 denotes the background refractive index, stemming from transitions far from the resonance line considered here). In (c) the absorption is changed by moving the absorption line by, e.g., applying an electric field, and (d) shows the corresponding index change. The dip in (d) is about -0.15. Such a large index change is difficult to use in practice, since the absorption is also very high for for the wavelengths in question. In practice, it is difficult to obtain single isolated lines like these in semiconductors. (II) Stark shift of absorption spectrum for different applied electric fields in an MQW (multiple quantum well) (100 Å GaAs/ 100 Å AlGaAs, $x = 0.4$). The electric field increases from right to left as 0, 30, 60, 90, 120, 150, 180 kV/cm (a); corresponding index changes (b); absorption spectra in bulk for different carrier densities in the current injection case. For decreasing absorption, the density varies as 0, 2, 4, 6, 8, 10 and $15 \cdot 10^{17}$ cm^{-3} (c); corresponding index changes (d).

LiNbO₃ AND SEMICONDUCTOR GUIDED WAVE OPTICS

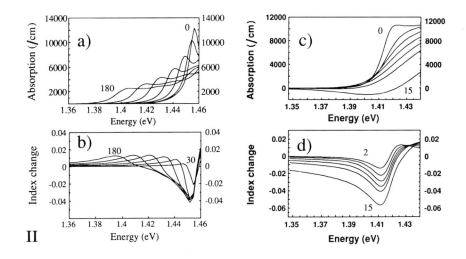

current results have been reported [80], with 4 mA switching current for a 2 mm long directional coupler giving 20-dB extinction ratio with a low insertion loss of a few decibels.

Switches relying on the quantum confined Stark effect (QCSE) index change in a multiple quantum well (MQW) structure have been reported [83,88]; as an example Figure 18 [88] shows a directional coupler switch requiring 20 V for a length of 170 μm, and with a crosstalk of 8 dB. The concept of a *voltage × length* product is in general not applicable here due to the nonlinear response, but it is interesting to note that the above numbers correspond to 0.34 V cm, which compares very favorably with LiNbO₃ (5 V cm; see above)

Figure 19 summarizes the status of semiconductor optical switches in terms of length versus drive voltage and current for the reverse- and forward-biased cases, respectively.

An issue of importance for semiconductor PICs and a distinguishing feature in relation to LiNbO₃ is the possibility of integration with *optical amplifiers*. As noted in Section II.B, the optical losses of the LiNbO₃ devices are a limiting factor in building large passive switching fabrics. In principle, one could envisage a hybrid optical network, where the optical losses in the LiNbO₃ devices are compensated by fiber amplifiers, but this would not warrant the advantages of integration, well established in integrated electronics. Hence, it is important to assess the features of integrated amplifiers, both from a performance and an applications point of view.

Concerning applications, semiconductor laser amplifiers can be used for detection (simultaneously with amplification), switching, or gating as well

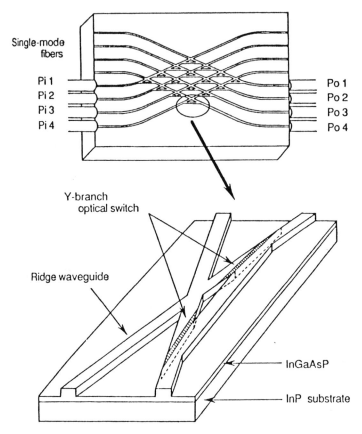

Figure 16 Schematic view of carrier injection switch element. Switching relies on total internal reflection in the shaded regions. These elements are used in a crossbar matrix with insertion losses ranging from 11 to 19 dB and extinction ratios from 9 to 25 dB [63].

as frequency translation, in addition to broadband amplification, making them *multifunctional* components. This is of special significance, since the possibility of using one generic type of device to perform a variety of functions alleviates the technological difficulties involved.

The optical amplifier is having a large impact on current research on optical networks, where the Er-doped fiber amplifier (EDF) is the prime candidate for creating a transparent optical, analog network (analog in the sense that no optical regeneration or pulse conditioning is envisaged). The EDF will most likely be used for preamplification, power boosting, and linear repeating because of its advantages of polarization insensitivity, ex-

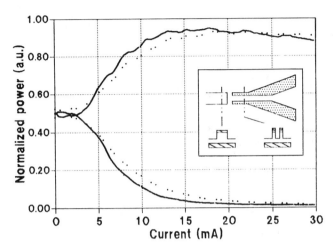

Figure 17 Switching curve for the TE (solid) as well as TM (dots) polarizations of a 3 mrad Y-junction digital switch in InP versus current. Waveguide width is 4 μm; switch length is 6.6 mm. Cf. [86].

tremely low "facet" reflectivity, good noise and saturation output power properties, and low crosstalk in multiwavelength systems. The EDF amplifier, however, is limited in versatility: in the semiconductor laser amplifier (SCLA), light amplification is effected by the coupling between the photon field and inverted electron population system, implying that the quasi-

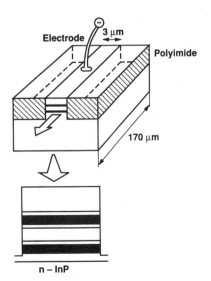

Figure 18 MQW-based vertical directional coupler switch [88].

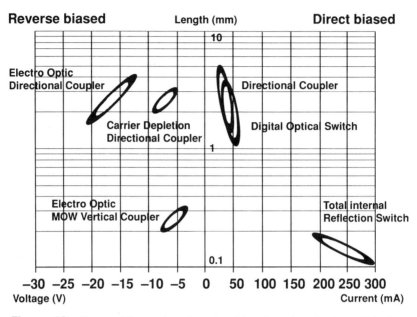

Figure 19 Status of forward- and reverse-biased semiconductor switch elements in the current–voltage–length plane. Note relation to a LiNbO$_3$ polarization-dependent switch with, e.g. 2 mm length and 25 V switch voltage. (From M. Ermau, Alcatel Alsthom Recherche, France.)

Fermi level separation provides an electronic means of monitoring the optical field. Further, since the spontaneous recombination time is of the order of nanoseconds, the optical field can be *detected* (simultaneously with amplification), *frequency converted*, and *gated* up to bandwidths of ≈ 1 GHz. In the EDF, there is no analogue to the electrooptic interface, and the spontaneous recombination time is on the order of milliseconds. This latter feature, however, gives the EDF its low crosstalk in multiwavelength systems. Hence, the SCLA offers unique possibilities, and all of the above functions can be implemented by one single generic structure that is integrable, in contrast to the EDF. But it should also be noted that the nanosecond time scale together with the homogeneous broadening can introduce severe crosstalk problems in multichannel systems.

The operation of the SCLA in the forward-biased detector mode of operation has been treated in a number of publications [90–93]. Recently, a quantum optics treatment was given [94]. From energy balance considerations, variations in the input power must be reflected in the quasi-Fermi level separation (when the load resistance parallel to the junction $R_L \to \infty$)

LiNbO₃ AND SEMICONDUCTOR GUIDED WAVE OPTICS

or in the injection current ($R_L \to 0$). In the latter case, for a traveling-wave amplifier,

$$i_{\text{junction}} = (G-1)p_{\text{in}}\frac{e}{\hbar\omega}\frac{\Gamma g_m}{g} \tag{13}$$

Here, g_m is the material gain coefficient, $g = \Gamma g_m - \alpha$ is the net gain coefficient, Γ the confinement factor, α the nonresonant losses, p_{in} is an input power change, and finally G the amplification factor. Figure 20 depicts calculated and measured responsivity and gain versus injection current [91], showing good agreement and the low-current transition from reverse-biased to forward-biased detector operation. The figure further shows the quantum optics theoretical predictions [94] and measurements of bit error rate (BER), showing reasonable agreement. The detection bandwidth in the latter case is $B = 380$ MHz, corresponding to a data rate in excess of 500 Mb/s. The theoretical sensitivity in this detector mode of operation is roughly comparable with an ordinary PIN FET receiver at these data rates. Systems experiments have been carried out up to 900 Mb/s [92], in general increasing the bias results in shorter carrier lifetimes and larger bandwidths [91,92].

Gating of an optical signal by an SCLA is treated theoretically and experimentally in a number of papers [95–98]; the speed is limited, however, which is a disadvantage in applications such as integrated "external" modulators. Higher input powers give higher modulation speed, but also mean less amplification; thus, there is a trade-off between amplification and speed. Faster modulation than in bulk SCLAs, a few Gb/s, can be attained with MQW SCLAs. Switches based on SCLAs are reported in different versions: theoretically [99], and experimentally [100] as directional couplers, and experimentally as gate switches [101–104]. Use of an SCLA-based directional coupler as a wavelength filter has been suggested [105]. In the directional coupler, it is mainly the real index change that effects switching, but the detailed crosstalk performance is determined by the imaginary part of the refractive index, i.e., the gain.

The same mechanism that can cause detrimental crosstalk in WDM systems can be employed for wide-range frequency conversion [106,107]. This nonlinear four-wave mixing (FWM) is based on beating between a signal (P_s) and a pump (P_{p1}), separated by less than 1 GHz; this beating creates a corresponding fluctuation in the carrier sea. A second pump (P_{p2}, where $\omega_{p2} - \omega_{p1} \gg 1$ GHz) can then copy P_s to a signal P_{s1} at the new frequency: $|\omega_{s1} - \omega_{p2}| = |\omega_s - \omega_{p1}|$. Such a wavelength converter, which retains all modulation formats, is highly desirable in multiwavelength photonic switching networks; in order to remove P_{p2}, however, filters with very steep frequency response are required.

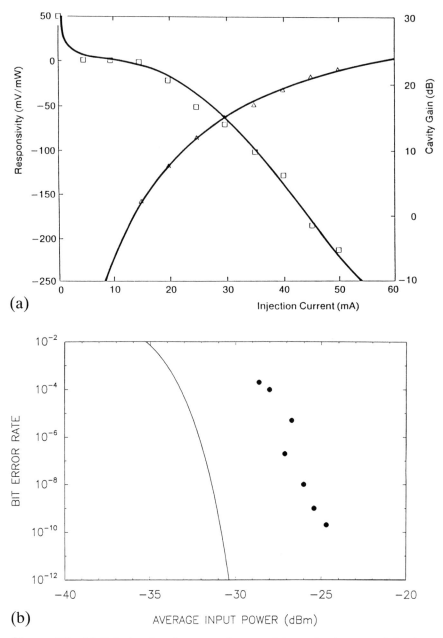

Figure 20 (a) Calculated and measured responsivity and gain versus injection current [91] in a 1.3-μm semiconductor laser amplifier, showing good agreement and the low-current transition from reverse-biased detector operation. (b) Quantum optics theory predictions [94] as well as measurements of BER, showing reasonable agreement. The detection bandwidth $B = 380$ MHz, corresponding to a data rate in excess of 500 Mb/s; notice sensitivity in relation to a PIN FET receiver.

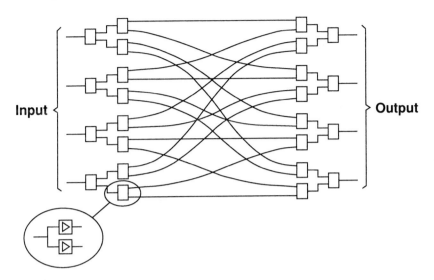

Figure 21 SCLA space switch architecture, with passive splitters, combiners, and SCLA gates. The structure is strictly nonblocking and permits broad- and narrowcasting. The same architecture has been used for LiNbO$_3$ switch arrays.

The gating function of the SCLA can be used to create switch arrays as in the suggested integrated photonic circuit of Figure 21 [108], where it performs routing (employing optical splitters and the following amplifier gates to determine the signal path), amplification, and can be used to monitor the status of the switching fabric with the above mentioned detector function. Such a switch array monolith could in principle be made quite large ($> 32 \times 32$ as an order of magnitude) in spite of the 3-dB losses inherent in each splitter or stage in the matrix. The size will be limited by noise and saturation due to amplified spontaneous emission (ASE) as well as by technological constraints. Due to the high extinction ratios (> 40 dB) obtainable with the SCLA gates, the crosstalk in this type of switch array is expected to be low. Figure 22 shows that comparatively high waveguide losses are allowed in this type of switch, due to the dominance of the 3-dB splitters from a loss point of view. Figure 23a shows a computer layout of the waveguides of an 8×8 switch array, and Fig. 23b shows the first reported 2×2 switch utilizing integrated SCLAs as gate switches (an integrated 4×4 SCLA gate switch array was reported in 1992) [109,109a]; facet-to-facet loss is as low as ≈ 3 dB, in spite of the intrinsic 6-dB losses in the splitters. From a noise point of view, it is more advantageous to use low-loss passive switches and booster amplifiers (e.g., EDFAs, erbium doped fiber amplifiers) at the inputs, especially if the input signal level is low.

It is obvious that this type of switch has potential to be superior to LiNbO$_3$ switches and would be useful in the fiber-optic transport network; as is the

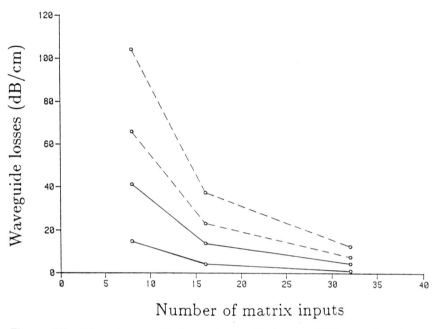

Figure 22 Allowed waveguide losses, for specified output SNRs, as a function of matrix size. The solid lines represent 1 μW input power with a required 20-dB output signal-to-noise ratio (SNR) for optical noise bandwidths of 10 and 40 nm, respectively. The broken lines represent 10 μW input power with a required output SNR of 30 dB for optical noise bandwidths of 10 nm and 40 nm, respectively [108].

case for $LiNbO_3$ devices, however, polarization independence is highly desirable. Figure 24 shows the measured amplification fiber to fiber in an integrated SCLA (i.e., comprising an amplifying section between passive waveguide sections) developed within the RACE R1033 OSCAR project (results from Thomson CSF). Excellent polarization independence is shown. Figure 25 shows another application of integrated SCLAs. Three interconnected SCLAs perform all the functions required in a bus node: detection/amplification and transmission. A hybrid systems experiment with such a configuration has been performed [110]. In an integrated version, detailed simulations show that transmit speeds of several Gb/s can be attained. It should be noted that the structure of Figure 25 could be used in a variety of ways: By reverse biasing one of the SCLAs, we get a high-frequency PIN detector; by introducing Bragg gratings, we get frequency selective feedback, etc.

The multifunctionality of the SCLA (detector, gate, wavelength shifter) as well as the possibility for integration make the SCLA an important building block in lightwave transmission and switching systems.

LiNbO₃ AND SEMICONDUCTOR GUIDED WAVE OPTICS

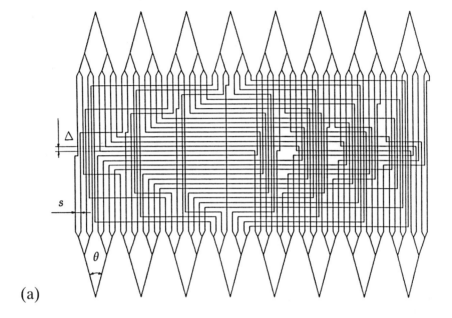

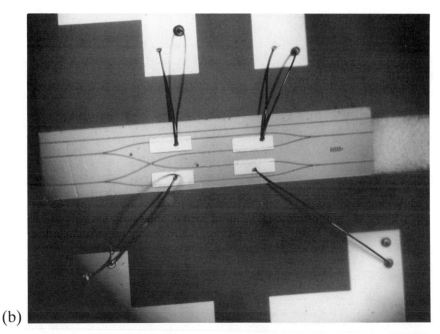

Figure 23 (a) Computer layout of a monolithic 8 × 8 semiconductor switch matrix, showing the waveguide pattern in the architecture of Fig. 21. (b) 2 × 2 switch array utilizing integrated SCLAs as gate switches [109].

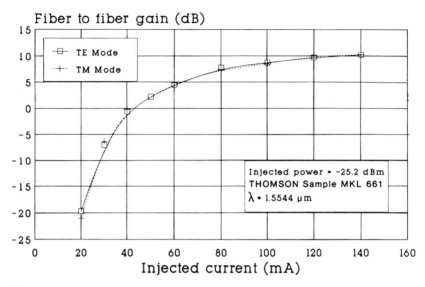

Figure 24 Amplification fiber to fiber for a three-section device comprising an amplifying section cascaded with two passive waveguiding sections, showing excellent polarization insensitivity (results of RACE 1033 OSCAR from Thomson CSF).

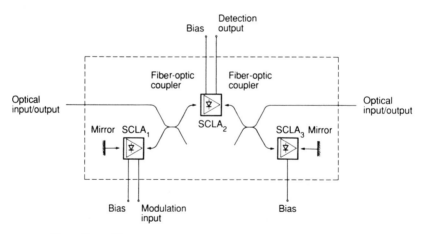

Figure 25 Three multifunctional semiconductor laser amplifiers connected as a node in a bus network. With only the center SCLA turned on, the node works in the amplifying and detection mode; with all amplifiers turned on, in the transmission mode. In the latter case, the SCLAs with feedback operate as a laser.

LiNbO₃ AND SEMICONDUCTOR GUIDED WAVE OPTICS

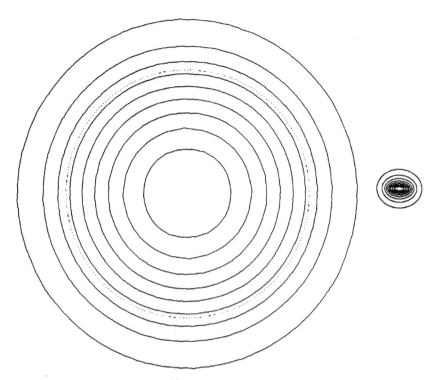

Figure 26 Near field of a representative InGaAsP/InP semiconductor waveguide as well as from a standard single-mode optical fiber, calculated with a finite element method (FEM) program. The semiconductor waveguide has dimensions 0.2×1 μm^2 and refractive indices 3.17 and 3.40 in the cladding and core, respectively. The wavelength is 1.55 μm. Note the mismatch in field profiles, giving in this case a coupling efficiency of well below -10 dB.

5. Coupling Efficiency Considerations

Figure 26 shows the near fields of a representative semiconductor waveguide and that of a single-mode fiber, exhibiting a large mismatch. In principle, using buried waveguides, the near field of the semiconductor waveguide could be well matched to the fiber by using a composition with an index step matching that of the fiber and the proper lateral dimensions. There are, however, various reasons for using larger index steps (much larger than in LiNbO₃) and the ensuing smaller waveguide dimensions, especially in the "depth" coordinate:

- Higher packing density with smaller bend radii of the bends interconnecting the various devices, as noted before.

- The existence of various "active" devices such as lasers and amplifiers, where a large thickness would mean large current densities (proportional to the thickness of the active region) to pump the device up to optical transparency; this necessitates the use of waveguides with thicknesses on the order of 0.1–0.2 μm.
- Carrier confinement in lasers and amplifiers requires potential barriers on the order of 0.2 eV; this implies large index steps, on the order of 0.2, and hence narrow structures in order to retain single-mode operation.

Typically, coupling efficiencies (overlap integrals) on the order of 0.1 are achieved between "laserlike" waveguides and standard optical fibers. The Fresnel reflections alone are responsible for 1.7 dB per interface in insertion loss. In most cases, antireflection coatings or special reflection suppression techniques have to be employed.

Whereas the coupling efficiency in the lateral dimension can be improved by tapering of the fiber and/or the semiconductor waveguide, the matching in depth is considerably more difficult, since "tapering" of the semiconductor wageguide in this dimension cannot be achieved by straightforward photolithography. The use of quantum wells has been demonstrated to spread out the optical field in the depth direction such that improved field matching is obtained [111], giving a coupling loss of 4.2 dB. As a comparison, field overlap contribution to the coupling efficiency between $LiNbO_3$ and optical fibers can routinely be made 0.5 dB, since the Ti indiffusion process by itself gives field shapes that are congruent with those of a standard single-mode fiber.

III. PHOTONIC SWITCHING IN THE CONTEXT OF GUIDED WAVES

A. Introductory Remarks

The area of photonic switching emerged as a natural outgrowth of the rapid development in fiber optics: Due to the anticipated total dominance of the optical fiber in the wirebound network, it was natural to investigate the potential of a deeper penetration of optics, in addition to the sheer point–point transmission, where the technology has proven successful in a short time span. Here, it is necessary to consider the unique properties of optics:

- The virtually unlimited bandwidth available: The carrier frequency is approximately 200 THz, and the available wavelength window, say 200 nm, corresponds to half a billion telephone channels or more than 300,000 high definition TV (HDTV) channels.
- The weak interaction between the information carriers, the photons, in transparent media. This is actually a condition for the success of the

LiNbO₃ AND SEMICONDUCTOR GUIDED WAVE OPTICS

optical fiber as a transmission medium, and it leads to some interesting consequences:
- Optical circuits can be crossed in a plane (no "short circuits," which has important consequences for the topologies of PICs and for optical interconnects, as outlined in Section IV).
- The difficulty of "controlling light by light."
- Strong interaction between light and material at the resonant bandgap wavelength, as outlined in Section II.C.

A comprehensive treatment of fully optical switching (i.e., where both the controlling and the controlled signals are optical) is outside the scope of this chapter; suffice it to say that at the present time, purely optical devices for logical or computational operations or RAM type storage are either far too power consuming, too slow in relation to electronics, or both [112].

Obviously, a successful development of photonic switching has to take the above conditions into consideration, and the aim of the research in the area has been to investigate the possibilities of increased capacity, flexibility, and functionality in the total network, bearing the future integrated broadband network in mind. Hence, we can expect photonic or optical switching to be introduced, starting from today's network (Fig. 27, top) in the way depicted in the middle figure of Figure 27, where the switching fabric is optical but still controlled by electronics; this can be labeled

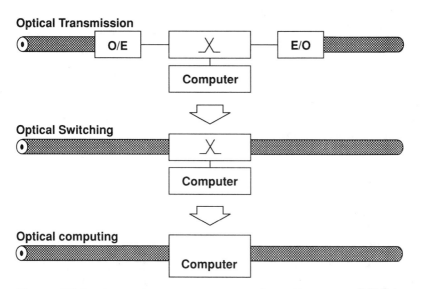

Figure 27 Introduction of photonic switching: From optical transmission and electronic switching and control (top) via electronic control of an optical switching fabric (middle) to the fully optical switching system (bottom).

optical interconnect under electronic control. This is the near-term vision of photonic switching, and the optical switching fabric can, e.g., be optical cross connects (OXCs); see below. Switching fabrics in the shape of asynchronous transfer mode (ATM) switches, communicative switching synchronous transfer mode (CS-STM) switches, etc., however, are expected to lie further in the future, for the same reason that the last stage, fully optical switching (Fig. 27, bottom), involving optical RAM type memories and optical logic and computing, is a long-range research item.

At this stage, it is illuminating to see what sort of capacity increases we are talking about in the future broadband network. In a current digital exchange handling 100,000 subscribers, we have a total aggregate capacity of, say, 10 Gb/s. If each subscriber, instead of having access to a narrowband 64 kb/s channel, was upgraded to broadband capacity of, say, 100 Mb/s, the aggregate capacity of the switch would be around 10 Tb/s, with an ensuing capacity increase in the transport network. Obviously, this is a quantum leap, and fiber optics can cope with this from a *transmission* point of view. The question asked in photonic switching is whether optics also has a wider role, and this can be discussed with reference to Figure 28. It is in the *transport and access layer* (in the shorter term) as well as in the *switching layer* (at a later stage) that the impact of photonic switching should assessed. Interestingly enough, the majority of the current photonic switching systems demonstrators pertain to the transport and access layer. Here, the compatibility with the transmission medium can be fully utilized to create a *transparent* network, i.e., a network virtually insensitive to the data rates and coding formats involved, as discussed in Section II.B, in marked contrast to electronic networks. This, of course, implies a considerable enhancement of the flexibility, capacity, and reliability of the total network. If some sort of temporal switching is employed, however, such as in the space stage in a time–space–time switch (TST switch) or in an ATM switch, the time domain is "broken" up, and the switching operations in general involve storage and logic operations for synchronization purposes, compromising the advantages of photonics. It is natural that any network based on time division will not exhibit the full transparency features, but space-division, as well as wavelength-division switching, retains the full transparency of the network. This does not necessarily say that time switching is not attractive in some applications; e.g., in the shape of time-multiplexed space switches [113–115], especially when combined with wavelength-division multiplexing (WDM); in either case, the packet speed can be much higher than the routing speed, giving "semitransparency"; however, electronic synchronization is still required, and guard bands must be inserted to allow time for optical switch reconfiguration. It should be noted that the distinction between switching and transmission implied in

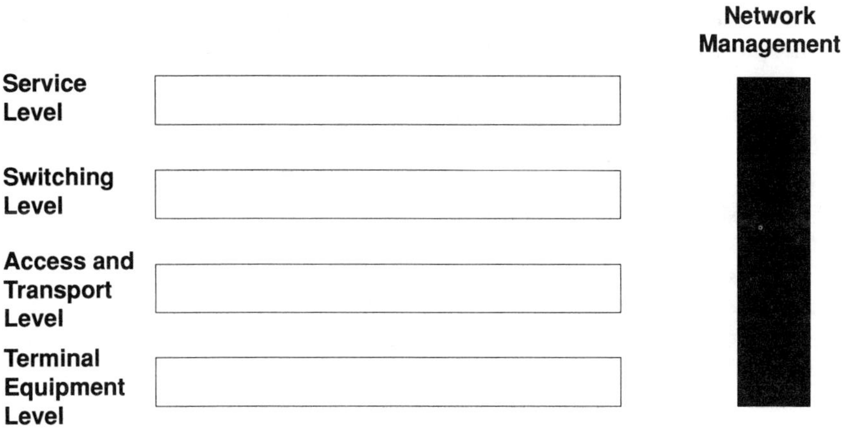

LiNbO₃ AND SEMICONDUCTOR GUIDED WAVE OPTICS

Figure 28 Model of the telecom network. Photonic switching is expected to have an impact on the two middle layers.

Figure 28 is becoming somewhat blurred: with the advent of digital cross connects (DXCs) and add/drop multiplexers, the transmission layer assumes some of the roles of the switching layer.

The first conference featuring a comprehensive approach to photonic switching as a research field was the OSA meeting in Incline Village, NV, in 1987. This meeting provided a forum for both systems and device research; an advantageous arrangement, which has been adhered to in subsequent conferences. At about the same time, programs devoted to photonic switching were started in Europe within the framework of COST (European Cooperation in the Field of Science and Technology Research) and RACE (Research for Advanced Communications in Europe, aimed at the introduction of broadband communications in Europe by 1995), as well as in the United States and in Japan. All these programs were to a degree "device" or technology driven, in the sense that no clear viable short-term applications for the technology available at the time could be identified. The original expectations of a rapid development in photonic switching, analogous to that of fiber-optic transmission, with its steady evolution from research to commercial systems, have not been met so far, and research in photonic switching has partly taken other directions, as discussed below.

Initially, the most obvious application of photonic switching was space-division circuit switching (in principle analogous to the old "crossbar" systems), where the device base was the strongest, in the shape of primarily LiNbO₃ switch arrays, as discussed in Section II.B. A fair amount of systems research was also aimed at replacing or complementing electronic switching by photonic switching in time-division switching systems (TDS), either by time switching of switch arrays [115,116], or, as in the latter part of the 1980s, by employing various types of bistable devices, which were considered prime candidates for applications as RAM type optical memories, e.g., for time slot interchange (TSI) type applications [117,118]. These devices have difficulty in competing with electronic devices, however, regarding speed, switching energy, and power consumption, and a systems study comparing photonic switching and electronic switching in a conventional time–space–time (TST) switch [119] showed the inadequacy from a performance and economical point of view of current and near-future photonic memory elements. The interest in bistable optical devices has lately been focused on device physics rather than on photonic switching applications. In addition, recent switching systems research has produced recommendations and standards for a new switching format, ATM (asynchronous transfer mode) aimed at the future broadband network: ATM is a combination of STM (synchronous transfer mode) and PTM (packet transfer mode) combining the advantages of the periodically recurring "packet" of STM with the flexibility of PTM [120]. This development points to an emerging mismatch between the switching level systems needs and the available (or foreseeable) photonic switching devices. Hence, much of recent research in photonic switching has been focused on transport network applications (where one sees a merging of transmission and switching as noted above, e.g., in the shape of optical cross connects, OXCs) and optical interconnects. Here, *space-division* and *wavelength-division multiplexing and switching* (*WDM* and *WDS*) play a key role in augmenting the capacity and flexibility of the total network. It should be noted that we are dealing with an *analog* network (transmitting digital information) in contrast to today's fully *digital* networks, which rely entirely on signal restoration. As noted above, this is a difficult task in the optical domain.

It is not clear whether the future network will be of pure ATM type; as an example, distributive video employing ATM might not be optimal from a systems point of view. Also, taking the continued rapid development of electronics into account, photonic switching should be regarded as a *complement to electronics* rather than a competitor. The treatment in this chapter pertains to applications where it appears that photonic switching will have an impact over the next 5–10 year period. That does not mean that more long-term approaches are uninteresting; as an example, the chapter by Dr. M. Islam in this volume (Chapter 4) gives an interesting example on how to utilize the extreme capacity of the optical medium in the temporal domain, by switching with ultrashort pulses. This is an approach that is analogous to todays heavily TDS-oriented systems.

B. Photonic Switching Systems

Several laboratories have constructed photonic switching systems for laboratory evaluation, and below a few representative examples are highlighted, again concentrating on the space switching type applications. Within the European RACE OSCAR project (R1033), several photonic switching *demonstrators* have been deviced. The word demonstrator is intended to mean that the ambition is higher than a laboratory "optical table" experiment in that the basic functions of the photonic switching applications in question should be shown employing a reasonably developed technology, making credible the *scalability* and *upgradability* to future full-scale systems. Hence, packaged, fiber compatible devices as well as practical optical and electronic interfaces are required.

We start by describing the first photonic switching demonstrator within the above mentioned RACE project: A broadband *access cross connect switch* [121]. Such a switch has the function to connect in a programmable and flexible manner high-speed multiplexed bit streams from groups of subscribers to any of the following: *service centers* such as

- A distributive services (DS) STM switch, for example, used for TV distribution
- A communicative services (CS) STM switch for different kinds of communicative services
- An ATM switch for all types of ATM services

or to

- The trunk network, via a digital cross connect (DXC)

In such a system, each subscriber could be equipped with a video telephone as well as with HDTV; in the demonstrator system (Fig. 29), three such subscriber sites are implemented, which are connected to network terminals (NTs) and subsequently via WDMs to a switching fabric of six 4 × 4 LiNbO$_3$ switches [48]. These switches are connected to the service centers listed above via line terminals (LTs), which are simulated in the actual demonstrator by 1550-nm receivers and 1300-nm transmitters coupled back to back. The use of WDMs permits bidirectional communication for the video telephone, with the wavelength 1550 nm used downstream and 1300 nm upstream. The transmission directions are demultiplexed in WDMs at the subscriber termination in order to make it possible to switch the different wavelengths separately. This is necessary due to the fact that the switch matrices are designed for a specific wavelength window (1300 *or* 1550 nm). The bit rate in the demonstrator is 140 Mb/s.

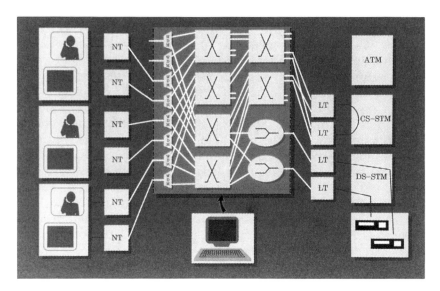

Figure 29 RACE OSCAR access cross connect demonstrator, involving 140 Mb/s-bidirectional video telephony as well as TV distribution through a space switch fabric [121]. See text.

In the demonstrator, the routing is centrally controlled, by specially developed software, run on a SUN workstation. The optical switch matrices are the major source of power loss in the optical signal path, and the power levels and the loss distribution in the optical access switch are shown in Figure 30. The crosstalk introduced from the various connections was also investigated, and it was confirmed that the switching network did not degrade bit-error-rate (BER) performance, even in the presence of crosstalk in a fully loaded system [121].

It is interesting to contrast the available $LiNbO_3$ array sizes with the systems requirements of Table 2 which shows that today's switch sizes in general only allow protection switching and digital cross connect (DXC, or in this context, optical cross connect, OXC) type applications; however, the latter will be very important in broadband networks. Taking a somewhat wider perspective, granting the cascading of $LiNbO_3$ chips and the employment of optical amplifiers, the situation can be described as in Table 3, which shows that $LiNbO_3$ can in principle meet all the requirements. Future technology pertains to improved material, e.g., by doping to control conductivity; there is nothing fundamental in *crosstalk* and *noise buildup* to preclude us from meeting the whole applications spectrum with $LiNbO_3$.

LiNbO₃ AND SEMICONDUCTOR GUIDED WAVE OPTICS

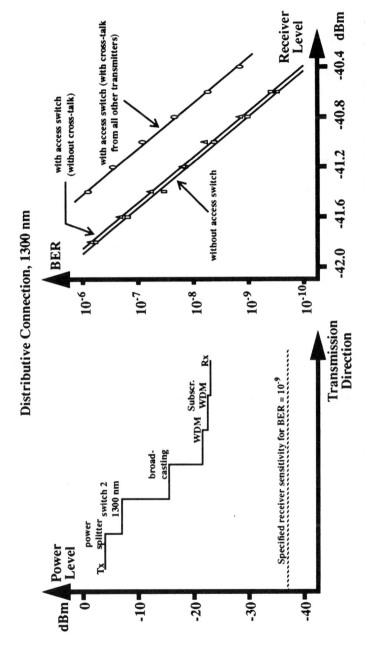

Figure 30 The power loss distribution for a distributive connection in the system of Fig. 29. The optical power level at the receiver is about the same as for a communicative connection. The system penalty originating from the optical access switch is < 0.1 dB, but the crosstalk degrades the system. The total penalty, however, is still < 0.5 dB.

Table 2 Requirements on Switch Sizes in Possible Applications of Optical Space Switching

Switch	Size	Bit rate	Slow/fast
Protective	(2–20) × (1–20)	620 Mb/s–2.4 Gb/s	Slow
MDF/DXCsub[a]	1000 × 500	620 Mb/s	Slow
Access switch	1000 × 500	140 Mb/s	Fast
DXC	(20–50) × (20–50)	620 Mb/s–2.4 Gb/s	Slow

[a]Main distribution frame/DXC at subscriber.
MDF = main distribution frame, DXC = digital cross connect, DXCsub = DXC at subscriber. "Bitrate" refers to routed data, "Fast" pertains to times on the order of call setup time, whereas "slow" signifies seconds up to months.

This can further be illustrated [122] as in Figure 31, which shows a 128-line space-division broadband switching system; the feasibility of such a system was demonstrated by BER measurements on a cascade of traveling-wave optical amplifiers and 8 × 8 polarization-insensitive $LiNbO_3$ switch matrices [47]. Recently, the same group reported results on four cascaded SCLAs in a network with such $LiNbO_3$ matrices [123].

Table 3 Applicability of $LiNbO_3$ Technology with Respect to Systems Requirements in Table 2

Device requirements derived from systems applications	LN Technology	
	Existing	Future
Size: One chip	1	+4
Cascade[a]	1,4	(+2,3)
Loss	1,4[b]	+2,3[b]
Crosstalk	1,4	(+2,3)
Polarization independence	1–4	
Blocking	1–4	
Bandwidth: Total	1–4[c]	
Instantaneous	1–4[d]	
Control complexity	1–4[c]	
Stability	?	1–4

[a]Requires optical interconnects.
[b]Depends on data rate; OAs will be required in most cases (doped LN?).
[c]With digital switches.
[d]Depends on data rate.
Key: 1—Protective; 2—DXCsub; 3—Access switch; 4—DXC.
Also compare Fig. 10 in Section II.B.

LiNbO₃ AND SEMICONDUCTOR GUIDED WAVE OPTICS

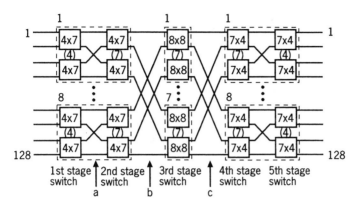

Figure 31 One hundred twenty-eight line photonic switching space-division switching network indicating the positions of optical amplifiers [122,123].

In order to implement reasonably practical full-scale systems of the type described in Table 2, however, developments in a number of technology areas are necessary. The size of monoliths has to be increased significantly. Here, larger monolithic matrices can be made in InP, integrating optical amplifiers and switches or using Er fibers as amplifiers. The higher integratability, stability, and flexibility of the semiconductor arrays make them the long-term choice in this area, as long as crosstalk problems in multiwavelength systems can be mastered. The sizes can further be increased by hybrid technique, in the shape of *optical interconnects*, e.g., involving optics on Si, to cascade switch chips.

Another important upgrade of space switching systems concerns using multiple wavelengths to increase the capacity *and* flexibility of the switch. The way the routing information is transferred to the switch is also important. In the case of using multiple wavelengths, the routing information could be provided to the switch on a specific wavelength, especially assigned for signaling, maintaining the transparency of the system.

Photonic switching utilizing a combination of space and wavelength switching can be discussed with reference to Figure 32 where Figure 32a shows a general network architecture with its proliferation of broadband cross connects, showing the role of *broadband switching*. Also shown (Fig. 32b) is a possibility to implement the cross connect function with a combination of an *electronic digital cross connect*, DXC, used for fine granularity switching in the temporal domain at lower bit rates and for signal restoration, as well as an *optical cross connect*, OXC, used for wideband switching of broadband signals. The total capacity of the switch is increased by the use of wavelength-division switching (WDS) techniques, as has been

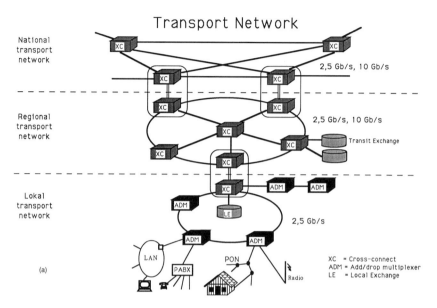

Figure 32 (a) Telecom network diagram, showing the role of broadband cross connects; (b) Possible implementation of the cross connects shown in (a) in the shape of combined optical (OXCs) and digital (DXCs) cross connects; the optical cross connect is for rerouting of broadband signals, the electronic one for switching at a higher level in the hierarchy as well as for signal regeneration, when necessary. The introduction of multiple wavelengths increases the capacity as well as the flexibility of the system.

demonstrated [59,124,125]. In addition, WDS techniques provide higher flexibility, since they allow switching with higher granularity in the optical layer. Here, a whole range of WDS devices could be employed: tunable sources, WDM devices, and tunable filters. In general, the use of *frequency converters* would be advantageous, since it would allow the construction of an analogue to the electronic TST switch, and the development of such a device remains an important research goal.

The network of Figure 32 can be considered to represent a near-term application of photonic switching, given that the required technology is available, and is an example of the combination of optics and electronics in their respective areas of strength to create a better systems solution.

Figure 33 shows a demonstrator system for optical cross connect (Ericsson exhibit at Telecom '91, Geneva, Switzerland, 1991), including four nodes, based on 4×4 $LiNbO_3$ switch matrices [48]. This system implements protection routing as well as protection switching, and general

OXC/DXC node

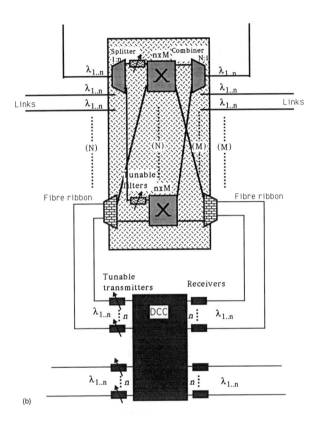

(b)

network reconfiguration features, under software control from a workstation. This demonstrator, the first of its kind, allowed an assessment of the basic optical cross connect features and the control issues and is a first step toward a transparent optical network.

Utilization of high-speed optical switching in the central space stage of a TST switch is shown in another RACE demonstrator. Here, a 16 × 16 LiNbO$_3$ switch array switches blocks (the bit rate within the block is 640 Mb/s for each of the 16 links) with a guard band of 5 ns in a synchronous fashion, giving an aggregate bit rate of 9.6 Gb/s. This is, as noted above, sufficient for a narrowband exchange. The temporal switching is done electronically, and the system permits the remote location of the T (time) stages in relation to the S (space) stages, although with ensuing problems in synchronization. In principle, the capacity (although not necessarily the

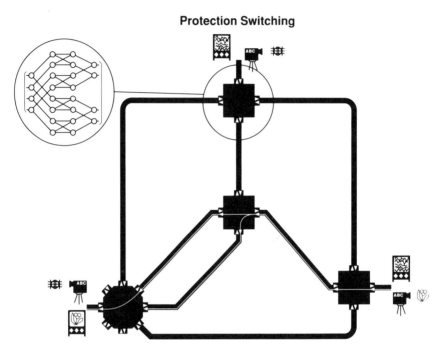

Figure 33 Demonstrator system for optical cross connect (Ericsson exhibit at Telecom '91, Geneva, Switzerland, 1991), including four nodes, based on 4 × 4 LiNbO$_3$ switch matrices [48]. The figure shows the graphic computer control interface, with the white lines indicating a configuration for protection switching ("hot standby"). Each port comprises a bidirectional fiber link. The system in general implements protection routing and protection switching, and general network reconfiguration features, under software control from a workstation.

switching granularity, i.e., the number of individually switchable channels) can be increased by adding WDM. This is an example of a system belonging to the second layer in Figure 28.

In another example in this area [115], time-division multiplexed switching on a word-by-word basis is demonstrated for a 4 × 4 LiNbO$_3$ switch matrix. The matrix is reconfigured in 250 ps (1-bit guard band) to switch 16-bit words at a 4-Gb/s line rate. Obviously, this requires good synchronization.

A further example is offered by photonic time-multiplexed permutation switching using the dilated slipped Banyan network [116], where a time-multiplexed LiNbO$_3$ switch is proposed as the time multiplexed hub of an active star local area network (LAN).

LiNbO$_3$ AND SEMICONDUCTOR GUIDED WAVE OPTICS

Photonic switching in a high-speed optical loop [126] is shown in Figure 34. Here, high-speed optical space switches in the passive node (i.e., the node does not comprise lasers) are used for tapping of information (in the listening state), switching the high-speed packets to the node when the node is addressed, as well as modulation of an incident CW signal from the hub when the node is transmitting. In addition, the optical switches perform synchronization in the optical domain.

IV. OPTICAL INTERCONNECT

A. Introductory Remarks

The area of optical interconnects can be subdivided in the following way:

- Fiber interconnects, to connect optical and/or electronic devices
- Free space (including microoptic) interconnect, to connect optical devices and/or electronics
- Planar guided-wave interconnect, to connect optical devices (such as the PICs described in Sections II.B and II.C in order to create large hybrid networks) and/or electronic devices

Fiber interconnects are the obvious short-term application for optical interconnects and are already being used to interconnect electronic equipment in telephone exchanges. They are further employed to expand the size of optical fabrics in a hybrid fashion, as in the photonic switching demonstrator described in Section III [121], Figure 29.

Free space (including microoptic) *interconnect* relates to the proliferating field of three-dimensional interconnects [127], where the addition of a third dimension offers the potential to outperform planar multilayer electronic interconnect. Representative of this area is the work on SEED-based switching fabrics [128]. (SEED: Self-electro-optic devices.)

Finally, the use of *planar guided wave optics* can be viewed as "integrating" the fiber optical interconnect in order to make possible the construction of large optical networks in a practical way. In fact, such a development of an "optical building practice," analogous to the long-established electrical one, has been identified as a key element in the implementation of photonic switching.

The interest in optical interconnects, as an alternative to the prevalent electronic means for interconnection, is due to a number of factors:

- The superior transmission capacity of optical waveguides, i.e., the same reasons that have led to the introduction of the optical fiber for long-distance transmission

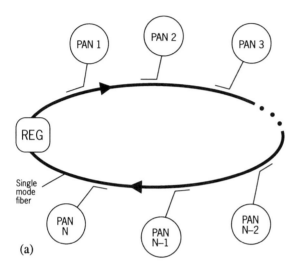

Figure 34 (a) High-speed optical loop with passive access nodes (PANs). The structure of the PAN is shown in (b): The AS (access switch) is used as a tap in the listening state, and the CS (correlation switch) is used together with the integrating receiver to perform optical correlation to assess when the node is addressed; in this case, the broadband receiver receives the high-speed data stream. The AS switch is used for high-speed modulation of a CW signal from the central node in the system. Here, both switches have to be high-speed (> 1 GHz) switches.

- The capability to all but eliminate crosstalk, EMI, etc., enabling high packing density even for extreme data rates, a notorious problem in electronics
- The possibility of crossing planar waveguides physically, which gives a significant topological leeway as compared with electronics (there are no optical "short circuits" when "connecting" two waveguides, if properly carried out); see Figure 11

On the other hand, there are no greater differences as far as the propagation speed is concerned; in the optical as well as in the electrical case, we are dealing with electromagnetic waves.

This section deals with the last of the three areas listed above, i.e., optical interconnects in the guided wave context and has in a way already been elucidated in Sections II.B and II.C. As an example, in the $LiNbO_3$

LiNbO₃ AND SEMICONDUCTOR GUIDED WAVE OPTICS

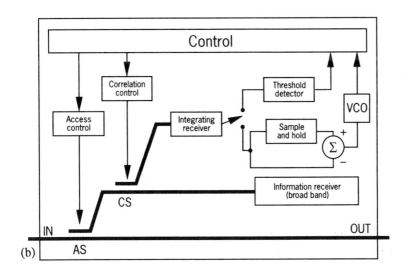

(b)

switch matrices described in Section II.B on-chip waveguides serve to interconnect the elemental switches (Fig. 11) so as to make possible a larger "monolithic" device, rather than joining the switches together in a hybrid fashion. Consequently, one could envisage LiNbO₃, LiTaO₃, InP, GaAs, etc., as being used for planar guided-wave interconnects. In general, however, we view interconnects employing these materials as endeavors to create larger monoliths, and the optical interconnects we are dealing with here as a means of *hybridizing* to achieve larger optical systems. The prevalent material systems for such passive waveguiding interconnect are *glass*, different types of *polymeric* materials and *silica*.

Below, we will concentrate on the silica based waveguide optical interconnect, since this material system is representative for the problems encountered, and it offers a number of advantages: In addition to providing a

- *Flexible waveguiding technology*

we have the possibility to utilize various techniques for

- *Physical positioning of lasers, fibers, etc.*

Furthermore, the material has

- *Excellent thermal and mechanical properties*

and offers

- *Mature processing in a well-established material system*

An addition asset could be integration of

- *Control electronics on the interconnect medium*

A number of devices have already been reported, as described below, and the possibility to introduce Er doping [129] opens up the possibility for a limited range of active devices, such as lasers.

B. Optics on Silicon

1. Waveguiding

Compatibility with the ordinary single-mode fiber and the need to align lasers, detectors, and fibers to the waveguide have made doped silica waveguides on a silicon substrate a most suitable material for the hybrid technology [130–133]. The adaption to single-mode fiber is obtained by using doped (e.g., GeO_2, TiO_2) silica for core material and pure silica for cladding layers. Dimensions and the refractive index step are moderate (typically 8×8 μm and 0.25 percent), which facilitates processing, minimizes birefringence, and accordingly decreases the polarization dependence. Excellent mode-matching to the fiber is achieved, resulting in connection losses about 0.05 dB [133]. Waveguide attenuation less than 0.04 dB/cm at 1.55 μm has been reported [134]. The mode field of a normal laser diode, however, does not correspond well to the mode fields of these waveguides. Hence, the coupling loss between a laser diode and waveguide is comparatively high (around 10 dB). This loss can only be partly compensated by manipulating (tapering) the waveguide geometry toward the laser diode.

Alternatively, the laser-to-waveguide coupling efficency can be maximized by choosing a core material with considerably higher refractive index, usually silicon nitride, exhibiting an index around 2.0 (the cladding layer is still silica), which reduces the size of the fundamental mode [130,135]. The problem is still not solved because the mode-mismatch problem is just transferred, and will appear at the waveguide-to-fiber coupling instead. In addition, the high index step gives some difficulties because of smaller dimensions, which affects produceability and increases birefringence, polarization dependence, and attenuation. Experiments have been performed, however, aiming at effective coupling between silicon nitride and doped silica waveguides by use of an adiabatic taper.

2. Mechanical Properties

The active positioning of active components to waveguides is a time consuming and expensive bottleneck for the manufacturer. If this problem is solved, the manufacturing cost will be significantly reduced. By adopting silicon as a substrate, its well-known etching and mechanical characteristics can be used to form V-grooves, trenches, holes, and guiding blocks, etc., as an aid to position lasers, detectors, fibers, etc., on a common, stable platform [130,131,136–139]. Additional reasons for the choice of silicon for substrate are its excellent thermal and electrical characteristics. Finally, a silicon substrate offers the potential advantage of housing electronics, e.g., laser diode driver circuits. This step toward higher integration between optics and electronics should allow for more compact packaging.

3. Fabrication

The methods that are used to manufacture waveguides on silicon show an obvious relationship to the processing of microelectronics. Naturally, there are differences; one is the large etching depth, which can be up to 100 μm in the multimode case; another is the demand for high definition of edges to keep guiding losses low, as well as the extreme (submicron) requirements on geometric tolerances for positioning laser diodes to the waveguide. The manufacturing steps are otherwise quite similar: deposition/growth, photolithography, etching, and in some cases, redeposition. The most commonly used deposition methods are PECVD (plasma-enhanced chemical vapor deposition) [140] and FHD (flame hydrolysis deposition) [132], Figure 35. The latter is able to produce very thick layers (200 μm).

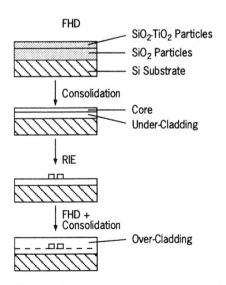

Figure 35 Fabrication process of silica-based waveguides by FHD and RIE [132].

The etching is usually performed by dry etching methods, e.g., RIE (reactive ion etching), to accomplish sharp edges and smooth sidewalls.

4. Devices

During the last few years a number of devices based on optics on silicon have been presented; a few of them are mentioned below; the survey is far from being complete.

1. Couplers and related devices for splitting and combining [133,141–143]. Building blocks are directional couplers as well as Y-branches, cascaded up to 64 output ports. For a 1 × 8-splitter the excess loss is typically 1.5 dB and the uniformity 0.5 dB.

2. Wavelength-division (de)multiplexers based on discrete filters, directional couplers, or a combination of the two [133,144]. In the case of directional couplers, insertion loss is less than 0.5 dB and crosstalk better than −20 dB. Discrete filters introduce an additional insertion loss, but improve crosstalk characteristics.

3. Bidirectional transmission modules (Fig. 36) incorporating splitters, WDM, detectors and light sources, e.g., for broadband subscriber access [131,145]. The combination of several elements into more complex devices points to the flexibility of this technology.

4. Interferometric devices [133,146]. By using 3-dB directional couplers as building blocks, Mach–Zehnder interferometers are obtained (Fig.

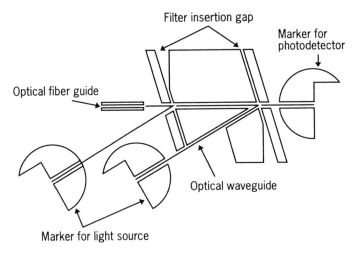

Figure 36 Schematic of multimode bidirectional transmission unit [145].

LiNbO$_3$ AND SEMICONDUCTOR GUIDED WAVE OPTICS

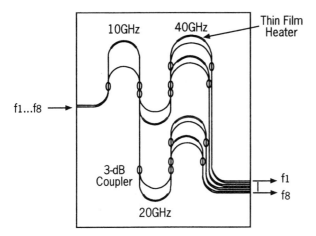

Figure 37 Configuration of a FDM with 10-GHz spacing between eight channels [133].

37). These are designed to various functions: WDMs, FDMs, and by applying a thermooptic phase shifter, wavelength/frequency tuning is realized. An optical switch can also be made, albeit with modest rise and fall time (0.8 ms).

5. Bragg reflection filters to accomplish optical feedback to laser diodes to obtain single-mode operation.

6. Adiabatic devices, e.g., tapers aiming at low-loss coupling between waveguides of different modal profiles, such as between silicon nitride and doped silica waveguides. A total loss of 3.1 dB, from laser diode to single-mode fiber, has been demonstrated [147].

Further, an optical microbench has been reported [130] (Fig. 38) giving the possibility for an optical building practice.

The hybrid technology described above is an interesting candidate for employment in future telecom applications, e.g., in interconnection links within local exchanges and for customer access modules in the distribution network, as well as in datacom applications, e.g., in links between and within computers.

V. SUMMARY

Initial applications of photonic switching are likely to be in the transport network for broadband routing involving devices for wavelength- and space-division switching. In the latter area, LiNbO$_3$ is currently the most

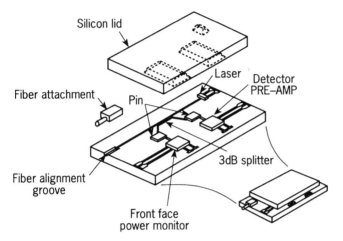

Figure 38 Exploded view of a transceiver made by hybrid assembly on silicon [130].

mature technology, but semiconductors offer the greatest potential and application spectrum, mainly due to the (current) unavailability of practical means for light amplification and generation in $LiNbO_3$. $LiNbO_3$ guided-wave devices, e.g., in the shape of space switches, are available for field trials and systems experiments today, and a number of such experiments have been reported. These are very important to assess the potential of photonic switching. The time scale of the actual deployment of PICs in networks, however, will mainly be governed by progress in PIC semiconductor manufacturing technology. In the area of photonic switching as well as in general high-speed, multigigabit per second systems, *optical interconnects* will play a key role, and in order to make optics practically useful in these applications, an optical "building practice," analogous to that long established in electronics, and admitting the scaling of optical systems to required sizes, is required. "Optics on silicon" was identified as a potential technology here.

ACKNOWLEDGMENTS

The author would like to acknowledge the assistance from the following colleagues within Ericsson, whose contributions were a material part in writing this chapter: P. Granestrand, M. Gustavsson, O. Sahlén, O. Steijer, M. Janson, B. Lagerström, P. Svensson, A. Djupsjöbacka, and B. Stoltz. Further thanks are due to L. Atternäs for computer assistance.

REFERENCES

1. T. Tamir, *Guided Wave Optoelectronics*, 2nd ed. (T. Tamir, ed.), Springer-Verlag, Berlin, 1985.
2. D. Hondros and P. Debye, *Ann. Phys.*, *32*:465 (1910).
3. A. Yariv and R. Leite, *Appl. Phys. Lett.*, *2*:55 (1963).
4. D. Nelson and F. Reinhart, *Appl. Phys. Lett.*, *5*:148 (1964).
5. S. E. Miller, *Bell System Techn. J.*, *48*:2059 (1969).
6. A. Djupsjöbacka, *J. Sel. Areas Comm.*, *6*:1227 (1988).
7. R. W. Keyes, *The Physics of VLSI Systems*, Addison-Wesley, Workingham, 1987.
8. M. A. Reed and W. P. Kirk, Nanostructure physics and fabrication, Proceedings of the International Symposium, College Station, Tx, 1989.
9. L. Thylén and O. Sahlén, *J. Quantum Electron.*, *28*: (1991).
10. R. V. Schmidt and I. Kaminow, *Appl. Phys. Lett*, *25*:458 (1974).
11. J. L. Jackel and C. E. Rice, *Appl. Phys. Lett.*, *41*:607 (1982).
12. G. L. Destefanis, J. P. Gailliard, E. L. Ligeon, S. Valette, B. W. Farmery, P. D. Townsend, and A. Perez, *J. Appl. Phys.*, *50*:7898 (1978).
13. L. Thylén, B. Lagerström, P. Svensson, A. Djupsjöbacka, and G. Arvidsson, Computer analysis and design of Ti: $LiNbO_3$ integrated optics devices and comparison with experiments, Proceedings of 9th ECOC, Geneva, Switzerland, 1983, p. 425.
14. M. D. Feit, J. A. Fleck, and L. McCaughan, *J. Opt. Soc. Amer.*, *73*:1296 (1983).
15. L. Thylén, *J. Lightwave Technol.*, *LT-6*:847 (1988).
16. I. P. Kaminow, *An Introduction to Electrooptic Devices*, Academic Press, New York, 1974.
17. P. Granestrand, B. Stoltz, L. Thylén, K. Bergvall, W. Döldissen, H. Heidrich, and D. Hoffmann, *Electron. Lett.*, *22*:816 (1986).
18. A. Neyer and W. Sohler, High-speed cut off modulator with Ti-diffused $LiNbO_3$ channel waveguide, Proceedings of 5th ECOC, Amsterdam, The Netherlands, 1979, paper 15.4.
19. R. V. Schmidt and H. Kogelnik, *Appl. Phys. Lett.*, *28*:503 (1975).
20. C. S. Tsai, B. Kim, F. R. El-Akkari, *IEEE J. Quantum Electron.*, *QE-14*:513 (1978).
21. A. Neyer, *Electron. Lett.*, *19*:553 (1983).
22. M. Papuchon and A. Roy, *Appl. Phys. Lett.*, *31*:266 (1977).
23. A. Neyer, W. Mevenkamp, L. Thylén and B. Lagerström, *IEEE J. Lightwave Technol.*, *LT-3*:635 (1985).
24. D. Marcuse, *Light Transmission Optics*, Van Nostrand, New York, 1972, Chapter 10.
25. J.-P. Weber, L. Thylén, and S. Wang, *IEEE J. Quantum Electron.*, *QE-24*:537 (1988).
26. W. K. Burns, A. B. Lee, and A. F. Milton, *Appl. Phys. Lett.*, *29*:790 (1976).
27. Y. Silberberg, P. Perlmutter, and J. E. Baran, *Appl. Phys. Lett.*, *51*:1230 (1987).

28. L. Thylén, P. Svensson, B. Lagerström, B. Stoltz, P. Granestrand, and W. K. Burns, Theoretical and experimental investigation of 1×2 digital switches, Proceedings of ECOC '89, Gothenburg, Sweden, 1989, p. 240.
29. W. K. Burns, *IEEE J. Quantum Electron.*, *QE-16*:446 (1980).
30. V. Ramaswamy, R. C. Alferness, and M. Divino, *Electron. Lett.*, *18*:30 (1982).
31. M. Fukuma, J. Noda, and H. Iwasaki, *J. Appl. Phys.*, *49*:3693 (1978).
32. M. Minakata, S. Saito, M. Shibata, and S. Miyazawa, *J. Appl. Phys.*, *49*:4677 (1978).
33. G. Arvidsson, K. Bergvall, and A. Sjöberg, *Thin Solid Films*, *126*:177 (1985).
34. S. Fouchet, A. Carenco, C. Daguet, R. Guglielmi, and L. Rivere, *J. Lightwave Technol.*, *LT-5*:700 (1987).
35. R. C. Alferness, *Appl. Phys. Lett.*, *35*:748 (1979).
36. L. McCaughan, *J. Lightwave Technol.*, *LT-2*:51 (1984).
37. H. F. Taylor, *J. Lightwave Technol.*, *LT-3*, 1277 (1985).
38. M. Kondo, Y. Tanisawa, Y. Ohta, T. Aoyama, and R. Ishikawa, *Electron. Lett.*, *23*:1167 (1987).
39. P. Granestrand, L. Thylén, and B. Stoltz, *Electron. Lett.*, *24*:1124 (1988).
40. W. K. Burns, Voltage–length product for modal evolution type digital switches, *IEEE J. Lightwave Technol.*, *8*:990 (1990).
41. K. Habara and K. Kikuchi, *Electron. Lett.*, *21*:631 (1985).
42. G. A. Bogert, 4×4 Ti: $LiNbO_3$ switch array with full broadcast capability, Proceedings of OSA Topical Meeting Photonic Switching, Incline Village, NV, 1987, paper ThD3.
43. R. A. Spanke, *IEEE Comm. Magazine*, *25*:42 (1987).
44. P. Granestrand, L. Thylén, and G. Wicklund, Analysis of switching employing a 4×4 switch matrix: Crosstalk requirements and system proposal, Proceedings of ECOC '89, Gothenburg, Sweden, 1989, paper WeA15-4.
45. P. Duthie and M. J. Wale, *Electron. Lett.*, *27*:1265 (1991).
46. T. O. Murphy, C. T. Kemmerer, and D. T. Moser, A 16×16 Ti: $LiNbO_3$ dilated benes photonic switch module, Proceedings of OSA Topical Meeting on Photonic Switching, Salt Lake City, UT, 1991, pp. 7–9.
47. H. Nishimoto, M. Iwasaki, S. Suzuki, and M. Kondo, *IEEE Photonics Technol. Lett.*, *2*:634 (1990).
48. P. Granestrand, B. Lagerström, P. Svensson, L. Thylén, B. Stoltz, K. Bergvall, J.-E. Falk, and H. Olofsson, *Electron. Lett.*, *26*:4 (1990).
48a. P. Granestrand et al., Tree structured 8×8 $LiNbO_3$ switch matrix with digital optical switches, Proceedings of the European Conference on Integrated Optics, Neuchâtel, Switzerland, 1993, pp. 10–12.
49. S. K. Korotky, E. A. J. Marcatili, J. J. Veselka, and R. H. Bosworth, *Greatly Reduced Losses for Small-Radius Bends in Ti*: $LiNbO_3$ *Waveguides* (H. P. Nolting and R. Ulrich, eds.), Springer-Verlag, Berlin, 1985, p. 207.
50. R. Brinkmann, W. Sohler, and H. Suche, Broadband gain in single mode Erbium diffused Ti: $LiNbO_3$ strip waveguides, Proceedings of 17th ECOC, Paris, France, 1991, pp. 157–160.

LiNbO₃ AND SEMICONDUCTOR GUIDED WAVE OPTICS 71

51. M. Fujiwara, S. Suzuki, and H. Nishimoto, Line capacity considerations for a photonic space division switching system with switch matrices and optical amplifiers, Proceedings of 1900 Int. Top. Meet. on Photonic Switching, Kobe, Japan, 1990, paper 13C-3.
52. R. V. Schmidt, P. S. Cross, and A. M. Glass *J. Appl. Phys.*, *51*:90 (1980).
53. G. Harvey, G. Astfalk, A. Feldblum, and B. Kassahun, *IEEE J. Quantum Electron.*, *QE-22*:939 (1986).
54. C. H. Bulmer, W. K. Burns, and S. C. Hiser, *Appl. Phys. Lett.*, 48:1036 (1986).
55. P. Skeath, C. H. Bulmer, S. H. Hiser, and W. K. Burns, *Appl. Phys. Lett.*, *49*:1221 (1986).
56. I. Sawaki, H. Nakajima, N. Seino, and K. Asama, Thermally stabilized z-cut Ti: LiNbO₃ waveguide switch, Proceedings of CLEO '86, 1986, paper MF2.
57. S. Yamada and M. Minakata, *Japan J. Appl. Phys.*, *20*:733 (1981).
58. R. A. Becker, *Opt. Lett.*, *10*:417 (1985).
59. H. J. Westlake, P. J. Chidgey, G. R. Hill, P. Granestrand, L. Thylén, G. Grasso, and F. Meli, Reconfigurable wavelength routed optical networks: A field demonstration, Proceedings of 17th ECOC, Paris, France, 1991, pp. 753–756.
60. J. Singh, I. Henning, M. Harlow, and S. Cole, *Electron. Lett*, 25:899 (1989).
61. P. Albrecht, W. Döldissen, U. Niggebrügge, H. P. Nolting, and H. Schmid, Waveguide mirror components in InGaAsP/InP, Proceedings of 14th ECOC, Brighton, UK, 1988, p. 235.
62. J. S. Osinski et al., *Electron. Lett.*, *23*:1156 (1987).
63. T. Kirihara, S. Kashimara, H. Inoue, H. Sano, Y. Sasaki, and M. K. Ishida: Insertion loss reduction of InP based carrier-injection type optical S3 switch (K. Tada and H. S. Hinton, eds.), Springer-Verlag, Berlin, 1990, p. 54.
64. P. J. Duthie, N. Shaw, M. J. Wale, and A. Moseley, 4 × 4 crossbar switch array using the electro-optic and carrier-depletion effects, Proceedings of Topical Meeting on Photonic Switching, Salt Lake City, UT, 1991, p. 197.
65. K. Komatsu, K. Hamamoto, M. Sugimoto, A. Ajisawa, Y. Kohga, and A. Suzuki, *J. Lightwave Technol.*, *LT-9*:871 (1991).
66. E. Lallier, A. Enard, D. Rondi, G. Glastre, R. Blondeau, M. Papuchon, and N. Vodjani, InGaAsp/InP 4 × 4 optical switch matrix with current injection turned directional couplers, Proceedings of 17th ECOC, Paris, France, 1991, part 3, pp. 44–47.
67. R. G. Hunsperger, *Integrated Optics: Theory and Technology*, Springer-Verlag, Berlin, 1982.
68. D. A. B. Miller, D. S. Chemla, T. C. Damen, A. C. Gossard, W. Wiegmann, T. H. Wood, and C. A. Burrus, *Phys. Rev. B*, *32*:1043 (1985).
69. J. H. Angenent, M. Erman, P. J. A. Thijs,'J. M. Auger, and R. Gamonal, *Electron. Lett.*, *25*:628 (1989).
70. S. Somekh, E. Garmire, A. Yariv, H. L. Garvin, and R. G. Hunsperger, *Appl. Opt.*, *13*:327 (1974).
71. F. J. Leonberger, *Appl. Phys. Lett*, *31*:223 (1977).

72. A. Carenco and L. Menigaux, *Appl. Phys. Lett*, *40*:653 (1982).
73. M. Fujiwara, A. Ajisawa, Y. Sugimoto, and Ohtaa, *Electron. Lett.*, *20*:790, (1984).
74. P. Buchmann, H. Kaufmann, H. Melchior, and G. Guekos, *Appl. Phys. Lett*, *46*:462 (1985).
75. R. A. Steinberg, and T. G. Giallorenzi, *Appl. Opt.*, *15*:2440 (1976).
76. M. Wegener, T. Y. Chang, I. Bar-Joseph, J. M. Kuo, and D. S. Chemla, *Appl. Phys. Lett*, *55*:583 (1989).
77. R. A. Smith, *Semiconductors*, 2nd ed., Cambridge University Press, 1979, Chapter 10.13.
78. O. Sahlén, L. Thylén, A. Karlsson, and U. Olin, *Bandfilling or Stark Effect for Photonic Switching: A Comparison* (K. Tada and H. S. Hinton, eds.), Springer-Verlag, Berlin, 1990 p. 42.
79. K. Ueki, Y. Kamata, and H. Yanagawa, *Polarisation and Wavelength Independent Four-Port Optical Routing Switch with Semiconductor Y-Junction Optical cross-points* (K. Tada and H. S. Hinton, eds.), Springer-Verlag, Berlin, 1990, p. 62.
80. G. Möller, L. Stoll, G. Schulte-Roth, and U. Wolff, *Electron. Lett.*, *26*:115 (1990).
81. T. Kikugawa, K. G. Ravikumar, K. Shimomura, A. Izumi, K. Matsubara, Y. Miyamoto, S. Arai, and Y. Suematsu, *IEEE Photonics Technol. Lett.*, *1*:126 (1989).
82. A. Ajisawa, M. Fujiwara, J. Shimizu, M. Sugimoto, M. Uchida, and Y. Ohta, *Electron. Lett.*, *23*:1121 (1987).
83. J. E. Zucker, K. L. Jones, M. G. Young, B. I. Miller, and U. Koren, *Appl. Phys. Lett.*, *55*:2280 (1989).
84. J. A. Cavailles, M. Erman, P. Jarry, J.-M. Auger, A. Goutelle, and J. H. Angenent, Cascaded carrier depletion optical switches based on InP/InGaAsP waveguides, Proceedings of ECOC '90, Amsterdam, The Netherlands, 1990, pp. 213–216.
85. M. Renaud, M. Erman, P. Jarry, C. Graver, and J. M. Auger, Single mode polarization insensitive GaInAsP/InP total internal reflection optical switch, Proceedings of ECOC '90, Amsterdam, The Netherlands, 1990, pp. 217–220.
86. J. A. Cavailles, M. Renaud, J. F. Vinchant, M. Erman, P. Svensson, and L. Thylén, *Electron. Lett.*, *27*:699 (1991).
87. T. C. Huang, T. Hausken, K. Lee, N. Dagli, L. A. Coldren, and D. R. Myers, *IEEE Photonics Technol. Lett.*, *1*:168 (1989).
88. M. Kohtoku, S. Baba, S. Arai, and Y. Suematsu, *IEEE Photonics Technol. Lett.*, *3*:225 (1991).
89. M. Cada, B. P. Keyworth, J. M. Glinski, C. Rolland, A. J. SpringThorpe, K. O. Hill, and R. A. Soref, *Appl. Phys. Lett.*, *54*:2509 (1989).
90. A. Alping, B. Bentland, and S. T. Eng, *Electron. Lett.*, *20*:794 (1984).
91. M. Gustavsson, A. Karlsson, and L. Thylén, *IEEE J. Lightwave Technol.*, *8*:610 (1990).
92. M. Gustavsson, L. Thylén, and A. Djupsjöbacka, *Electron. Lett.*, *25*:1375 (1989).

93. K. T. Koai, R. Olshansky, and P. M. Hill, *IEEE Photonics Technol. Lett.*, 2:926 (1990).
94. L. Thylén, M. Gustavsson, T. K. Gustafson, I. Kim, and A. Karlsson, *IEEE J. Quantum Electron.*, 27:1251 (1991).
95. R. Rörgren, P. A. Andrekson, K. Bertilsson, and S. T. Eng, Gain switching modulation of a semiconductor laser optical amplifier, Proceedings of IEEE/LEOS Topical Meet on New Semiconductor Laser Devices and Applications, Monterey, CA, 1990, paper SCF3.
96. G. Eisenstein, U. Koren, T. L. Koch, G. Raybon, R. S. Tucker, B. I. Miller, A multiple quantum will optical amplifier/modulator integrated with a tunable DBR laser, Proceedings of IOOC '90, Kobe, Japan, 1989, paper 19C2.
97. A. F. Alrefaie, H. Izadpanah, and A. Alhamdan, 8 Gb/s current modulation of semiconductor optical amplifiers, Proceedings of ECOC '90, Amsterdam, The Netherlands, 1990, p. 625.
98. L. Gillner, Studies of semiconductor lasers and laser amplifiers, Doctoral dissertation, Royal Institute of Technology, Stockholm, Sweden (1991).
99. C. J. Setterlind and L. Thylén, *IEEE J. Quantum Electron.*, QE-22: 595 (1986).
100. D. Mace, M. Adams, J. Singh, M. Fisher, I. Henning, and W. Duncan, *Electron. Lett.*, 25:987 (1989).
101. S. Oku, K. Yoshino, M. Ikeda, M. Okamoto, and T. Kawakami, Design and performance of monolithic LD optical matrix switches, Proceedings of 1990 Int. Top. Meet. on Photonic Switching, Kobe, Japan, 1990, paper 13C-17.
102. W. Idler, M. Schilling, G. Laube, K. Wünstel, and O. Hildebrand, High speed wavelength and spatial switching with a Y-coupled cavity laser (YCCL), Proceedings of Topical Meeting Photonic Switching, Salt Lake City, UT, 1991, pp. 224–227.
103. S. Lindgren, M. G. Öberg, J. Andros, S. Nilsson, B. Broberg, B. Holmberg, and L. Bäckbom, *J. Lightwave Technol.*, LT-8:1591 (1990).
104. I. H. White, J. J. S. Watts, J. E. Carroll, C. J. Armistead, D. J. Moule, and J. A. Champelovier, Demonstration of a compact InGaAsP active crosspoint switch operating at a 1.5 micron using novel reflective Y-coupler components, Proceedings of ECOC '90, Amsterdam, The Netherlands, 1990, pp. 209–212.
105. L. Thylén, *IEEE J. Quantum Electron.*, QE-23:1956 (1987).
106. K. Inoue, T. Mukai, and T. Saitoh, *Appl. Phys. Lett.*, 51:1051 (1987).
107. G. Grosskopf, R. Ludwig, and H. G. Weber, *Electron. Lett.*, 24:1106 (1988).
108. M. Gustavsson and L. Thylén, Switch matrix with semiconductor laser amplifier gate switches: A performance analysis, Proceedings of OSA Topical Meeting Photonic Switching, Salt Lake City, UT, 1989, paper FE5.
109. M. Janson, L. Lundgren, A. C. Mörner, M. Rask, B. Stoltz, M. Gustavsson, and L. Thylén, Monolithically integrated 2 × 2 InGaAsP/InP laser amplifier gate switch arrays, Proceedings of 17th ECOC, Paris, France, 1991, part 3, pp. 28–31.
109a. M. Gustavsson et al., Monolithically integrated 4 × 4 InGaAsP/InP laser amplifier gate switch arrays, *Electron. Lett.*, 28:2223 (1992).

110. M. Gustavsson and L. Thylén, Multifunctional semiconductor laser amplifiers: Optical amplifier, detector, gate switch and transmitter, Proceedings of OSA Topical Meeting Integrated Photonics Research, Hilton Head, SC, 1990, paper WI2.
111. U. Koren, T. L. Koch, G. Eisenstein, M. G. Young, M. Oron, C. R. Giles, and B. I. Miller, Tapered waveguide InGaAs/InGaAsP multiple quantum-well lasers, Proceedings of OSA Topical Meeting Integrated Photonics Research, Hilton Head, SC, 1990, paper WD2.
112. For a discussion of some fundamental constraints of optical switching, see R. L. Fork, *Phys. Rev. A*, 26:2049 (1982) and P. W. Smith, *Bell System Technol.*, pp. 1975–1993 (1982). Further reading: H. M. Gibbs, *Optical Bistability: Controlling Light with Light,* Academic Press, New York, 1985, and H. Haug (ed.), *Optical Nonlinearities and Instabilities in Semiconductors,* Academic Press, San Diego, 1988.
113. N. Whitehead et al., Photonic switching systems experiments in the European RACE program, Proceedings of OSA Topical Meeting Photonic Switching, Salt Lake City, UT, 1989, paper WA2.
114. J. J. Veselka et al., Low-voltage low-crosstalk 8×8 Ti: $LiNbO_3$ switch for a time-multiplexed switching system, Proceedings of Optical Fiber Commun. Conf., Houston, TX, 1989, paper THB2.
115. S. Korotky et al., Experimental synchronized optical network using a high-speed time-multiplexed Ti: $LiNbO_3$ switch, Proceedings of Optical Fiber Commun. Conf., Houston, TX, 1989, paper THL3.
116. R. A. Thompson, Photonic time multiplexed permutation switching using the dilated slipped Banyan network, Proceedings of OSA Topical Meeting Photonic Switching, Salt Lake City, UT, 1991, paper WE6.
117. S. Suzuki, T. Terakado, K. Komatsu, K. Nagashima, A. Suzuki, and M. Kondo, *J. Lightwave Technol.*, *LT-4*:894 (1986).
118. K. Tada and H. S. Hinton (eds.), *Photonic Switching II,* Springer-Verlag, Berlin, 1990, pp. 378–381.
119. P Amrén, Initial investigation of an optical time–space–time (TST) switch, MSc Thesis, University of Uppsala, Dept. of Technology (1987).
120. There are a large number of references in this area; see, e.g., M. de Prycker, *Asynchronous Transfer Mode Solution for Broadband ISDN*, Ellis Horwood, New York, 1990.
121. M. Lindblom, P. Granestrand, and L. Thylén, Optical access switch—first photonic switching demonstrator within the RACE program, Proceedings of OSA Topical Meeting Photonic Switching, Salt Lake City, UT, 1991, paper PD1.
122. C. Burke, M. Fujiwara, M. Yamaguchi, H. Nishimoto, and H. Honmou, Studies on a 128 line photonic space division switching network using $LiNbO_3$ switch matrices and optical amplifiers, Proceedings of OSA Topical Meeting Photonic Switching, Salt Lake City, UT, 1991, paper FA4.
123. M. Fujiwara, S. Suzuki, C. Burke, H. Sakaguchi, H. Honmou, H. Nishimoto, and M. Yamaguchi, Studies on an optical digital cross-connect system

LiNbO₃ AND SEMICONDUCTOR GUIDED WAVE OPTICS 75

 using photonic switching matrices and optical amplifiers, Proceedings of ECOC '91, Paris, France, 1991, p. 97.
124. P. J. Chidgey, I. Hawker, G. R. Hill, and H. J. Westlake, The role for reconfigureable wavelength multiplexed networks and links in future optical networks, Proceedings of OSA Topical Meeting Photonic Switching, Salt Lake City, UT, 1991.
125. G. R. Hill, *IEE Proc.*, 77 (1989)
126. T. M. Martinson, Synchronization of passive access nodes in very high speed optical packet networks, Proceedings of ECOC '90, Amsterdam, The Netherlands, 1990, p. 473.
127. J. E. Midwinter, Communications, VLSI, optoelectronics, and self routing switches, Proceedings XIIIth International Switching Symposium, Stockholm, Sweden, 1990, vol. III, pp. 37–41.
128. For a discussion of SEED devices, see A. L. Lentine, L. M. F. Chirovsky, L. A. d'Asaro, C. W. Tu, D. A. B. Miller, *IEEE Photonics Techol. Lett.*, 1:129 (1989). For a discussion on obtainable parallelism, see F. B. McCormick, A. L. Lentine, R. L. Morrison, S. L. Walker, L. M. F. Chirovsky, and L. A. d'Asaro, *IEEE Photonics Technol. Lett.*, 3:232 (1991).
129. T. Kitagawa, K. Hattori, M. Shimizu, Y. Ohmori, and M. Kobayashi, *Electron. Lett.*, 27:334 (1991).
130. C. H. Henry, G. E. Blonder, and R. F. Kazarinov, *IEEE J. Lightwave Technol.*, 7:1530 (1989).
131. T. Miyashita, S. Sumida, and S. Sakaguchi, *SPIE Integrated Optical Circuit Engineering VI*, 993:288 (1988).
132. N. Takato and A. Sugita, Silica-based single-mode waveguides and their applications to integrated-optics devices, Proceedings of Mat. Res. Soc. Symp, vol. 172, 1990, pp. 253–264.
133. N. Takato, M. Kawachi, M. Nakahara, and T. Miyashita, Silica-based single-mode guided-wave devices", *SPIE Integrated Optics and Optoelectronics*, 1177:92 (1989).
134. T. Kominato, Y. Ohmori, H. Okazaki, and M. Yasu, *Electron. Lett.*, 26:327 (1990).
135. C. H. Henry, R. F. Kazarinov, H. J. Lee, K. J. Orlowsky, and L. E. Katz, *Appl. Opt.*, 26:2621 (1987).
136. Y. Yamada and M. Kobayashi, *J. Lightwave Technol.*, 5:1716 (1987).
137. K.-A. Steinhauser and E. Hörmann, Single-mode laser module in silicon packaging technology, Proceedings of ECOC '90, Amsterdam, The Netherlands, 1990, p. 597.
138. C. A. Armiento et al., Passive coupling of an InGaAsP/InP laser array and single-mode fibers using silicon waferboard, Proceedings of OFC '91, San Diego, CA, 1991, p. 124.
139. Y. Yamada, M. Kawachi, M. Yasu, and M. Kobayashi, *J. Lightwave Technol.*, 4:277 (1986).
140. N. Nourshargh, E. M. Starr, and J. S. McCormack, *IEE Proc. Pt. J*, 133:264 (1986).

141. S. Kobayashi, T. Kitoh, Y. Hida, S. Suzuki, and M. Yamaguchi, *Electron. Lett.*, *26*:707 (1990).
142. N. Noursharg, E. M. Starr, and T. M. Ong, *Electron. Lett.*, *25*:981 (1989).
143. Y. Yamada, T. Miya, M. Kobayashi, and T. Miyashita, *SPIE Fiber Optic Datacom and Computer Networks*, *991*:4 (1988).
144. M. Kawachi, Y. Yamada, M. Yasu, and M. Kobayashi, *Electron. Lett.*, *21*:314 (1985).
145. S. Sumida, Y. Yamada, M. Yasu, and M. Kawachi, *Trans. IECE Jpn.*, *E69*:352 (1986).
146. N. Takato, K. Jinguji, M. Yasu, H. Toba, and M. Kawachi, *J. Lightwave Technol.*, *6*:1003 (1988).
147. Y. Shani, C. H. Henry, R. C. Kistler, K. J. Orlowsky, and D. A. Ackerman, *Appl. Phys. Lett.*, *55*:2389 (1989).

2
Wavelength-Division Multiplexing Technology in Photonic Switching

Masahiko Fujiwara and Shuji Suzuki

NEC Corporation, Kawasaki, Japan

I. INTRODUCTION

Broadband networks providing various kinds of services, such as video telephony and video broadcasting, have been receiving increasing attention in recent years. To realize such networks, both broadband transmission and switching systems are required. Application of photonic technologies is promising in achieving such broadband transmission and switching systems. One of the most attractive features of photonic technologies is the capability of making full use of the wavelength domain. More than 30 THz bandwidth can be utilized by making full use of the low-loss region of single-mode optical fibers, which extends from 1.2 to 1.6 μm. By directly utilizing this wavelength region, wavelength-division-multiplexing (WDM) transmission can increase transmission capacity without any limitation imposed by electronic circuit operation speed. WDM technologies are promising also for achieving large-scale photonic switching systems. There are three possible types of photonic switching networks: space-division (SD), wavelength-division (WD), and time-division (TD) switching networks [1]. The photonic SD switching network is most practical and will be introduced into local area networks at an early stage [2]. In constructing large-capacity photonic switching networks, however, the photonic TD and WD switching networks are attractive because they do not require large numbers of photonic switching devices in order to expand line capacity. Compared with the TD switching network, the WD switching network has two main advantages. Firstly, because there is a bit-rate independency for individual wavelength-division (WD) channels, various speed broadband signals can be exchanged without difficulty. The other advantage is that there is no necessity for high-speed operation in switching control circuits. In the TD switching network, ultra-high speed operations are required, not only for the switching network but also for control circuits, in order to exchange a time-division multiplexed broadband signal. For example, if 32-channel, 156-Mb/s signals are time-division multiplexed, operation speed of the control circuits, such as control memories, will reach as high as 5 Gb/s. On the other hand, high-speed operation for the control system is not required in the WD switching network for a circuit switching application. Moreover, the WD switching system has potential capability for extension to a wide-area network in partnership with the WDM transmission systems. The WDM transmission also has bit-rate independency for individual channels. Therefore, this wide-area network will be able to provide optical bit-rate independent connection between subscribers. From this viewpoint, an optical broadband network architecture using photonic WD switching systems and WDM optical transmission systems was proposed [3,4]. So far, various photonic WD switching systems have already been proposed and demonstrated.

This chapter describes the photonic switching networks utilizing WDM technologies. In Section II, the basic structure of photonic WD switching networks is described and possible line capacity numbers are discussed. Following this, the present status of photonic functional devices for the photonic WD switching networks is overviewed in Section III. System experiments on photonic WD switching are explained in Section IV.

II. PHOTONIC WAVELENGTH-DIVISION SWITCHING NETWORK

A. Basic Network Structure

Three general architectures have been proposed for WD switching networks, namely, the wavelength switch [3], the WDM passive star network [5,9], and the wavelength routing network [10–12].

A wavelength switch (λ switch), which can interchange wavelengths for an input WDM signal, is a basic component in the WD switching network, as a time switch is a basic component in the TD switching network. The λ switch consists of tunable wavelength filters and wavelength converters. There are two kinds of λ switches, as shown in Figure 1 [4]. In the λ switch shown in Figure 1a, an input WDM signal is split and parts thereof are led to individual tunable wavelength filters, each of which extracts a specific wave-

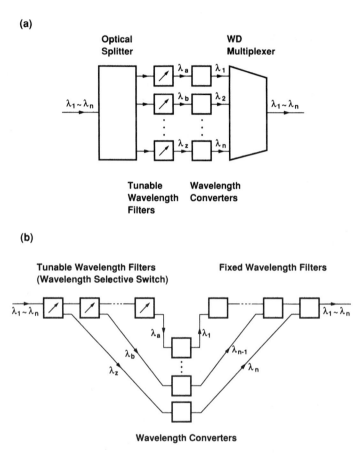

Figure 1 Wavelength (λ) switch structure.

length signal from the input signal. The output optical signal from each tunable wavelength filter is then applied to a wavelength converter, which results in wavelength interchange. Any kind of one-to-one wavelength interchange is possible. Moreover, a multicast function can be achieved by controlling all tunable wavelength optical filters to select the same WD channel. If the multicast function is not necessary, optical attenuation caused by optical signal splitting in the λ switch, shown in Figure 1a, can be avoided by using another type of wavelength filter (or wavelength selection space switch), which can separate a specific wavelength signal from the WDM signal, as shown in Figure 1b.

Other important photonic WD switching networks are the WDM passive star network and the wavelength routing network. The structure of

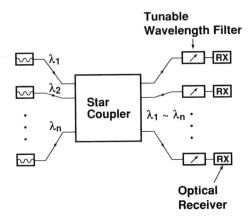

Figure 2 WDM passive star network with fixed wavelength transmitters and tunable wavelength optical receivers.

the WDM passive star network is illustrated in Figure 2, where each node has a fixed-wavelength optical transmitter and an optical receiver with a tunable-wavelength optical filter. This network is based on the "broadcast-and-select" principle. The transmitted signals from the nodes are wavelength-division multiplexed and divided at the central star-coupler and are again sent back to the nodes. At each node, the desired signal is selected using an optical filter. The multicast connection is possible in this type of network. There exists another possibility, where the transmitters' wavelengths are tunable and receivers are set to select the fixed wavelengths. In this case, wavelengths are used for output port addressing, and only one-to-one connections are possible. If both the transmitters and receivers are tunable, a reduction in number of wavelengths required may be possible. In this case, however, the simultaneous $n \times n$ connection is not possible.

In the wavelength routing approach, a wavelength-selective network is employed instead of the simple $n \times n$ star-coupler, shown in Figure 2. The path where the signal goes through the network is uniquely determined by the wavelength of the signal. The signal routing can be obtained by tuning the transmission wavelength at each port. Tunable-wavelength filters are not required at the nodes because the wavelength selection is accomplished at the central network.

There are two ways to expand the line capacities of the photonic WD switching networks. One method is to expand the number of WD channels, and the other is to use multistage switching networks. An increase in the number of WD channels mainly depends on improvements in photonic

WAVELENGTH-DIVISION MULTIPLEXING TECHNOLOGY

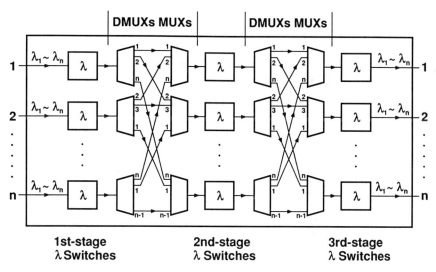

Figure 3 λ^3 switching network (from [4]).

functional devices. On the other hand, use of multistage switching networks is a system-oriented line-capacity expansion scheme and is common in the conventional electronic switching networks. In the WD switching networks, a large-capacity switching network can be constructed with multistage connections of λ switches, with fewer WD channels. Figure 3 shows a three-stage photonic WD switching network (λ^3 switching network), using wavelength multiplexers and demultiplexers (MUXs, DMUXs) in interstage networks [3]. The interstage network can provide connectivity between each stage λ switch. A nonblocking switching network equivalent to a multistage Clos network [13] using $n \times n$ switch matrices can be constructed with this multistage arrangement. Utilizing the wavelength interchange function of the λ switches, this kind of network allows the reuse of WD channels in different switching stages. This leads to a large-capacity switching system with few WD channels. Consequently, all WDM input lines to the switching network can use the same WD channels, $\lambda_1, \lambda_2, \ldots, \lambda_n$. Line capacity for a single λ switch is expressed as n, which is the same for the WDM passive star network, and the line capacities for λ^3 and λ^5 switching networks are expressed as n^2 and n^3, respectively [4]. Therefore, extremely large line capacity can be achieved using multistage connection of λ switches, even with a limited number of WD channels. The complexity in the structure shown in Figure 3 may cause a problem in constructing a large line capacity switching network. Introduction of photonic integrated circuits (PICs), however, would solve the problem. Especially, the development of multiwavelength laser diode arrays will significantly help the large line capacity switching network construction.

B. Feasible Number of WD Channels in Photonic WD Switching Networks

Realizing a large line capacity photonic switching system would be difficult without the introduction of photonic integrated circuits. InGaAsP/InP is expected to be the most suitable material for photonic integrated circuits, because it allows the integration of active and passive elements in the low-loss wavelength region of silica optical fibers.

In photonic wavelength-tunable devices based on the InGaAsP/InP system, wavelength tuning is generally accomplished by refractive index change. The relative change in the center wavelength $\Delta\lambda/\lambda$ is proportional to the refractive index change $\Delta n/n$. The maximum of $\Delta n/n$ is about 1% for InGaAsP/InP materials, and the maximum wavelength tuning range $\Delta\lambda/\lambda$ is expected to be the same. Therefore, a tuning range of about 15 nm is expected at the 1.5-μm wavelength region. Because of the temperature dependency of the refractive index, the wavelength of an InGaAsP/InP photonic device as a wavelength–temperature dependency of about 0.1 nm/K. Although a change in temperature causes a shift in the wavelength tuning range, about 10-nm wavelength range can arbitrarily be used even allowing 50 K atmosphere temperature change. Dense WDM systems with WD channel spacing of 0.1 nm and even less have been demonstrated using both direct and coherent optical detection techniques [14]. Therefore, with future improvements in InGaAsP/InP photonic devices, about 100 WD channels can be used within the wavelength tuning ranges of photonic

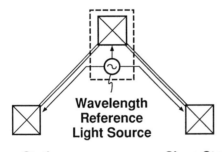

Figure 4 Basic model of subject synchronized WD communication network (from [4]).

WAVELENGTH-DIVISION MULTIPLEXING TECHNOLOGY

functional devices. In that case, the λ^3 switching network can provide a line capacity of 10,000.

C. Wavelength Network Synchronization

To achieve multistage WD switching networks like that shown in Figure 3, individual wavelengths for input and output WDM signals of each λ switch should be exactly the same. Such a condition is also required when constructing a large-scale WDM communication network by linking individual WDM networks together. The introduction of wavelength network synchronization has been proposed to solve these problems [3].

An example of a basic model of a subject synchronized WD communication network can be seen in Figure 4. A wavelength reference light source is mounted in the network, and internal channel wavelengths in all WD switching systems are synchronized according to the reference light supplied from the light source. Recently, several kinds of wavelength channel separation locking systems have been reported [15–17]. Among them, the "reference pulse method" [17] is easy to achieve for use in the wavelength network synchronization. Figure 5 illustrates wavelength synchronization using this method. In this case, the wavelength reference light source consists of a wavelength swept laser and a Fabry–Perot resonator (F–P). Direct output

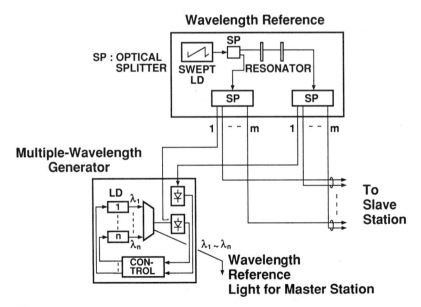

Figure 5 Wavelength synchronization using the *reference pulse method* (from [4]).

and output via the F–P from the wavelength swept laser were distributed to all photonic WD switching systems. Each photonic WD switching system has lasers with controlled output wavelengths, $\lambda_1, \ldots, \lambda_n$, respectively. The laser output lights are combined to form a WDM wavelength reference light for each switching network. Wavelengths are controlled so that the generation time for the beat pulses between the transmitter laser lightwaves and the swept laser lightwave coincides with the occurrence time for the F–P output pulses. As a result, WDM wavelength reference lights in all WD switching systems are locked to F–P resonance wavelengths. Through this process, wavelength synchronization is achieved. The wavelength for every controlled LD in every photonic WD switching system can be stabilized absolutely when one of the resonant wavelengths for the optical resonator is stabilized to an absolutely wavelength-stabilized LD.

In this wavelength synchronization scheme, the reference pulse sequence and wavelength swept light can be distributed by one or two optical fibers. When two optical fibers are used, there may be an optical path length difference between these two optical fibers. In this case, the arriving reference pulse sequence generation time does not indicate the exact wavelength. This can cause wavelength deviation from the resonator's resonance wavelength. For a 2-ms wavelength swept period and about 4.4-nm wavelength sweep range, the optical fiber length difference leads to a stabilized wavelength deviation of about 14 MHz per cm. Therefore, an optical fiber length difference of several centimeters is small enough to retain wavelength deviation to less than several tens of megahertz. This wavelength deviation can also be avoided by transmitting reference pulse sequences and wavelength swept light with only one optical fiber.

Mutual and independent synchronization can also be used in the WD communications networks, analogous to the clock synchronization in conventional time-division digital networks. It has already been shown that mutual wavelength network synchronization using an optical phase-locked loop technique may be possible [18]. The crucial point in developing the optical phase-locked loop is to expand the pull-in range, which is now restricted by photodiode bandwidth. The independent wavelength network synchronization method requires absolutely wavelength-stabilized reference light sources [19] for every photonic switching system. This is seen as an advantage in regard to network reliability as compared with subject and mutual network synchronization methods.

D. Photonic Hybrid Switching Networks

In an electrical TD switching system, large electronic digital switching systems make use of a time switch (T switch) in combination with a time-multiplexed space switch. Analogous to the conventional electronic switching systems, the practical photonic switching system will be a hybrid switching system, using two or three types of photonic switching systems. Several photonic hybrid switching networks using a photonic WD switching network have been proposed. The recently proposed WD/TD hybrid switching system [20] is a good example of such hybrid photonic switching systems. A wavelength and time hybrid switch (W&T switch) is a basic component in WD&TD hybrid switching networks. In this type of hybrid switching system, fast wavelength switching is required. The basic structure of the W&T switch is shown in Figure 6a. Both WD- and TD-multiplexed input optical signals are split into k parts and sent to tunable wavelength filters. The selected wavelength of a tunable filter output signal changes every time slot. The output signal from the filter is then applied to a photonic T switch, from where it is then sent to a wavelength

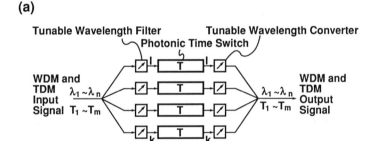

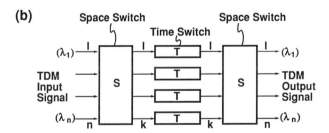

WDM : Wavelength-division multiplexed
TDM : Time-division multiplexed

Figure 6 Photonic WD/TD hybrid switching system structure (a) and equivalent time-division switch representation (b) (from [20]).

converter, which changes its output wavelength on a time slot basis. Finally, wavelength-converted output signals are combined to form a WD&TD multiplexed output. This W&T hybrid switch is equivalent to a space–time–space (STS) three-stage time-division switch with n input and output time-division highways, each of which corresponds to an individual-wavelength optical signal, as shown in Figure 6b. Therefore, this W&T hybrid switch can be strictly nonblocking, in case of $k = 2n - 1$. In the W&T hybrid switch, the total multiplexity value is given by the product of WD and TD multiplexities. Therefore, a large multiplexity value can be achieved, even with current WD and TD multiplexity values. A WD multiplexity value of eight has already been demonstrated, and TD multiplexity values of four or eight for the STM-1 signal (155 Mb/s) are easy to obtain with both electronic and photonic TD switching technologies. Therefore, even with current technologies total multiplexity values of 32 or 64 are possible. For a near-term implementation, it is the most practical way to use opto electronic converters, and wavelength-tunable lasers instead of photonic T switches and wavelength converters. Future progress in high-speed electronic T switches will benefit this W&T hybrid switch. Furthermore, this W&T hybrid switch will be integrated in some LSIs, with progress in optoelectronic integrated circuits (OEICs).

Required wavelength switching time depends both on the TDM signal highway speed and the multiplexing format. A byte-interleaved or bit-interleaved TD multiplexed signal format would be used in the photonic WD&TD hybrid switching system. TD multiplexity for a bit-interleaved return-to-zero (RZ) signal format is directly limited by attainable wavelength switching time. On the other hand, in byte-interleaved signal format, guard time for wavelength switching can be inserted between channel bytes. Large TD multiplexity is allowable for byte-interleaved format because of its capability for more flexible guard time setting. For example, to achieve a TD multiplexity of four for 100-Mb/s signals, using bit-interleaved signal format, wavelength switching time must be reduced to below 1.25 ns. This requirement can be reduced to lower than 2 ns, using byte-interleaved signal format with 10% guard time [21].

Another promising hybrid switching network utilizes space-division and wavelength-division switching networks. Akiyama and Miyagi have proposed a photonic SD/WD hybrid switching network with multiplexed link connection [22]. It has the structure of link connection between photonic SD switches, which have lattice switch function with input branching circuits and output combining circuits. A wavelength conversion laser is attached at each input, and also a tunable wavelength filter is attached at each output. The links of the network carry multiplexed wavelengths. This switch has the advantages of simple structure with minimized link stage number and switching elements. It can provide broadcast-type multiple connection also. A photonic SD/WD switching network based on a multiple access bus structure also has been proposed [23]. The optical bus has a number of parallel spatial paths. By using coherent WDM technology in each of the spatial paths, it is possible to build up an interconnection network of astonishing capacity. Using unidirectional dual bus, both the local oscillator and transmitter wavelengths are derived directly from a common reference source. A switching network is achieved by using photonic space switches to select the desired spatial channel and reference wavelength.

E. Photonic Packet Switch Using WD Switching Technologies

The future broadband network should support a variety of services, such as voice, high-speed data, high-speed still picture, and video services. Such a wide variety of services can be supported by a high-speed packet switching system. A broadband circuit switching system would still be attractive for supporting broadband/high-speed communication services, however, even in the future broadband integrated services digital networks [24]. The WD switching systems described above are attractive for applying to such a broadband circuit switching system. In this case, fast wavelength switching is not required, and some kind of computer controlled wavelength tuning scheme [25,26] can be applied for random access for WD channels. In addition to these circuit switching applications, WD switching technologies can be applied to photonic packet switching systems [27,28] by making use of a fast wavelength tuning technique.

Figure 7 shows the structure of an example of such a photonic packet switching system; the HYPASS reported by Arthurs et al. [28]. This system consists of two WDM passive star networks, one for transport and another for control. The control protocol of the HYPASS is based on an input-buffered/output-controlled arbitration. The packet to be transported is temporarily stored at a buffer in the input port until a poll is received from the destination output port. When output ports become available, they send polls over the control network to request packets from the input ports, using fixed-wavelength transmitters, and tunable-wavelength receivers at the input ports. After receiving a poll, the packet is transmitted over the transport network using the specific wavelength corresponding to the desired output port. The time overhead required by the wavelength tuning as well as electronic control, however, would limit the throughput. In the design considerations for HYPASS, a peak throughput of about 200 Gb/s is estimated, allowing total overhead time of 100 ns and about 10 ns for the

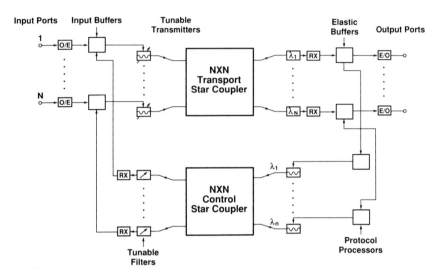

Figure 7 The HYPASS structure (from [28]).

wavelength tuning. An application of fast wavelength switching techniques to a parallel computer network has also been proposed [29].

III. PHOTONIC FUNCTIONAL DEVICES FOR WD SWITCHING SYSTEMS

As can be seen in the previous section, the photonic WD switching system requires a variety of photonic functional devices. This section reviews the present status of these devices. The number of WD channels is determined by the maximum wavelength tuning ranges of the wavelength tunable devices and the required WD channel separation for suppressing interchannel crosstalk. Therefore, serious consideration must be given to the wavelength tuning range. For the tunable-wavelength filter, a measure of the crosstalk characteristic is important in realizing practical functional devices. Finally, in the case of TD and packet switching applications, wavelength tuning speed should be investigated.

A. Wavelength Tunable Lasers [30,31]

Wavelength tunable lasers can be classified into two categories: hybrid tunable lasers and monolithic tunable lasers. A hybrid tunable laser has an external wavelength-selective filter section, such as a grating mirror, with a semiconductor gain section in the cavity. In monolithic tunable lasers, all sections are made of semiconductor materials. In the practical sense, monolithic tunable lasers are, of course, superior to the hybrid tunable lasers. The larger tunable wavelength ranges have been reported for hybrid tunable lasers, however.

1. Hybrid Tunable Lasers

The most popular hybrid tunable laser is the grating-type external cavity laser. This laser consists of an InGaAsP/InP active region, a grating, and a coupling lens system. The active section is usually a normal Fabry–Perot semiconductor laser with an antireflection coating on one facet. The center wavelength for light reflected from the grating is tuned by rotating the grating mechanically. Discrete tuning as wide as 55 nm centered at 1.5 μm has been reported, where a cavity 20 cm long was used [32]. The tuning range was limited by the gain bandwidth of the semiconductor material. A continuous tuning range of 15 nm centered at 1.26 μm has also been achieved by simultaneous control of the angle and placement of the grating [33]. The spectral linewidth is very narrow (< 100 kHz) because of long cavity length. Hybrid tuning lasers with electrical control of mode-selective filters have also been developed. Both an electrooptic (EO) and an acoustooptic (AO) filter have been employed. The hybrid tunable laser using a LiNbO$_3$ electrically tunable TE–TM convertor and a polarization filter has been demonstrated [34]. A discrete tuning range as wide as 7 nm, centered at 1.52 μm, has been achieved by applying a 100 V voltage swing. Continuous tuning over 1 GHz was performed by changing the applied voltage to the polarization filter. The EO external cavity laser has a potential for high-speed tuning of > 1 GHz. The hybrid tunable laser with an AO filter also has been reported at the 0.8 μm wavelength region [35]. A discrete tuning range of 35 nm has been reported. This was achieved by changing the acoustic frequency from 37.5 to 38.1 MHz. A hybrid tunable laser using an AO modulator has been reported in 1.3-μm wavelength region, using a fiber ring laser structure [36]. A tuning range of 80 nm has been obtained. The switching time may be on the order of microseconds.

2. Monolithic Tunable Lasers

Figure 8 shows typical examples of monolithic tunable lasers, which are classified into three types: distributed Bragg reflector (DBR), distributed feedback (DFB), and tunable twin guide (TTG) lasers. For all types of monolithic tunable lasers, wavelength tuning is carried out by changing the injection currents into the active section or passive section.

The three-section DBR lasers [37,38] shown in Figure 8a are the monolithic versions of the grating-type external cavity lasers. As the result of the free carrier plasma effect, injection currents into the phase-control and

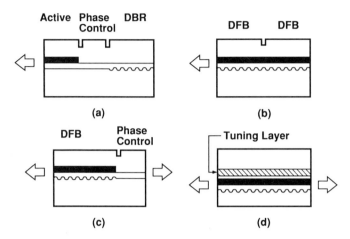

Figure 8 Typical examples of monolithic wavelength tunable lasers. They are classified into three types: DBR lasers, DFB lasers, and TTG lasers. (a) Three-section DBR laser; (b) phase-controlled DFB laser; (c) two-section DFB laser; and (d) TTG DFB laser.

DBR sections change the refractive index in those sections, resulting in a change in the light phase and the center wavelength. Figure 9 shows the structure and tuning characteristics of a three-section tunable DBR laser with double channel planar buried-hetero (DC-PBH) structure [38]. A maximum continuous tuning range of 4.4 nm (550 GHz) at 1 mW output power has been obtained. Continuous tuning was carried out by simultaneous changes in the DBR section current, I_d, and the phase-control section current I_p. The maximum quasicontinuous tuning range was 10.0 nm (1.25 THz). In quasicontinuous tuning, one of the resonant modes was selected by changing the DBR section current I_d, and narrow-range continuous tuning is done by changing the phase-control section current I_p in the vicinity of the mode. A two-section DBR laser, which has no phase-control section, has also been reported [39–41]. The continuous tuning range, however, is restricted to within the amount of frequency separation between the longitudinal modes; that is, about 1 nm at 1.55 μm.

Tunable DFB lasers reported to date are classified into two types; multielectrode DFB lasers [42–44], and phase-controlled DFB lasers [45,46]. The two-electrode DFB laser shown in Figure 8b has a simple structure; the two sections have the same layer structures and only the electrode is divided into two parts along the cavity, to control the injection currents of those sections independently. Thus, the fabrication procedure is relatively simple, nearly the same as that of normal DFB lasers. The tuning

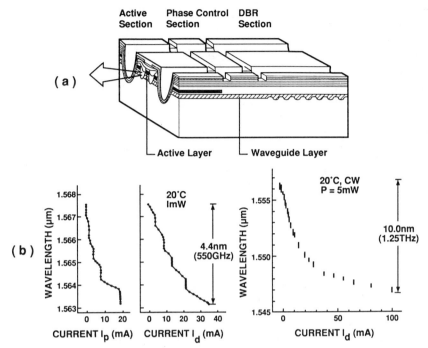

Figure 9 Structure and tuning characteristics of three-section DBR lasers with DC-PBH structure (from [38]).

principle, however, is more complicated than that of a tunable DBR laser, because the gain and the center wavelength of the mode-selective filter change simultaneously when current is injected. The lasing wavelength is usually tuned by changing the current ratio of the two sections, which results in changes in the resonant condition through changes in the distribution of the refractive index along the cavity. For a 1.55-μm two-electrode DFB laser, a continuous tuning range of 3.3 nm (410 GHz) has been obtained [42]. For the phase-controlled DFB laser shown in Figure 8c, the wavelength is tuned by changing the current in the phase-control section, which is added to one end of (or within) the DFB cavity. The tuning current changes the effective grating phase and results in a change in the resonant wavelength of the cavity. The wavelength and output power may be controlled almost independently by changing the phase-control and active-section currents, respectively. Thus, wavelength control may be easier than with a multielectrode DFB laser. A continuous tuning range of 1.2 nm centered at 1.54 μm has been reported [45]. Another type of 1.55-μm phase-controlled DFB laser, with a phase-control section in the center of the cavity, has also been reported [46]. The continuous tuning range was 0.9 nm at 2 mW output power. The maximum tuning range of tunable DFB lasers reported to date is narrower than DBR-type counterparts. The reported spectral linewidths of the tunable DFB lasers, however, were slightly narrower than those of tunable DBR lasers.

Tunable twin guide (TTG) lasers (Fig. 8d) have a passive tuning layer extending over the whole device length, in addition to a grating layer and an active layer [47]. The injection current to the tuning layer can be controlled independently of that to the active layer. It has been shown theoretically that the useful continuous tuning range of the TTG DFB laser is 30% larger than the three-section DBR laser [48]. The TTG structure requires only a single control current for wavelength tuning, which seems to be

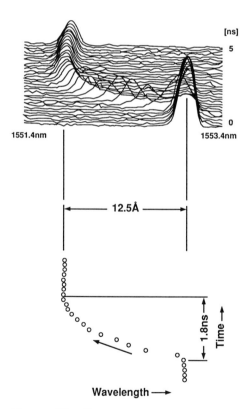

Figure 10 Time-resolved spectra for the DBR laser diode (LD) output (from [21]).

WAVELENGTH-DIVISION MULTIPLEXING TECHNOLOGY

advantageous. The TTG laser has the widest continuous tuning range—as large as 7 nm (875 GHz) at the 1.55-μm wavelength region [47].

As explained above, the development of tunable lasers with wide tuning range is progressing. In addition, a wider tuning range is expected by using several tunable lasers, each of which covers a different wavelength region. Recently, a photonic integrated circuit containing three quantum well tunable lasers, an optical combiner, and an optical amplifier has been demonstrated [49]. This approach appears to be very effective in expanding the wavelength tuning range.

Nanosecond-order wavelength switching experiments have been demonstrated using both tunable DBR and DFB lasers [21,50–52]. One example of an experimental result is shown in Figure 10 [21]. Here, a time-resolved spectra for a three-section DBR laser output, driven with a 400 ps rise time switching signal is shown. Wide-range wavelength switching of 1.25 nm was accomplished with as short as 1.8 ns switching time, which would be limited by the carrier lifetime. The switching speed for the multielectrode DFB laser would be expected to be higher than that of passively controlled lasers, because the carrier lifetime is shorter due to the stimulated emission in the gain section. The small signal frequency modulation bandwidths of multielectrode DFB lasers have been reported to exceed 10 GHz [53,54].

B. Tunable-Wavelength Optical Filters

A summary of the typical characteristics of tunable wavelength filters is outlined in Table 1 [55]. Two types of multichannel systems are foreseen from Table 1. One is the WDM with coarse (low density) wavelength setting, and the other is a dense packing system. AO (acoustooptic) and EO (electro optic) filters in Table 1 are suitable for coarse WDM systems. For the dense WDM networks considered in this chapter, a semiconductor resonant optical amplifier, a tunable-wavelength filter using coherent detection, Mach–Zehnder interferometer, Fabry–Perot filter, and stimulated Brillouin amplifier can be applied.

The wavelength tunable lasers described in Section III.A can be used as tunable-wavelength filters with injection current biased below their threshold values [56–60]. In this case, they act as wavelength-selective optical amplifiers. The features of this type of filter are optical gain and high WD channel selectivity. The activities for developing these tunable filters have been toward the production of broadly tunable narrow passband width, and constant gain filters. The structure and characteristics of the phase-shift controlled DFB filter [59] are shown in Figure 11. A constant gain of about 25 dB, throughout the almost 1-nm (120-GHz) tuning range was obtained. With an optical crosstalk value of about −13 dB, these character-

Table 1 Characteristics of Typical Tunable Wavelength Filters

Filter	Tuning method (speed)	Tuning range (nm)	Filter bandwidth (nm)	Loss	Crosstalk	Polarization dependence	Temperature dependence	Power dissipation
Semiconductor amplifier	Injection current (ns)	~10	~0.1	20 dB gain	−20 dB	△	1 Å/C	~50 mA
Coherent	Injection current (ns)	~10	Arbitrary	Receiver sensitivity improvement	Very low	△	1 Å/C	~20 mA
Fabry–Perot	PZT (ms)	Can be large	Arbitrary	~1.5 dB	−20 dB	○	0.1 Å/C	~10 W
Mach–Zehnder	Thermal or PZT (ms)	Can be large	Arbitrary	~2 dB	−20 dB	△	0.1 Å/C	~0.6 W
Brillouin amp.	Pump wavelength (ns)	~10	~100 MHz	~10 dB gain	−10 dB	○	1 Å/C	~10 mW
AO filter	RF (μs)	1.3–1.5 μm	1–2	~4 dB	−15~−20 dB	○	7 Å/C	~100 mW
EO filter	Voltage (ns)	~10	1–2	~8 dB	−15~−20 dB	○	1 Å/C	±100 V

Source: [55].

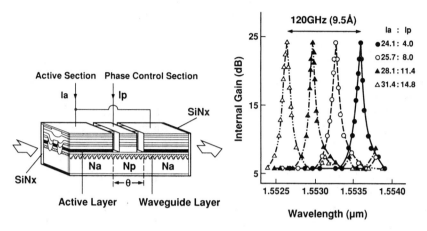

Figure 11 Phase-shift controlled DFB LD filter (from [59]).

istics allow for successful selection from 16 WD channels. Recently, a tuning range of 4.2 nm has been demonstrated at the 1.55-μm region using a $\lambda/4$-shifted DFB laser diode (LD) filter with TTG structure [60]. Using this filter, about 50 WD channels may be accessible. The wavelength switching times of these filters, which are limited by carrier lifetime, are expected to be a few nanoseconds.

A coherent optical receiver with a tunable wavelength local oscillator (LO) can be used as a tunable wavelength filter [55,61–63]. In this case, WD channel selection is accomplished in intermediate frequency (IF) signal level, not in optical signal level, utilizing IF filters with steep cutoff characteristics and high outband rejection. Therefore, low crosstalk switching for dense WDM signals is possible. The tuning range of the filter is determined by the wavelength tuning range of the lasers used as an LO. In this case, linewidth as well as tuning range is important. Spectral linewidths of less than several tens of megahertz have been obtained for wavelength tunable lasers throughout the tuning range. These values are narrow enough for an optical frequency shift keying (FSK) heterodyne detection application. About 100 WD channels would be expected for the coherent tunable filter. Wavelength switching time is expected to be nanoseconds or less, depending on the structure of the tunable laser used as an LO.

The Fabry–Perot filter also is an attractive means for tunable-wavelength selection in dense WDM networks. The spectral response of a Fabry–Perot etalon can be tuned by varying the optical path length in the Fabry–Perot cavity. The required pass length change to obtain spectral response shift of a free spectral range is only a half wavelength. The changes in the optical path length can easily be created mechanically, thermally, or electrooptically. A microoptic Fabry–Perot filter with passband widths of 5 nm and a free spectral range (FSR) of 180 nm at the 1.5-μm wavelength region has been demonstrated [64]. The finess (F) was about 37. Several types of tunable fiber Fabry–Perot filters also have been demonstrated [65]. The obtained finess values were about 200. In both types of filters, the cavity length change was performed mechanically using piezoelectric translators. Another means of tuning wavelength is rotating the Fabry–Perot filter using a step motor [66]. Recently, micromechanical Fabry–Perot devices using Si micromachining technology have been reported [67]. The tuning of cavity length was obtained electrostatically; therefore, no additional translator is required. The electrooptic implementation of tunable Fabry–Perot devices has also been demonstrated using liquid crystal [68]. The filter has a narrow passband of 0.4 nm and a tuning range greater than 60 nm with a low drive voltage of 1–5 V. Although a large tuning range is possible for a Fabry–Perot filter, the WD channel arrangement is restricted within one free spectral range due to the periodic spectral response. For Fabry–Perot filters the possible number of WD channels is expressed approximately as $F/6$, where F is the finess value of the Fabry–Perot filter [69]. As the reported F value is around 200, estimates of the number of WD channels obtainable using Fabry–Perot filters were around 30. It has been also reported that an effective overall finess of about 3000 is possible by using a two-stage Fabry–Perot filter structure. In this case, the possible WD channel number can be increased to 1000, because the WD channel number for a two-stage filter is expressed as $F/3$ [69]. Recently, a connectorized cascaded fiber Fabry–Perot filter with 5.2-dB insertion loss and finess greater than 3500 has been demonstrated [70]. This filter could separate 1000 WD channels in a 40-nm Er^{3+}-doped fiber optical amplifier gain bandwidth.

The Mach–Zehnder tunable filter is based on a silica-based integrated optic asymmetric interferometer [71]. It consists of two 3-dB couplers and two waveguides with a different connecting length between each set. The phase-delay difference of the waveguides determines the wavelength spacing of the periodic filter. The wavelength tuning is conducted through the phase shift in one of the interferometer arms, caused by temperature control using thin-film heaters. It is possible to increase the number of WD channels by serially connecting filters, each of which has a different pass difference between two arms of the interferometer. This filter has been demonstrated with 100 WD channels separated by 10 GHz in optical frequency with seven-stage construction [72]. Insertion loss of the filter was 6.9 dB. Tunable range was 1280 GHz. The total crosstalk level was less than -13 dB for any selected channel. The measured tuning time was on the order of milliseconds.

A Brillouin optical amplifier also can be used as a tunable wavelength filter, because its gain bandwidth is narrow and its gain peak wavelength can be tuned by changing the pumping wavelength [73]. The Brillouin amplification occurs due to a stimulated scattering process. The pump wavelength is only 11 GHz away from the optical signal to be amplified. It has been shown that 45-Mb/s signals can be detected error-free with an interfering channel spaced only 140 MHz away in the 1.5-μm wavelength region. Its passband width (less than 100 MHz), however, seems to be too narrow for practical application. Another problem with this type of filter would be a relatively large noise figure.

The acoustooptic (AO) [74,75] and electrooptic (EO) [76] filters are based on the wavelength-selective polarization transformation. The orthogonal polarizations of waveguide are coupled together at a specific tunable wavelength. Wavelength tuning is carried out by changing the applied voltage or by changing the frequency of the acoustic driver for EO and AO filters, respectively. A passband width is about 1–2 nm for both filters. This characteristic is compatible with direct intensity modulation of a DFB LD, where spectral broadening of several angstroms due to chirping may occur. Therefore, there is no need to use external modulators, and hence a simple and practical system can be constructed. In addition, strict WD channel wavelength control is not necessary in the network using AO or EO filters, because of the relatively large WD channel spacing. The AO filter has a much broader tuning range than the EO counterpart. Another attractive feature of the AO filter is the possibility of constructing a wavelength-selective space switch function [77]. A three-channel λ switch experiment has been reported using AO deflectors [78]. The EO filter has the potential of high-speed wavelength switching exceeding 1 GHz, while the switching speed of an AO filter is on the order of megahertz.

C. Wavelength Converters

Wavelength converters are used together with tunable-wavelength filters in constructing a λ switch. In most of the photonic WD switching experiments, the wavelength conversion function was achieved using optoelectronic (OE) and electrooptic (EO) converters. To achieve a large-scale λ switch, development of wavelength converters with optoelectronic integrated circuit (OEIC) format is desirable. Recently, two kinds of all-optical wavelength converters have been demonstrated.

In the wavelength conversion scheme using nearly degenerated four-wave mixing (NDFWM) proposed by Grosskopf et al. [79], the third lightwave (converter wave, E_c) is introduced in addition to the signal lightwave E_s and the pump lightwave E_p, which are conventional for the

NDFWM process. All three light waves are coupled to a semiconductor optical amplifier. The pump lightwave wavelength is set close to the signal lightwave (< 5 GHz) for efficient NDFWM interaction. The converter lightwave wavelength is set far away from E_s and E_p, to avoid direct NDFWM interaction between E_c and E_s or E_p. The NDFWM process between E_s and E_p creates a replica of the signal as one of the two sidebands of E_s and E_p. With the converter wave E_c, sidebands appear also at each side of E_c. These sidebands represent both replicas of the input signal wave. By tuning of E_c wavelength, wavelength conversion with wide tuning range, is almost same as the optical amplifier gain bandwidth, is possible. As one of the drawbacks of this scheme is the complicated output signal spectra, an optical filter may be required.

Another possibility for achieving all-optical wavelength conversion is to use wavelength conversion lasers [80,81]. The wavelength conversion laser is basically a bistable laser with a multisection DBR structure. Wavelength conversion within about a 3-nm output wavelength range has been demonstrated [81]. In this type of laser, the spectral broadening of the wavelength converted signal may limit the maximum WD channel numbers.

Another proposal for achieving a wavelength converter is using an optically controlled optical modulator [3,4]. In this way, a preassigned wavelength continuous wave (CW) lightwave is modulated by the optically controlled optical modulator according to the input optical signal whose wavelength is to be converted to the preassigned wavelength. This process is similar to that in an electronic transistor. The optically controlled modulator can be realized using optical nonlinear devices, such as nonlinear etalons [82] and S-SEED devices [83].

IV. PHOTONIC WD SWITCHING EXPERIMENTS

This section reviews some of the reported photonic WD switching experiments using different technologies, and examines possibilities and practical limitations of the current photonic WD technologies.

A. Photonic Wavelength-Division Switching Experiments Using Tunable DFB Filters [4,84]

An eight-channel photonic WD switching system has been reported using DFB filters. Figure 12 shows a block diagram of the experimental photonic WD switching system considered in this section. This system consists of a wavelength multiplexer, a λ switch, and a wavelength demultiplexer. In the setup shown, the wavelength synchronization was accomplished by distributing reference lights both to the wavelength multiplexer and to the λ

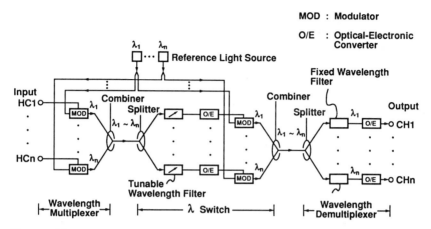

Figure 12 Block diagram of experimental photonic WD switching system (from [4]).

switch. The wavelength multiplexer consists of laser diode (LD) modulators [85], which intensity-modulate light carriers supplied from wavelength reference light sources according to input channels 1–n signals, and an optical combiner. In the λ switch, an input WDM signal is split and parts are led to individual tunable-wavelength filters, each of which extracts a specific wavelength signal. The output signal from the tunable-wavelength filter is then converted to an electronic signal by an optical–electronic converter. A preassigned-wavelength light carrier is intensity-modulated by a LD modulator according to the electronic signal from the optical–electronic converter. The phase-shift controlled DFB LD filters were used as tunable-wavelength filters in the experiments.

Experimental bit error rate characteristics, as a function of the filter input power, measured under no-crosstalk conditions for each wavelength channel, are shown in Figure 13. The solid line shown represents the theoretical value where optical gain is 16. Good agreement with the theoretical value has been obtained. Figure 13 also shows error rate characteristics, when the phase-shift controlled DFB filter selects the λ_4 signal from the five-channel WDM input signal, whose wavelengths are $\lambda_2, \ldots, \lambda_6$. The measured power penalty is at a very low level (about 0.5 dB), which agrees well with the calculated value, shown with a dotted line.

The number of WD channels n in this photonic switching network is mainly determined by the tunable filter performance. The phase-shift controlled DFB LD tunable filter has advantages of wide wavelength tuning range and narrow transmission bandwidth, whose gain and bandwidth can

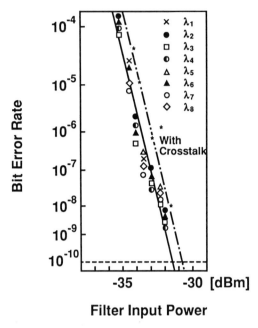

Figure 13 Experimental bit error rate characteristics, as a function of the filter input power, measured under no-crosstalk conditions for each WD channel (from [4]).

be constant over the tuning range. Tunable filter optical gain and spontaneous emission noise characteristics determine the required filter input power. Calculated filter input peak power values, taking the filter spontaneous emission into account, are -31.5 and -38.5 dBm, to satisfy the 10^{-10} bit error rate for a 200-Mb/s signal, on condition that tunable filter optical gains are 15 and 75, respectively. Assuming 3-dBm modulator output peak power and 5-dB coupling loss between optical devices and fibers, 21 and 28 dB, respectively, for 15 and 75 optical gains, could be assigned to the combiner and splitter maximum loss, while the operation margin is 3.5 dB. Therefore, it is concluded that the n values can reach 8 and 16, respectively, where 0.5-dB single-stage optical coupler excess loss in the combiner and splitter is counted. With improvement in filter optical gain and replacement of the optical combiner by a low-loss wavelength multiplexer, n can increase. For example, required filter input peak power will be reduced to -43 dBm, using a filter with 200 optical gain. As a result, a 32.5 dB loss value can be assigned to the wavelength multiplexer and optical splitter.

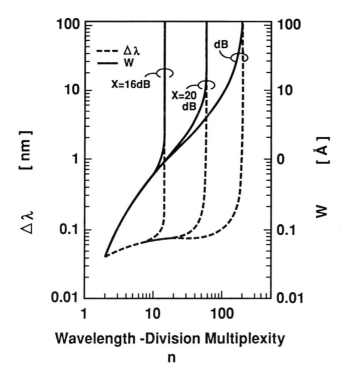

Figure 14 The relationship between required channel separation $\Delta\lambda$, wavelength tuning range $W = (n - 1)\,\Delta\lambda$, and n (from [4]).

Therefore, n can reach 100 by using the wavelength multiplexer with 10-dB loss and the optical splitter with 22.5-dB loss.

Tunable filter selectivity and tuning range also bound the number of WD channels, n. The relationship between the required channel separation $\Delta\lambda$, wavelength tuning range $W = (n - 1)\,\Delta\lambda$, and the n value are calculated as shown in Figure 14. $\Delta\lambda$ was determined to be the minimum value that gave a 1-dB power penalty due to the crosstalk caused by $(n - 1)$-channel WDM signals. The power penalty was determined as the required increase in filter optical input power to satisfy an error rate of 10^{-11} with crosstalk. With $X = 20$ dB and $W = 1.2$ nm, the n value can reach 16. With future improvements in characteristics of the tunable filter, n may reach 100 while $X = 24$ dB and $W = 10$ nm. Moreover, replacing optical combiners with low-loss wavelength multiplexers, such as Mach–Zehnder interferometers, will enable us to fabricate the λ^m multistage switching network using many λ switches.

Fast wavelength switching of less than 6.7 ns has also been demonstrated using the same filter [86].

B. Coherent Photonic Wavelength-Division Switching System [61–63]

Figure 15 shows the structure for the coherent photonic WD switching system, which consists of a wavelength multiplexer, a wavelength demultiplexer, and a coherent λ switch. In this system, the optical FSK modulation format is considered because an FSK system is most practical regarding laser linewidth and modulation. In Figure 15 lightwave signal paths are indicated by solid lines and electrical signal paths by dashed lines. In the coherent λ switch, an input WDM signal is split, and parts thereof are led to an individual coherent optical receiver with tunable wavelength local oscillator (LO). At every coherent optical receiver, the desired signal channel can be selected and demodulated by tuning the LO wavelength with a control signal from the switch controller. Each demodulated electrical signal from the coherent optical receiver is then applied to a single wavelength laser diode (DFB or DBR laser) to create an output optical FSK signal. The wavelengths of the lasers are preassigned also to be $\lambda_1, \lambda_2, \ldots, \lambda_n$, as shown in Figure 15. By this process wavelength interchange can be accomplished.

In addition to high wavelength channel selectivity, the receiver sensitivity improvement with coherent optical detection can increase the allowable loss value for the λ switches and transmission lines. Therefore, a large-scale network is possible with the coherent photonic WD switching system. Design considerations have shown that 32 WD channels are possible with a coherent λ switch, and a broadband metropolitan area network with a line capacity of over 1000 is possible using a multistage connection in the coherent λ switches.

Figure 16 shows a diagram of the experimental coherent WD switching system. This switching system consists of a wavelength multiplexer, a coherent λ switch, and a wavelength controller for wavelength synchronization. The *reference pulse method* was applied for wavelength synchronization. Here, 1.55-μm wavelength tunable DBR lasers were used as the transmitters, LOs, and as the sweep laser. Transmitted and switched signals were 280-Mb/s optical FSK signals with frequency deviation of 1 GHz. Frequency separation for the WDM signals was set to 8 GHz, which is sufficient to avoid interchannel crosstalk at bit rates up to 400 Mb/s [87]. Coherent optical receivers in the λ switch consist of balanced receivers and FSK single-filter detection systems with 400 MHz–1.4 GHz passbands. Beat spectral widths, between two DBR LDs, were around 30 MHz, which is sufficiently narrow for FSK single-filter detection [88]. The continuous wavelength tuning range was 2.06 nm (257 GHz) for the DBR lasers that

WAVELENGTH-DIVISION MULTIPLEXING TECHNOLOGY

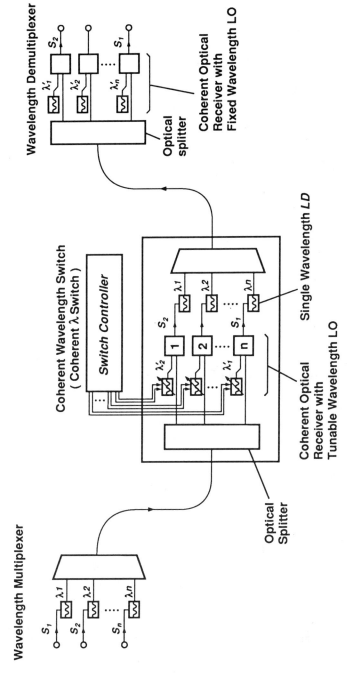

Figure 15 The structure for the coherent photonic WD switching system (from [63]).

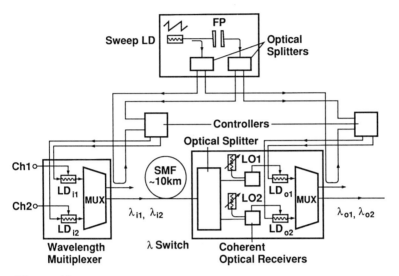

Figure 16 Experimental coherent WD switching system (from [63]).

were used as LOs. Therefore, 32 WD channels with 8-GHz separation can be selected using this LO.

Optical spectra of the input and output WDM signals for the coherent λ switch, measured using a scanning Fabry–Perot interferometer, are depicted in Figure 17. In this case, only the LD_{i1} was modulated, and LO2 was tuned to select the channel 1 signal. Switching between synchronized WD channels is clearly observed. In the wavelength switching experiments, no bit error rate degradation due to optical crosstalk has been observed in each coherent optical receiver, even under simultaneous two-channel operating conditions.

The fast wavelength capability has been demonstrated in coherent optical detection. Demodulated signals, using the high-speed wavelength switching LD, with LO wavelength switching, are shown in Figure 18 [21]. Wavelength switching was conducted after every ninth bit. As shown in this figure, the wavelength switching was accomplished completely during a 1 bit duration of the 500-Mb/s signal with this 1 bit duration (2ns) assumed to be guard time. This result shows that coherent detection, which follows fast and wide-range LO wavelength switching, was successfully accomplished.

C. Star-Coupler Based System Experiments

The star-coupler based WDM network experiments were started around 1985. British Telecom Research Laboratories reported WDM passive star network with broadcast-select mode operation, using an 8 × 8 star-coupler

WAVELENGTH-DIVISION MULTIPLEXING TECHNOLOGY 105

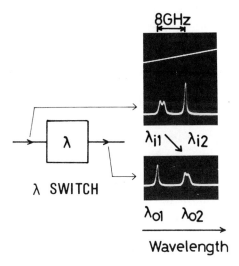

Figure 17 WDM signal spectra of input and output WDM signals for a λ switch (from [63]).

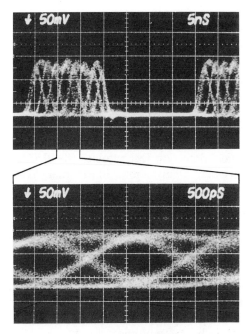

Figure 18 Demodulated signals with LO wavelength switching (from [21]).

[89]. The tunable-wavelength filters used in the experiment were mechanically tunable dichromatic gelatin holographic reflection filters. The tunable range and passband width were 400 and 10 nm, respectively.

A series of experiments have been demonstrated by Bellcore for WDM passive star networks using direct detection technology, called the LAMBDANET [6,7]. This network has an array of N receivers at each receiving node, using a grating demultiplexer to separate the different wavelength channels. The LAMBDANET experimental demonstration system consists of 18 DFB LDs whose wavelengths were separated by about 2 nm in the 1527–1561 nm wavelength region. For precise wavelength tuning, the temperature was adjusted and then stabilized to within 0.01 deg. The receiver sensitivity at 10^{-9} bit error rate was -34.5 dBm at 1.5 Gb/s. At 1.5 Gb/s, 18 channels were transmitted over 57.8 km of the single-mode optical fiber with zero dispersion at 1.3 μm. The received power levels ranged from -28.1 to -32.5 dBm for a bit error rate of 10^{-9}. This result indicates up to a 6-dB system power penalty due to fiber dispersion, reflection noise, and imperfect extinction ratio. Thus, a bandwidth–distance product of as high as 1.56 Tb/s · km has been demonstrated.

In 1986, the Heinrich Hertz Institute reported the first WDM passive star network using coherent optical detection technology [90]. The experimental system had 10 WDM channels spaced at 6 GHz, each carrying a 70-Mb/s video signal. The possibility of using a 128 × 128 star-coupler was demonstrated experimentally. Since then, there have been many reports on WDM passive star networks using the coherent optical detection as tunable-wavelength filters, especially for the video signal distribution application [91]. The experimentally demonstrated number of WD channels has been increasing. Recently, a 1.244-Gb/s, 32-channel experiment has been demonstrated [92].

AT&T Bell Labs reported WDM network experiments using 45-Mb/s FSK modulated WD channels with channel separation as small as 300 MHz using both direct and coherent optical detection technologies [8,69]. The transmitters used in the experiments were highly stabilized external cavity lasers. In the direct detection experiments, fiber Fabry–Perot tunable filters were employed as both a tunable filter and an FSK-to-ASK (Amplitude Shift Keying) converter. These experiments have shown that almost the same WD channel arrangement is possible for both direct and coherent optical detections.

As described in Section II.D, the use of some switching technologies to form a hybrid switching system is a usual technology in the conventional electronic switching systems. Use of WD/TD hybrid multiplexing technologies in WDM passive star networks has also been proposed [93–95]. Figure 19 shows the structure of the WD/TD hybrid multiplexed network. This

WAVELENGTH-DIVISION MULTIPLEXING TECHNOLOGY

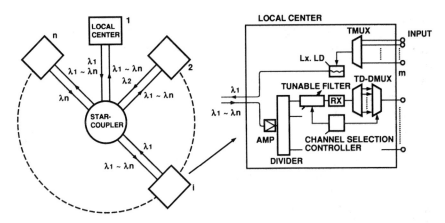

Figure 19 Wavelength-division and time-division hybrid multiplexed network (from [95]).

network was developed especially for the digital video signal distribution/ selection application in a broadcasting studio. Digital video signals are time-division multiplexed into a high-speed digital signal at each local center. Each local center transmits the TDM signal using an optical transmitter whose wavelength corresponds to the preassigned WD channel. For the WD channel selection, tunable wavelength filters are utilized. After WD channel selection and optoelectronic conversion, the desired video signal is selected using a TD demultiplexer. In the experiments, eight 155.5 Mb/s signals were multiplexed into a 1.244-Gb/s signal for direct intensity modulation of 1.53–1.54 μm DFB LDs. Six DFB LDs were used as transmitters. The wavelength spacing of the transmitter LDs was set at around 2.7 nm. AO filters with passband width of 2.2 nm were employed as tunable-wavelength filters. The WD and TD signal selection from 48 multiplexed NTSC digital video signals have been demonstrated. The possible number of video signals that can be handled in this network has been estimated to be several hundred.

V. CONCLUSION

Photonic wavelength-division (WD) switching networks have been discussed, and the present status of the photonic functional devices and photonic WD switching system experiments has been reviewed. The extremely high frequency of a lightwave carrier enables photonic WD switching systems to exhibit excellent features, namely, bit-rate independency for individual channels and also the suitability for large line-capacity switching systems. The WD switching system, using the λ switches and multistage switching networks, will be able to provide large line capacity, with an estimated possibility of 100 WD channels, based on InGaAsP/InP photonic integrated circuits. The wavelength network synchronization introduction will enable the network to utilize such a large number of WD channels without wavelength misalignment and drift. As a result, the photonic WD switching network is the most suitable for developing future large-scale broadband communication networks.

The most important issue in developing photonic WD switching networks is the development of large-scale integrated photonic devices. As can be seen in this chapter, photonic WD switching networks require many kinds of photonic functional devices, such as tunable-wavelength lasers, filters, and wavelength converters. Although the development of these devices is progressing, they are now realized in discrete device format. Present digital switching system expansion is mainly due to the advancement in Si integrated circuit technology. Semiconductor photonic device technology seems to be following the path of Si integrated circuit technology, but lags behind by about 20 years. The development of a practical photonic switching system similarly depends upon the photonic integrated circuit technology. Also significantly important for achieving a practical photonic WD switching system is the optical interconnection technology for achieving a multistage switching network. Even after obtaining integrated photonic functional devices, the interconnection between these devices cannot be achieved effectively by conventional optical fiber based interconnection technologies. New kinds of interconnection technologies utilizing optical waveguides or free-space interconnection would be required. With future advances in these technologies, a practical large-scale photonic WD switching network will be possible.

REFERENCES

1. M. Sakaguchi and H. Goto, High speed optical time-division and space-division switching, Proceedings of ECOC/IOOC '85, Venezia, Italy, 1985, 2, pp. 81–88.
2. S. Suzuki, K. Nagashima, and B. Hirosaki, Evolutional optical switching systems for broadband communications networks, Proceedings of ICC '87, Seattle, Washington, 1987, 3, pp. 1565–1569.
3. S. Suzuki and K. Nagashima, Optical broadband communication network architecture utilizing wavelength-division switching technology, Proceedings of Topical Meeting on Photonic Switching, Nevada, 1987, pp. 21–23.
4. S. Suzuki, M. Nishio, T. Numai, M. Fujiwara, M. Itoh, S. Murata, and N. Shimosaka, *IEEE J. Lightwave Technol.*, 8:660 (1990).
5. D. B. Payne and J. R. Stern, *IEEE J. Lightwave Technol.*, LT-4:864 (1986).
6. M. S. Goodman, H. Kobrinski, and K. W. Loh, Application of wavelength

division multiplexing to communication network architecture, Proceedings of ICC '86, Toronto, Ont., 1986, 2, pp. 931–933.
7. H. Kobrinski, R. M. Bully, M. S. Goodmann, M. P. Vecchi, C. A. Brackett, L. Curtiss, and J. L. Gimlett, *Electron. Lett.*, *23*:824 (1987).
8. B. S. Glance, K. Pollock, C. A. Burrus, B. L. Kasper, G. Eisenstein, and L. W. Stulz, *IEEE J. Lightwave Technol.*, *6*:67 (1988).
9. K. Y. Eng, M. Santoro, T. L. Koch, W. W. Snell, and J. Stone, An FDM coherent optical switch experiment with monolithic tunable laser covering a 1,000 GHz range, Postdeadline Paper of Topical Meeting of Photonic Switching, Salt Lake City, UT, 1989, paper PD10.
10. N. A. Olsson and W. T. Tsang, *IEEE J. Quantum Electron.*, *QE-20*:332 (1984).
11. H. Kobrinski, *Electron, Lett.*, *23*:974 (1987).
12. G. R. Hill, A wavelength routing approach to optical communications networks, Proceedings of IEEE Communications Society INFOCOM '88, New Orleans, LA, 1988, pp. 354–362.
13. C. Clos, *Bell System Techn. J.*, *32*:407 (1953).
14. K. Nosu, H. Toba, and K. Iwashita, *IEEE J. Lightwave Technol.*, *LT-5*:1301 (1987).
15. E. J. Bachus, R. P. Brawn, W. Eutin, E. Grossmann, H. Foisel, K. Heimes, and B. Strebel, Bidirectional four-channel coherent transmission, Proceedings of OFC '86, Atlanta, GA, 1986, pp. 88–90.
16. B. Glance, P. J. Fitzgerand, K. J. Pollck, J. Stone, C. A. Burrus, and L. W. Stulz, *Electron. Lett.*, *23*:750 (1987).
17. N. Shimosaka, K. Kaede, M. Fujiwara, S. Yamazaki, S. Murata, and M. Nishio, *IEEE J. Lightwave Technol.*, *8*:1078 (1990).
18. L. G. Kazovsky and D. A. Atlas, *IEEE Photonics Technol. Lett.*, *1*:395 (1989).
19. Y. C. Chung and C. B. Roxlo, *Electron. Lett.*, *24*:1045 (1988).
20. S. Suzuki, M. Fujiwara, and S. Murata, Photonic wavelength-division and time-division hybrid switching networks for large line-capacity broadband switching system, Proceedings of GLOBECOM '88, Hollywood, FL, 1988, pp. 933–937.
21. N. Shimosaka, M. Fujiwara, S. Murata, N. Henmi, K. Emura, and S. Suzuki, *IEEE Photonics Technol. Lett.*, *2*:301 (1990).
22. M. Akiyama and M. Miyagi, Photonic switching network utilizing wavelength multiplexing, Tech. Dig. 1990 International Topical Meeting on Photonic Switching, Kobe, Japan, 1990, pp. 176–178.
23. D. W. Smith, P. Healey, and S. A. Cassidy, Extendible optical interconnection network, Tech. Dig. Topical Meeting on Photonic Switching, Salt Lake City, UT, 1989, pp. 95–97.
24. H. Suzuki, T. Takeuchi, and T. Yamaguchi, Very high speed and high capacity packet switching for broadband ISDN, Proceedings of ICC '86, Toronto, Canada, 1986, pp. 749–754.
25. S. Yamazaki, M. Shibutani, N. Shimosaka, S. Murata, and M. Shikada, *Electron. Lett.*, *25*:507 (1989).

26. B. Glance, T. L. Koch, O. Scaramucci, K. C. Reichmann, L. D. Tzeng, U. Koren, and C. A. Burrus, *Electron. Lett.*, 25:883 (1989).
27. K. Y. Eng, A photonic knockout switch for high-speed packet networks, Proceedings of GLOBECOME '87, Tokyo, Japan, 1987, pp. 1861–1865.
28. E. Arthurs, M. S. Goodman., H. Kobrinski, and M. P. Vecchi, *J. Sel. Areas Comm.*, 6:1500 (1988).
29. E. Arthurs, J. M. Cooper, M. S. Goodman, H. Kobrinski, M. Tur, and M. P. Vecchi, *Electron. Lett.*, 24:119 (1986).
30. K. Kobayashi and I. Mito, *IEEE J. Lightwave Technol.*, 6:1623 (1988).
31. S. Murata and I. Mito, *Optical and Quantum Electron.*, 22:1 (1990).
32. R. Wyatt and W. J. Devlin, *Electron. Lett.*, 19:110 (1983).
33. F. Favre, D. Le Guen, J. C. Simon, and B. Landousies, *Electron. Lett.*, 22:795 (1986).
34. F. Heismann, R. C. Alferness, L. L. Buhl, G. Eisenstein, S. K. Korotoky, J. J. Veselka, L. W. Stulz, and C. A. Burrus, *Appl. Phys. Lett.*, 51:164 (1987).
35. G. A. Coquin and K. W. Cheung, *Electron. Lett.*, 24:599 (1988).
36. S. Oshiba, T. Kamijoh, and Y. Kawai, Electrically accesible wavelength tuning using acousto-optic modulator in fiber ring laser, Tech. Dig. 1990 International Topical Meeting on Photonic Switching (PS '90), Kobe, Japan, 1990, pp. 202–204.
37. Y. Tohmori, S. Suematsu, H. Tsusima, and S. Arai, *Electron. Lett.*, 19:656 (1983).
38. S. Murata, I. Mito, and K. Kobayashi, *Electron. Lett.*, 24:577 (1988).
39. Y. Thomori, H. Oohashi, T. Kato, S. Arai, K. Komori, and Y. Suematsu, *Electron. Lett.*, 22:138 (1986).
40. S. Murata, I. Mito, and K. Kobayashi, *IEEE J. Quantum Electron.*, QE-23:835 (1987).
41. B. Broberg and S. Nilsson, *Appl. Phys. Lett.*, 52:1285 (1988).
42. Y. Yoshikuni, K. Oe, G. Motosugi, and T. Matsuoka, *Electron. Lett.*, 22:1153 (1986).
43. M. Okai, S. Sakano, and N. Chinone, *Trans. IECE Jpn.*, OCS88-65 (1988) (in Japanese).
44. Y. Yoshikuni and G. Motosugi, *IEEE J. Lightwave Technol.*, LT-5:516 (1987).
45. S. Murata, I. Mito, and K. Kobayashi, *Electron. Lett.*, 23:12 (1987).
46. T. Numai, S. Murata, and I. Mito, *Electron. Lett.*, 24:1526 (1988).
47. S. Illek, W. Thulke, C. Schanen, H. Lang, and M.-C. Amann, *Electron. Lett.*, 26:46 (1990).
48. M.-C. Amann and W. Thulke, *J. Sel. Areas Comm.*, 8:1169 (1990).
49. U. Koren, T. L. Koch, B. I. Miller, G. Eisenstein, and R. H. Bosworth, *Appl. Phys. Lett.*, 54:2056 (1989).
50. H. Kobrinski, M. P. Vecchi, T. E. Chapuran, J. B. Georges, C. E. Zah, C. Caneau, S. G. Menocal, P. S. D. Lin, A. S. Gozdz, and F. J. Favire, Simultaneous fast wavelength-switching and direct data modulation using a 3-section DBR laser with a 2.2 nm continuous tuning range, Proceedings of OFC '89, Houston, TX, 1989, paper PD3.
51. T. E. Chapuranm, H. Kobrinski, M. P. Vecchi, C. E. Zah, C. Caneaau, M. G.

Menocal, P. S. D. Lin, A. S. Gozdz, and F. J. Favire, Proceedings of OFC '90, San Francisco, CA, 1990, p. 20.
52. J. M. Cooper, J. Dixon, M. S. Goodman, H. Kobrinski, M. P. Vecchi, M. Tur, S. G. Menocal, and S. Tsuji, *Electron. Lett.*, 24:1237 (1988).
53. N. Takachio, K. Iwashita, K. Nakanishi, and S. Koike, Chromatic dispersion equalization in an 8Gbit/s 202km optical CPFSK transmission experiment, Proceedings of IOOC '89, Kobe, Japan, 1989, 5, pp. 36–37.
54. S. Ogita, Y. Kotaki, M. Matsuda, Y. Kuwahara, H. Onaka, H. Miyata, and H. Ishikawa, *IEEE Photonics Technol. Lett.*, 2:165 (1990).
55. K. Emura, M. Shibutani, S. Yamazaki, I. Cha, I. Mito, and M. Shikada, Coherent optical FDM broadcasting system with optical amplifier, Proceedings of the Fourth Tirrenia International Workshop on Digital Communications, Tirrenia, Italy, 1989.
56. T. Numai, Semiconductor tunable wavelength filters, Proceedings of OEC '90, Chiba, Japan, 1990, 140–141.
57. K. Kobrinski, M. P. Vecchi, E. L. Goldstein, and R. M. Bully, *Electron. Lett.*, 24:959 (1988).
58. K. Magari, H. Kawaguchi, K. Oe, Y. Nakano, and M. Fukuda, *Appl. Phys. Lett.*, 51:1974 (1987).
59. T. Numai, S. Murata, T. Sasaki, and I. Mito, *Appl. Phys. Lett.*, 54:1859 (1989).
60. K. Tanaka, T. Inoue, M. Matsuda, T. Yamamoto, H. Kobayashi, K. Wakao, and T. Mikawa, A wide-wavelength-tunable active filter with a $\lambda/4$-shift DFB structure, Proceedings of Topical Meeting on Photonic Switching, Salt Lake City, UT, 1991, pp. 17–20.
61. M. Fujiwara, S. Suzuki, K. Emura, M. Kondo, K. Manome, I. Mito, K. Kaede, M. Shikada, and M. Sakaguchi, Optical switching in coherent lightwave systems, Proceedings of Topical Meeting on Photonic Switching, Incline Village, NV, 1987, pp. 27–29.
62. M. Fujiwara, S. Suzuki, K. Emura, M. Kondo, I. Mito, K. Kaede, M. Shikada, and M. Sakaguchi, *Trans. IECE Jpn.*, E78:55 (1989).
63. M. Fujiwara, N. Shimosaka, M. Nishio, S. Suzuki, S. Murata, and K. Kaede, *IEEE J. Lightwave Technol.*, 8:416 (1990).
64. S. R. Mallinson, *Electron. Lett.*, 21:121 (1985).
65. J. Stone and L. W. Stulz, *Electron. Lett.*, 23:781 (1987).
66. A. Frenkel and C. Lin, *Electron. Lett.*, 24:159 (1988).
67. S. R. Mallinson and J. H. Jerman, *Electron. Lett.*, 23:103 (1987).
68. M. W. Maeda, J. S. Patel, C. Lin, and R. Spicer, Novel electrically tunable filter based on a liquid-crystal Fabry–Perot etalon for high-density WDM systems, Proc. ECOC '90, Amsterdam, The Netherlands, 1990.
69. I. P. Kaminow, P. P. Iannone, J. Stone, and L. W. Stulz, *IEEE J. Lightwave Technol.*, 6:1406 (1988).
70. C. M. Miller, Low-loss cascaded fiber Fabry–Perot filter with finess greater than 3500, Tech. Dig. ECOC/IOOC '91, Paris, 1991, pp. 141–144.
71. N. Takato, K. Jinguji, M. Yasu, H. Toba, and M. Kawachi, *IEEE J. Lightwave Technol.*, 6:1003 (1988).

72. H. Toba, K. Oda, K. Nakanishi, N. Shibata, K. Nosu, N. Takato, and M. Fukuda, 100-channel optical FDM transmission/distribution at 622Mb/s over 50km, Proc. OFC '90, San Francisco, CA, 1990, postdeadline paper PD1.
73. A. R. Chraplyvy and R. W. Tkach, *Electron. Lett.*, 22:1084 (1986).
74. B. L. Heffner, D. A. Smith, J. E. Baran, A. Yi-Yan, and K. W. Cheung, *Electron. Lett.*, 24:1562 (1988).
75. D. A. Smith, J. E. Baran, J. J. Johnson, and K. W. Cheung, *J. Sel. Areas Comm.*, 8:1151 (1990).
76. F. Heismann, W. Warzanskyj, R. C. Alferness, and L. L. Buhl, *Electron. Lett.*, 23:572 (1987).
77. N. Goto and Y. Miyazaki, Wavelength-multiplexed optical switching system using acousto-optic switches, Conf. Proceedings of GLOBECOM '87, Tokyo, 1987 2, pp. 1305–1309.
78. Y. Shimazu, S. Nishi, and N. Yoshikai, *IEEE J. Lightwave Technol.*, LT-5:1742 (1987).
79. G. Grosskopf, R. Ludwig, and H. G. Weber, *Electron. Lett.*, 24:1106 (1988).
80. S. Yamakoshi, M. Kuno, K. Kondo, Y. Kotaki, and H. Imai, An optical-wavelength conversion laser with tunable range of 30A, Postdeadline papers of OFC '88, New Orleans, LA, 1988, paper PD-10.
81. H. Kawaguchi, K. Magari, H. Yasaka, M. Fukuda, and K. Oe, *IEEE J. Lightwave Technol.*, 24:2153 (1988).
82. N. Peyghambarian and H. M. Gibbs, *J. Opt. Soc. Amer. B*, 2:1215 (1989).
83. A. L. Lentine, H. S. Hinton, D. A. B. Miller, J. E. Henry, J. E. Cunningham, and L. M. F. Chirovsky, *IEEE J. Quantum Electron.*, QE-25:1928 (1989).
84. M. Nishio, T. Numai, S. Suzuki, M. Fujiwara, M. Itoh, and S. Murata, Eight-channel wavelength-division switching experiment using wide-tuning range DFB LD filters, Proceedings of ECOC '88, Brighton, UK, 1988, pp. 49–52.
85. M. Fujiwara, S. Murata, T. Numai, and H. Honmoh, *Trans. IECE J.*, E71:972 (1988).
86. M. Nishio, S. Suzuki, N. Shimosaka, T. Numai, T. Miyakawa, M. Fujiwara, and M. Itoh, An experiment on photonic wavelength-division and time-division hybrid switching, Proceedings of the Second Topical Meeting on Photonic Switching, Salt Lake City, UT, 1989, pp. 98–100.
87. K. Emura, S. Yamazaki, S. Murata, and M. Fujiwara, Design consideration for a coherent FDM-FSK dual filter detection system with wavelength tunable DBR LD, Proceedings of OFC '88, New Orleans, LA, 1988, p. 54.
88. K. Emura, S. Yamazaki, M. Shikada, S. Fujita, M. Yamaguchi, I. Mito, and K. Minemura, *IEEE J. Lightwave Technol.*, LT-5:469 (1987).
89. D. B. Payne and J. R. Stern, Single mode optical local network, Proc. GLOBECOM '85, Houston, TX, 1985, paper 39.5.
90. E.-J. Bachus, R.P. Braun, C. Caspar, E. Grossman, H. Foisel, K. Hermes, H. Lamping, B. Strebel, and F. J. Westphal, *Electron. Lett.*, 22:1002 (1986).
91. For example, S. Yamazaki, M. Shibutani, N. Shimosaka, S. Murata, T. Ono, M. Kitamura, K. Emura, and M. Shikada, *IEEE J. Lightwave Technol.*, 8:396 (1990).

92. H. Tsushima, S. Sasaki, K. Kuboki, S. Kitajima, R. Takeyari, and M. Okai, 1.244-Gb/s 32-channel 121km transmission experiment using shelf-mounted CPFSK optical heterodyne system, Tech. Dig. ECOC/IOOC '91, Paris, 1991, pp. 397–400.
93. A. Oliphant, R. P. Marsden, and J. T. Zubruzycki, *SMPTE J.*, *96*:660 (1987).
94. R. P. Marsden, J. J. Allen, G. J. Cannel, J. P. Laude, and H. Mulder, A multi-Gigabit/s optical business communication system using wavelength and time division multiplexing techniques, Proc. ECOC '90, Amsterdam, The Netherlands, 1990, pp. 779–786.
95. N. Shimosaka, M. Fujiwara, H. Nishimoto, C. Burke, T. Kajitani, and M. Yamaguchi, A photonic wavelength-division and time-division hybrid multiplexed network using tunable wavelength filters for a broadcasting studio application, Tech. Dig. ECOC/IOOC '91, Paris 1991, pp. 545–548.

3
Time-Division Optical Microarea Networks

Paul R. Prucnal

Princeton University, Princeton, New Jersey

Raymond K. Boncek, Steven T. Johns, Mark F. Krol, and John L. Stacy

Rome Laboratories, Griffiss Air Force Base, New York

I. INTRODUCTION

The speed of VLSI circuits is limited primarily by the low bandwidth–distance product of electrical interconnects [1]. Furthermore, the integration density of VLSI chips is limited by the small number of electrical pin-outs that can be accommodated on the perimeter of a chip without crosstalk. It is well known that this electrical I/O bottleneck can be remedied with optical interconnects, which provide immunity from crosstalk and an enormous bandwidth–distance product [2].

The optical interconnection of N VLSI processors physically distributed over a very small area (transmission distances $\ll 1$ km, as in chips or boards) has been referred to as a microarea network (μAN) [3–8]. Though it is possible to construct such a μAN with point-to-point links by simply providing a dedicated link between each processor and every other processor, the number of links required in such a fully connected topology would grow quadratically with the number of processors N. If the number of processors is large, then the number of links required for full interconnection, $N(N - 1)/2$, may be impractical to implement. In addition, not all of the links in a fully connected network can be used simultaneously. If, at a given time, every processor is either transmitting or receiving, then the number of links required is between $N/2$ and $N - 1$, and the fraction of unused links is between $(N - 2)/(N - 1)$ and $(N - 2)/N$. Not only is it impractical to provide full interconnection in a large μAN, but the resources would not be utilized efficiently.

To increase the efficiency and decrease the complexity of interconnections in a μAN, fewer links can be provided at the expense of requiring a means of sharing these links. For example, in a partially connected network with a star, bus, or ring topology, the number of links grows only linearly with the number of users. To provide connectivity in such a network, switching is required to establish connections as they are needed, and to release them when they are no longer needed. Consequently, optical μANs require, in addition to point-to-point fiber-optic links, some form of switching.

Switches can be broadly classified into two categories, in terms of how connections are established. Those switches in which the end-to-end network connection is established in advance and maintained during the entire transmission are called *circuit switches*. Those switches for which the transmission is segmented into short blocks of data preceded by an address (packets) and piecewise connections are temporarily established "on the fly" as packets propagate through the network, are called *packet switches*. Circuit switches have traditionally been used in telecommunications networks, because the duration of the call is usually much longer than the amount of time required to set up the circuit in advance of a call. Packet switches were developed for data networks in which the traffic is bursty (short transmissions at random intervals) and where the duration of the data burst is not much longer than the amount of time required to set up an end-to-end circuit. Since it would be inefficient to spend excessive time setting up a dedicated circuit only to use it for a brief transmission, the connection is established by the packet itself as it propagates. Standards are already being developed for packet-switched networks such as asynchronous transfer mode (ATM) in which the packet comprises 48 bytes of data and a 5-byte header containing address, priority, and error-check information.

Clearly, the potentially large throughput in fiber-optic links, as well as the large number of packets per second, will place severe demands on the required transmission bandwidth and reconfiguration speed of future switches for μANs. Although electronic switching technology has already achieved high switching speeds, it is quite possible that it will be difficult for electronic switches to match the transmission bandwidths that fibers can provide. Of course, it is possible to argue that the bandwidth of any individual channel will never exceed the reconfiguration speed or transmission bandwidth of an electronic switch, but many channels are usually multiplexed on the fiber, producing a high aggregate data rate. It is then essential that the transmission bandwidth of the switch match the aggregate bandwidth of the fiber. It is unlikely that this requirement can be met by electronic switches.

Thus, photonic switches may be needed to provide a transmission bandwidth that matches the aggregate bandwidth carried by the optical μAN. If packet switching is used, then the reconfiguration speed of the switch must also be fast. Another advantage of using a photonic switch is that optoelectronic conversion is avoided, which may reduce the cost and increase reliability. Because the photonic switching fabric is transparent to the transmitted optical signal, phase and frequency information are preserved, which may be important in coherent as well as frequency- and wavelength-division multiplexed systems

Photonic switching architectures can be classified in two categories, interchanger and shared medium. In the interchanger architecture, switching is performed centrally by a mapping operation. In a space-interchanger (crossbar) switch, signals entering the switch on physically disjoint input paths are spatially mapped to physically disjoint output ports. In a time-, wavelength-, or frequency-interchanger switch, a set of multiplexed signals on a shared input path are mapped by interchanging time, wavelength, or frequency slots. In the shared-medium architecture, a single transmission medium (an optical fiber) is shared by all nodes; a multiaccess protocol performs the switching function. The topology of the shared medium may be a star, bus, ring, or other local area network configuration. Access to the shared medium may be provided by time-, code-, wavelength-, or frequency-division multiple access (T-, C-, W-, or FDMA). Access may also be provided by a random access protocol such as carrier-sense multiple access with collision detection (CSMA/CD). The task of switching is distributed among the tunable transmitters and/or receivers. Either the shared-medium or the interchanger switch architecture may be used in an optical μAN. The appropriate choice of architecture will depend, in part, on the insertion loss, maximum size of N, internal blocking characteristics, required transmission bandwidth, and required switching speed. Because the number of processors to be interconnected in an optical μAN may be very large, and the maximum dimension of interchanger switches can be severely limited by technological considerations, only the shared-medium switch architecture will be considered further in this chapter.

We have just seen that optical microarea networks (μANs) can provide flexible communications among VLSI processors and eliminate electrical I/O bottlenecks. In Section II, shared-medium multiple access protocols will be discussed that can avoid the access delays associated with statistical multiple access protocols (which are unacceptable in multiprocessor applications) and increase the throughput, at the expense of wasting optical bandwidth. Time-division multiple access (TDMA, Section II.B.3) may be more practical to implement in a μAN than other shared-medium multiple access protocols such as frequency division (Section II.B.1) or code division (Section II.B.2). Since the total throughput of TDMA is given by the inverse of the optical pulsewidth, the throughput can be increased by making the pulsewidth small. Accomplishing this goal requires avoiding the use of low-bandwidth electronics in the portion of the μAN that directly processes these short pulses. Instead, optical processing can be used in those portions of the network. The architecture of a TDMA μAN, which uses optical multiple access processing and is self-clocking, is described in detail in Section III.A. Experimental demonstrations of key subsystems for optically generating, delaying, modulating, and correlating short optical pulses are presented. The feasibility of a variable-integer-delay line, which provides rapid tuning, wide tuning range, and high precision, is demonstrated and described in Section III.B. A transmitter consisting of a mode-locked laser with an external modulator is considered in Section III.C, since arbitrarily short pulses can be controlled with a modulator that need only operate at the bit rate, which translates into extremely high total throughput. The use of an off-chip optical source and an on-chip modulator also has circuit-integration advantages over a laser diode, including lower electrical drive power, electrical drive power that is independent of output optical power, greater ease of hybrid integration on silicon substrates, and less required real estate on the chip. Experimental measurements of the modulation depth, excess transmission loss, and required electrical drive power of a multiple quantum well (MQW) modulator are presented in Section III.C and are related to system performance. In Section III.D, an optical correlation receiver is demonstrated using 2-ps optical pulses and a two-gap photoconductive AND gate. The sensitivity, rise time, and fall time of the device are measured. In Section IV, the demonstration of a 5-GB/s optical TDMA μAN is presented, and its performance is analyzed in Section IV.B. Finally, the power budget of the TDMA μAN is calculated in Section IV.C, and it is determined that a 1000-node μAN is feasible.

II. μAN ACCESS CONTROL

A. Classical Access Control Methods

When more than one user shares any medium, an access control method is needed to determine when a user is allowed to begin to transmit [9]. Centralized control methods, as the name implies, delegate this decision to a central processor. Though centralized control can guarantee fairness in the decision process, the transmission of control information between the individual users and the control processor may be slow, limiting the total throughput in the network. Also, failure of the central processor will cause the entire network to fail. Distributed control methods, in which each user is responsible for using a protocol that governs access to the shared channel, can avoid some of these problems. Examples of early distributed control methods, developed for copper-based local area networks, include statistical multiple access protocols such as carrier-sense multiple access with collision detection (CSMA/CD) and token passing. Statistical multiple access protocols, as the name implies, only guarantee access to the network on a random basis. Users must contend with one another for access to the network; any given user must wait for the channel to become free before attempting to transmit. With statistical multiple access protocols, the delay in gaining access to the network increases as the number of users attempting to transmit increases. Also, because collisions occur when more than one user attempts to transmit simultaneously, the total throughput on the network decreases as the data rate increases (for a fixed transmission distance and fixed average transmission time). Clearly, statistical multiple access protocols are not suitable for high-data-rate transmission channels such as optical fibers. Furthermore, the access delays associated with statistical multiple access protocols would not be acceptable in many processor interconnect applications, where instantaneous access to the network must be guaranteed.

B. Multiple Access with High Throughput and Low Delay: Shared-Medium Protocols

To overcome the problems associated with statistical multiple access protocols, another class of distributed protocols, called *shared-medium* multiple access protocols, has been developed for fiber-optic networks. Because the capacity of the fiber is so large, shared-medium protocols do not attempt to share the fiber's bandwidth efficiently. In fact, shared-medium protocols waste the fiber's bandwidth by dividing the channel into many lower capacity channels, each of which is dedicated to a user. In this way, some of the contention problems inherent in statistical multiple access protocols are avoided. Shared-medium protocols do not utilize the channel as efficiently as do statistical protocols, since each user may only require access to its dedicated channel on a sporadic basis. Therefore, with shared-medium multiple access protocols, efficiency of bandwidth utilization is traded for reduced access delay and increased total throughput.

A block diagram of an optical microarea network (μAN) using a shared-medium multiple access protocol is shown in Figure 1. A passive star topology is chosen because the number of users (N) allowed can be larger than with other passive network topologies. The μAN consists of N optical

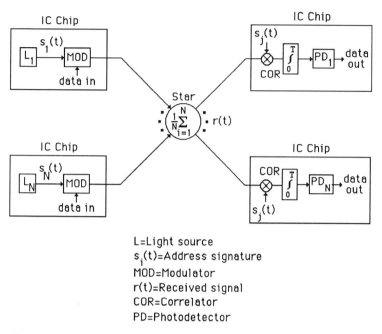

Figure 1 Block diagram of an $N \times N$ shared-medium microarea network architecture using optical orthogonal multiple access. Each light source L_i produces an orthogonal (in time, code, wavelength, or frequency) address signature $s_i(t)$, which is modulated by the optical data stream at input port i and broadcast by the passive star coupler to all output ports; at any receiver, to tune to the jth input port the received signal $r(t)$ is optically correlated with the address signature $s_j(t)$.

transmitters and N optical receivers on IC chips. Each transmitter consists of a light source producing a unique address signature $s_i(t)$, $i = 1, \ldots, N$. All address signatures are orthogonal, as defined by the expression

$$\int_0^T s_i(t)s_j(t)\,dt = \begin{cases} 0, & i \neq j \\ 1, & i = j \end{cases} \tag{1}$$

The address signature may be generated by driving the light source directly, by using an external modulator, or internally by the light source itself. Examples of sets of orthogonal address signatures include distinct optical carrier frequencies (or wavelengths), radio-frequency subcarriers, time-hopped code sequences, time slots, or spatial positions. We will not consider the last example further here, since it corresponds to an $N \times N$ photonic switch (an active star), which is limited in size to about $N = 20$

TIME-DIVISION OPTICAL MICROAREA NETWORKS 121

with present technology. The address signature is then modulated by a data stream. This modulation can take place either by driving the light source directly or by using an external modulator. In the system shown, the light source internally generates the address signature, which is then externally modulated by a data stream. The modulated address signatures are combined by a passive star-coupler and uniformly broadcast to all receivers. At any receiver, to tune to the jth transmitter, the received signal $r(t)$ is correlated with the address signature $s_j(t)$ by performing the operation

$$\int_0^T r(t)s_j(t)\, dt \qquad (2)$$

Note that the signature inserted at the receiver must be precisely in phase (coherent) with the corresponding signature portion of the received signal. The output of the correlator is then detected and compared with a threshold to recover the data from the jth transmitter. Though the correlation process can be performed electronically after photodetection (as in the case of radio-frequency subcarrier multiple access), the correlation in Figure 1 is performed optically before photodetection. As will be discussed later, with shared-medium multiple access the number of address signatures (the size of the network) increases with the channel bandwidth used. Thus, the use of an optical correlator, which has higher bandwidth than an electronic correlator, allows the size of the network to be larger.

Note that the network considered here assigns fixed address signatures to transmitters, and therefore can perform broadcasting. To establish communication, a (tunable) receiver must first poll all transmitters to identify an incoming transmission. To avoid the need for polling, fixed address signatures could be assigned to receivers, and communication would then be established by simply tuning the transmitter to the desired receiver address. The fixed-receiver assignment scheme, however, requires a reservation protocol to avoid collisions (i.e., multiple simultaneous transmissions to the same receiver), which may be difficult to implement.

Particular examples of shared-medium multiple access protocols that may be used in the optical μAN shown in Figure 1 include frequency-division multiple access (FDMA), code-division multiple access (CDMA), and time-division multiple access (TDMA). These are illustrated in Figure 2.

1. FDMA

In the first example illustrated in Figure 2, FDMA, the orthogonal address signatures are given by $s_i(t) \alpha \cos \omega_i t$, $i = 1, \ldots, N$, corresponding to N distinct carrier frequencies [10]. If the transmitted data has baseband bandwidth $1/T$, then the power spectral density $R(\omega)$ of the received signal

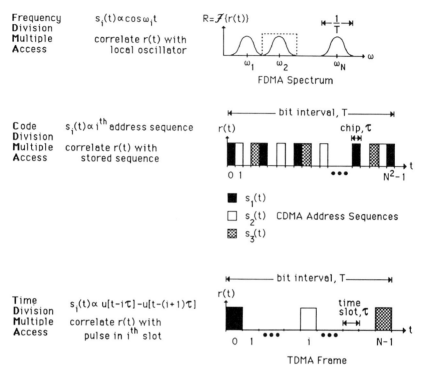

Figure 2 Examples of optical orthogonal multiple access schemes.

corresponds to a set of shifted versions of the baseband spectra, centered around the carrier frequencies ω_i, $i = 1, \ldots, N$. These spectra must be separated by a guard band of sufficient width to prevent interchannel crosstalk. Optical correlation is performed at the receiver by mixing the received signal with the output of an optical local oscillator. FDMA systems therefore require N laser sources with N selected stable (temperature controlled) center frequencies and narrow linewidths (e.g., distributed feedback (DFB) lasers), as well as laser local oscillators with stable center frequency, narrow linewidth, wide tuning range, and rapid tunability. In addition, some form of frequency registration is needed to establish a common frequency reference throughout the network. The baseband bandwidth required in the transmitter and receiver circuitry is $1/T$, and the total spectral bandwidth required in the optical fiber is at least N/T. In addition to the technological problems presently associated with the integration of laser diodes of specified frequencies on integrated circuit (IC) substrates, the integration of tunable local oscillators and coherent mixers is even a

TIME-DIVISION OPTICAL MICROAREA NETWORKS

more difficult task. If instead the laser sources and local oscillators are isolated from the IC chip, then modulators with frequency-independent absorption must be integrated on the chip. Another option, which relaxes the stringent linewidth and frequency stability constraints, and avoids the use of coherent local oscillators, is to widely separate the carrier frequencies $(\omega_{i+1}) - \omega_i \gg 1/T$, for $i = 1, \ldots, N - 1$, as in wavelength-division multiple access (WDMA) [11,12], and use a tunable bandpass filter with direct detection at the receiver. Though most of the above FDMA constraints are relaxed, they are nevertheless present in WDMA, and the required tuning range of the receiver is increased as well. For these reasons, it appears that it will be difficult to implement FDMA or WDMA in a μAN.

2. CDMA

In the second example illustrated in Figure 2, CDMA, the orthogonal address signatures correspond to a set of intensity-modulated waveforms $s_i(t)$, $i = 1, \ldots, N$, of duration T (the bit interval). Each waveform represents a code sequence, which in the case of prime code sequences [13–16], for example, consists of N^2 chips of duration $\tau = T/N^2$. Optical generation and correlation of code sequences using fiber-optic delay lines permits the duration of τ to be very short and the number of code sequences to be large, as has been previously demonstrated [13–16]. For prime code sequences, the baseband bandwidth required in the transmitter and receiver is N^2/T (though the use of optical code sequence generation and correlation techniques can reduce the bandwidth required in the optical source and detector to $1/T$), and the total spectral bandwidth required in the optical fiber is also N^2/T. Thus, CDMA requires an extremely large expansion of the channel bandwidth, and a similar increase in transmitter and receiver bandwidths unless optical coding and correlation is used. One advantage of CDMA is that like FDMA it allows asynchronous access to the network.

CDMA has been extensively studied in the areas of satellite and mobile radio communications [17–22]. In recent years, the use of CDMA in local area networks (LANs) has also been investigated [13–16,23–34]. CDMA offers several advantages over conventional multiple-access schemes, such as carrier-sense multiple access with collision detection (CSMA/CD) and ALOHA [21], in LANs. First, CDMA makes efficient use of the channel by allowing asynchronous access to each user. Since the traffic in LANs is typically bursty, asynchronous multiplexing schemes, which allow multiple users to share the entire channel simultaneously, are more suitable than synchronous multiplexing schemes (e.g., TDMA), where a portion of the channel is dedicated to each bursty user. For CDMA, system performance depends on the number of simultaneous users, which may be much smaller than the number of users actually subscribing to the network. Second, no scheduling is required in CDMA and new users can be easily added to the network. Third, CDMA is more suitable for networks with high traffic loads and with high signal rates, compared with conventional protocols (i.e., CSMA/CD and ALOHA), which are designed to handle low traffic loads and relatively low signal rates (e.g., 10 Mb/s). The number of collisions between packets increases and delays accumulate when these contention protocols are used at higher speed or at high traffic loads. Fourth, CDMA permits multiple users to simultaneously access the channel with no waiting time, in contrast to other asynchronous techniques, where each user must wait for the channel to become idle before gaining access. In addition, the waiting time with CSMA/CD and ALOHA increases with transmission distance, which is not the case with CDMA.

CDMA is based upon the assignment of orthogonal codes to the address of each user, which substantially increases the bandwidth occupied by the transmitted signal [20,35–44]. Therefore, to provide many subscribers and simultaneous users in a CDMA system, a high bandwidth expansion ratio (e.g., long code sequences) is needed, which implies that very narrow pulses have to be used. A CDMA μAN therefore requires a wide bandwidth channel such as an optical fiber; coaxial cable and twisted pairs are not suitable transmission media. In principle, a single-mode optical fiber can operate at a bandwidth–distance product of 1 THz · km and can offer an attenuation factor as low as 0.157 dB/km for optical signals at a wavelength of 1.55 μm [45]. Fiber-optic CDMA systems suffer a severe restriction on the usable bandwidth by the speed of the electronics required for signal processing and optoelectronic interfaces at the input and output of the optical fiber, however, resulting in a data-flow bottleneck and restricting the aggregate channel bandwidth.

Nevertheless, these limitations can be overcome by introducing optical signal processing in fiber-optic CDMA networks at points where high-speed electronics and high-bandwidth optoelectronic and/or electrooptic conversion would otherwise be required [13–15,24,27,31–34]. For example, to avoid high-speed current modulation of laser diodes, mode-locked lasers are used to generate narrow optical pulses, representing the transmission of data bits, and these pulses are then gated by integrated-optic modulators for further optical processing. In addition, the electronic bottleneck associated with optical code sequence generation and correlation and the use of high-bandwidth photodetectors can be eliminated by using optical encoders and decoders to generate and correlate address sequences all-optically. In general, optical signal processing techniques in optical CDMA can be classified as incoherent or coherent, depending on whether the intensities or fields of optical pulses are processed.

In coherent optical signal processing, the optical pulses in each code sequence generated at each optical encoder are phase coherent, and the optical fields of all these coherent pulses are then superimposed at an optical decoder [21,31–33,46,47]. Research has shown that these coherent optical encoders and decoders can be made of, for example, diffraction gratings and lenses with a phase mask in between [27] or by fiber-optic ladder networks [21,31–33,46,47]. Because of using phase information, coherent optical processing allows the use of bipolar (-1, $+1$) sequences, thus utilizing those extensively well-studied orthogonal code sequences, such as maximal length codes and Gold codes, used in spread spectrum satellite and mobile radio communications [20]. In general, this coherent technique offers a natural discrimination against interference from other orthogonal code sequences as well as high contrast ratios in the correlation process (i.e., more simultaneous users) at the expense of increased system complexity and sensitivity to environmental changes. In fact, the interference discrimination capability at each optical decoder depends on various parameters of the coherent pulses of each code sequence received, such as their relative phase shifts, polarization states, and amplitudes. Coherent correlation will only take place if the delays at the decoder match those of the corresponding encoder within the coherence time of the laser source. In addition, the polarization vectors and phases of the individual pulses in a pulse train have to be aligned. Otherwise, incoherent superposition and strong interference result. Therefore, the coherence time of the source is important in determining the required precision of the matching between the delays of the encoder and decoder. Using a laser source with the full width at half-maximum (FWHM) of 2 nm, corresponding approximately to a coherent time of 3.99 ps for an operation wavelength of 1.55 μm, the accuracy of the delays required matching to better than 798 μm in a single-mode fiber, which may be difficult to implement in fiber-optic systems [31]. In addition, polarization maintaining fibers may be needed to provide polarization and phase control of the coherent pulses.

On the other hand, incoherent optical processing allows only optical pulses at each encoder with intensity levels corresponding to light ON (i.e., 1) or light OFF (i.e., 0); the intensity of these incoherent pulses in each code sequence are then superimposed at the decoder. These optical encoders and decoders can be made of, for example, fiber-optic delay-line correlators [14,15,24] or fiber-optic ladder networks [46,47]. Since this kind of incoherent optical system can only accommodate unipolar (0, $+1$) sequences, orthogonal codes intended for communication systems in which both positive and negative levels are available are not optimal here. Therefore, pseudoorthogonal codes that have small code weight to sequence length ratio are needed to reduce undesired interference, thus supporting an acceptable number of simultaneous users. Extensive research has been investigated to find optical (unipolar) pseudoorthogonal codes, such as optical orthogonal codes (OOCs) [25,27,39], 2^n codes [43], Alberta codes [41], quasiprime codes [42], and prime sequence codes [13,34,36], that have good correlation properties.

Incoherent optical CDMA based on unipolar sequences can allow only a limited number of subscribers, and even fewer simultaneous users, before a rapid deterioration of the system performance occurs, due to the nonscheduled transmissions in a random access network. Synchronous code-division multiple access (S/CDMA) [16,24] for fiber-optic LANs using all-optical signal processing can increase the number of subscribers and simultaneous users. This abbreviation is chosen to stress the fact that, in contrast to CDMA, S/CDMA requires that network access among all users be synchronized.

3. TDMA

The third example illustrated in Figure 2, TDMA, is a special case of CDMA where the orthogonal address signatures are again given by a set of intensity-modulated waveforms $s_i(t)$, $i = 1, \ldots, N$, of duration T (the bit interval). Each waveform represents a code sequence, which in the case of TDMA consists of a single pulse in one of the N chip positions (called time slots) of duration $\tau = T/N$. In the case of fixed-transmitter assignment TDMA, the address signature $s_i(t)$ of the ith transmitter consists of a single pulse in the ith time slot. The receiver correlates the received TDMA frame, consisting of the time-interleaved data streams from all N transmitters, with the address signature of the desired transmitted data. Note once again that the signature inserted at the receiver must be precisely in phase with the corresponding signature portion of the received signal. Therefore, not only does TDMA require phase synchronization throughout the network, but it also requires narrow pulse generation and modulation, time delays with rapid tuning, wide tuning range and high precision, and high-speed correlation at the receiver. The baseband bandwidth required in the transmitter and receiver circuitry is N/T (though, as discussed below, this constraint is reduced to $1/T$ if optical encoding and correlation is used), and the total spectral bandwidth required in the optical fiber is also N/T. Thus, the baseband and spectral bandwidth requirements of TDMA are a factor N smaller than CDMA, at the expense of requiring network phase synchronization. Since the total throughput of the network with TDMA is $N/T = 1/\tau$ (N channels of baseband bandwidth $1/T$), the throughput can be increased by making τ small. The architecture of a TDMA μAN with extremely high throughput will be described below. The experimental demonstration of subsystems for generating, modulating, delaying, synchronizing, and correlating optical pulses of very short duration τ will be presented, as well as the demonstration of the full network architecture at 5 Gb/s.

III. HIGH-THROUGHPUT µANs WITH OPTICALLY PROCESSED TDMA

A. Optical TDMA µAN Architecture

A fiber-optic µAN using TDMA can achieve extremely high total network throughput if τ is made very short. Accomplishing this goal requires avoiding the use of low-bandwidth electronics in the portion of the network that directly processes these short pulses. Instead, optical processing can be used in those portions of the network [48,49].

The architecture of a TDMA µAN with dimension $(N - 1) \times N$ is shown in Figure 3. In this system, optical processing is used for multiple access. A train of optical clock pulses of duration τ, energy E_L, and repetition rate $1/T$ is generated by a mode-locked laser. To increase the total pulse energy, an array of N synchronous mode-locked lasers may be used instead of a single laser. Using fiber-grating pulse compression techniques, subpicosecond optical pulses can routinely be produced, resulting in a total network throughput of greater than 1 THz. With fixed transmitter assignment TDMA, the clock period T is divided into a time frame comprised of $N = T/\tau$ time slots, where the ith slot in the frame corresponds to the ith transmitter address. The system shown in Figure 3 is designed to be self-clocking, so the 0th slot is always reserved for a framing pulse. The output of the pulse compressor is distributed by a $1 \times N$ splitter to $N - 1$ fixed optical delays $i\tau, i = 1, \ldots, N - 1$; the remaining output of the splitter is connected directly to the star-coupler and provides a framing pulse for synchronization. The fixed optical delay $i\tau$ moves the optical clock pulse into the ith time slot, yielding a waveform that corresponds to the ith address signature, which is then input to the ith electrooptic modulator integrated on an IC chip. The modulator is driven by an electrical data stream of rate $1/T$ that modulates the address signature. In this way, arbitrarily short optical pulses can be modulated and the bandwidth of the modulator need only be as large as the baseband bandwidth of the data. As will be discussed below in detail, modulators offer circuit-integration advantages over laser diodes, but suffer a system performance penalty due to their typically lower modulation depth. The modulated address signatures from all $N - 1$ transmitters and the framing pulse are combined in the star-coupler and evenly distributed to the receivers on N chips. To maintain timing synchronization throughout the network, it is important that all optical path lengths between the pulse compressor output and the receivers, except for intentional address signature delays, be an integral multiple of τ, with an error of less than 0.1τ. The TDMA frame

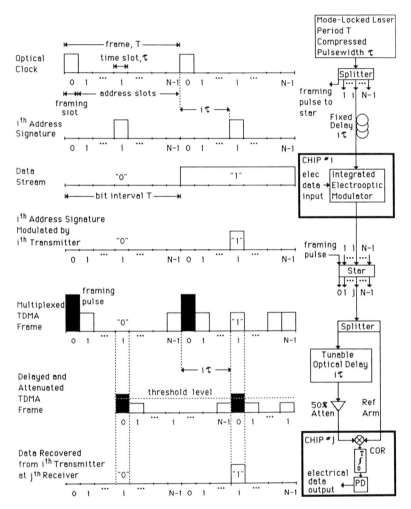

Figure 3 Self-clocking optically processed TDMA shared-medium microarea network architecture; optical clock pulses of duration τ and repetition rate $1/T$ are generated by a mode-locked laser; the clock period represents a time frame with N slots, where the ith slot corresponds to the ith input port address; a framing pulse is always in slot 0; the optical clock is distributed via a splitter and an optical delay $i\tau$ to the ith input port, modulated by the data stream at input port i, and broadcast by the passive star-coupler to all output ports; at the jth output port, to tune to the ith input port the received time frame is optically correlated with a copy to the received time frame delayed by $i\tau$; in this way, the framing pulse is aligned with the desired input address slot.

comprised of the superposed $N - 1$ modulated address signatures and the framing pulse is illustrated in Figure 3. Because the framing pulse has bypassed the insertion loss of the modulator, its amplitude is assumed to be at least 3 dB greater than that of the modulated address signatures. At each correlation receiver, the TDMA frame is divided by a 1×2 splitter. One arm of the splitter carries a reference version of the TDMA frame. To sample the ith slot of the reference TDMA frame, the other output of the splitter is delayed by $i\tau$ with respect to the reference TDMA frame, so that the framing pulse is in phase with the ith slot in the reference TDMA frame. In order to tune the receiver to any one of the time slots, a variable optical time delay is required. To select among a large number of time slots, the variable time delay must have wide tuning range and high precision. For rapid access to desired channels, the variable time delay must also be tunable at high speed. The demonstration of a tunable optical time delay with these characteristics will be described in detail below. The output of the tunable delay is attenuated by 3 dB, so that the amplitude of the data falls below a specified threshold level, whereas the amplitude of the framing pulse exceeds the threshold. The framing pulse can then be used to carry out self-clocked optical correlation of the ith slot in the reference TDMA frame. Because τ is small, the optical correlation must be performed at high speed. Though high-speed optical correlation is difficult to do with low-energy pulses, promising experimental results using a photoconductive AND gate will be presented below, as well as the development of an experimental 5-Gb/s optical TDMA network and calculated bit error rate results. First, the demonstration of a tunable optical delay line and performance measurements of an integrated modulator will be described.

B. Variable-Integer-Delay Line Demonstration

A variable-integer-delay line implementation for the optical TDMA coder, shown in Figure 4, allows a large number of delays with fast reconfiguration time. The feed-forward structure consists of $\log_2 N$ delay stages $k = 1, \ldots,$ $\log_2 N$ and an output stage. Thompson has shown analytically that a feed-forward structure requires fewer stages than a feedback structure [50]. Each delay stage consists of a 2×2 optical switch, a connection to the next stage at one output, and a fixed optical delay in excess of the "connection" delay at the other output. The value of the fixed excess delay for the kth stage is $T/2^k$. Only one input to the first stage is used. The output stage consists of a 2×2 optical switch, where only one output is used. Each optical switch can be set in either the bar or the cross state. The state of a switch is set by the electrical control input, where a 0 at the control sets the 2×2 switch in the cross state, whereas a 1 sets the 2×2 switch in the bar state.

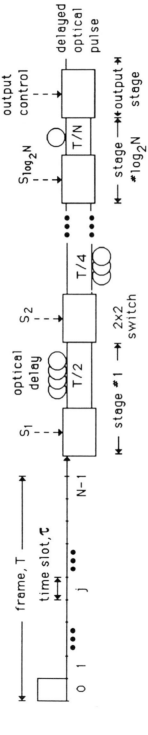

Figure 4 Tunable optical delay producing N possible delays; feed-forward structure consists of $\log_2 N$ delay stages and an output stage; the kth stage consists of a 2×2 optical switch with a connection to the next stage at one output, and a delay $T/2^k$ in excess of the connection delay at the other output; to delay the input pulse to the jth slot, successive bits in the binary representation of the number j are used to control each stage. The input is a time frame of duration T comprised of N slots of duration τ. The state of the encoder is set by a control sequence $(s_1, s_2, \ldots, s_{\log N})$, where $s_k = 0$ (1) sets the kth switch in the cross (bar) state.

The state of the coder is set by a control sequence $(s_1, s_2, \ldots, s_{\log N})$, where control bit s_k sets the state of the kth stage, and the output stage is set equal to 0 if the parity of the control sequence is odd, or to 1 if the parity of the control sequence is even. The output stage serves only to ensure that the delayed pulse always exists at the chosen output of the 2×2 optical switch. The control sequence for the jth slot is generated from the binary representation of the integer j, $(b_1, b_2, \ldots, b_{\log N})$, where b_1 is the most significant bit, according to the rule $s_1 = $ {the complement of b_1}; and for $i = 2, \ldots, \log_2 N$, $s_i = 0$ if $b_i = b_{i-1}$; otherwise $s_i = 1$. After the control sequence has set the coder, the sampled data will be delayed by an amount $j\tau$ in excess of the reference delays, accomplishing the desired time-division encoding operation.

An experimental demonstration of an optical TDMA variable-integer-delay line coder is presented for 64 100-Mb/s channels [51]. Here $T = 10$ ns and $\tau = 156.25$ ps. The complete coder requires six delay stages $k = 1, \ldots, 6$ with time delays $D_k = 10/2^k$ ns, corresponding to fiber lengths $L_k = 2.052/2^k$ m, where the index of refraction of the fiber core is $n_f = 1.462$. If the error in time delay is required to be less than 10% of a time slot, then the aggregate positioning error must be less than 3.2 mm, or 0.53 mm per stage, requiring careful trimming of the fiber lengths.

In this demonstration, only stages 1, 2, and 6 are implemented, demonstrating the longest ($D_1 = 5$ ns), intermediate ($D_2 = 2.5$ ns), and shortest ($D_6 = 156.25$ ps) delays, and allowing access to eight times slots: 0, 1, 16, 17, 32, 33, 48, 49. The experimental setup of the coder, comprised of three LiNbO$_3$ polarization-independent directional couplers (DC) with 50 MHz bandwidth and one passive coupler (C1) interconnected by single-mode fiber (SMF) delays, is shown in Figure 5. Use of a passive coupler at the output stage, rather than another directional coupler, introduces 3 dB of additional loss. The fiber delays were trimmed by carefully measuring and cutting the required lengths. Low-loss connections between the fiber delays and the DC pigtails were made using a fusion splicer. The state of the coder is set by applying a bias voltage of 0 volts (cross state) or 20 volts (bar state) to each of the directional couplers. The reconfiguration time of the coder is determined by the switching speed of the directional couplers.

The various delays produced by the coder were measured using 1.3-μm wavelength, 100-ps optical pulses generated by a mode-locked Nd: YAG laser with 100-MHz repetition rate. As shown in Figure 5, the output of the laser is split by coupler (C), so that part of the optical pulse propagates through the coder and the remainder through a reference delay. The output of the coder and the reference delay are combined by a coupler (C2), detected by a 20-GHz-bandwidth photodetector (PD) and displayed on a sampling oscilloscope.

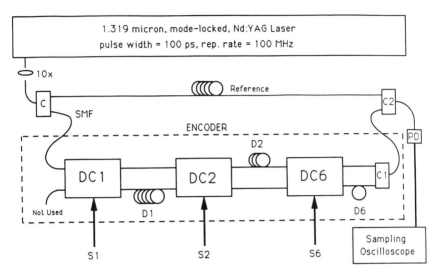

Figure 5 Experimental setup of OTDM encoder for 64 100-Mb/s channels. Heavy lines represent coaxial cable; light lines represent single-mode fiber (SMF). Though the complete encoder requires six stages, only the longest (D_1 = 5 ns), intermediate (D_2 = 2.5 ns) and shortest delays (D_6 = 156.25 ps) are implemented. DC1, DC2, and DC6 are directional couplers biased at either 0 volts (cross state) or 20 volts (bar state) by the control signals s_1, s_2, and s_6. C, C1, and C2 are passive couplers. Encoded delays are measured relative to a reference delay using 1.3-μm wavelength, 100-ps optical pulses generated by a mode-locked Nd : YAG laser with 100-MHz repetition rate, a photodetector (PD), and a sampling oscilloscope.

Shown in Figure 6 are the reference pulses (higher peaks) and encoded pulses (lower peaks), corresponding to the following time slots: (a) slot 0 (ideal delay 0 ns); (b) slot 1 (ideal delay 156.25 ps); (c) slot 16 (ideal delay 2.5 ns); (d) slot 32 (ideal delay 5 ns); (e) slot 49 (ideal delay 7.65625 ns). Slots 17, 33, and 48 were also encoded, but the results are not shown here. The time delay relative to slot 0, measured with the oscilloscope's cursor function, is shown for each case. The maximum aggregate error in measured delay is 2.5% of a time slot. This could be due to measurement error (positioning of the cursor on the oscilloscope) or an error in the length of fiber in stage 2.

The measured input and output powers of the coder are 399 and 1 μW, corresponding to approximately a 26-dB total insertion loss. This includes a 3-dB splitting loss in C1, and insertion losses of 3.5 dB in DC1, 4.5 dB in DC2, and 15 dB in DC6. These insertion losses are primarily due to fiber-to-waveguide coupling at the input and output of the directional couplers.

TIME-DIVISION OPTICAL MICROAREA NETWORKS 133

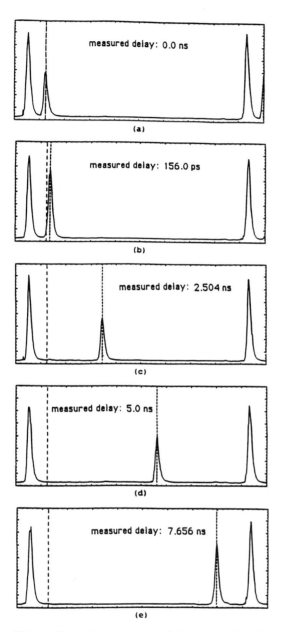

Figure 6 Reference pulses (higher peaks) and encoded pulses (lower peaks) for the following time slots (ideal delays): (a) 0 (0 ns); (b) 1 (156.25 ps); (c) 16 (2.5 ns); (d) 32 (5 ns); (e) 49 (7.65625 ns). The time interval between a cursor centered on slot 0 and a cursor centered on the delayed pulse is indicated.

In general, the most suitable technology for implementing the optical TDMA coder is determined by the number of channels N and the bit rate per channel $1/T$: $T/2$ determines the maximum delay required, whereas T/N determines the minimum delay as well as the precision required. For delays $T/2$ longer than approximately 50 ps ($1/T < 10$ Gb/s), fiber-optic delays are suitable due to the long path length required (> 1 cm). For delays shorter than 50 ps, integrated-optic waveguides are suitable, since lithographic techniques can routinely yield a precision of less than 1 μm (5 fs delay).

In situations where N is large, both long and short delays, using fibers and integrated optics, may be required. Here the main difficulty would be in trimming the fiber length with an error of less than, say, $T/10N$. For example, if $1/T = 1$ Gb/s and $N = 1000$ (10 stages), then the total error in delay should be less than 20 μm. This is easily achieved with integrated-optic waveguides, which would be used for the last seven stages of the coder, ranging in length from 1.25 cm ($T/16$) to 195 μm ($T/1028$). This precision could not easily be achieved with fiber-optic delays, however, which would be used for the first three stages of the coder, ranging in length from 10 cm ($T/2$) to 2.5 cm ($T/8$). A small static adjustment in the fiber length could be made, for example, by stretching the fiber at high temperature.

The maximum number of stages $\log_2 N$ that can be used is limited by the insertion loss. The insertion loss can be minimized in the integrated-optic delay stages by integrating all of the stages together, including the directional couplers and the waveguide delays, on a single substrate. The insertion loss in the fiber-optic delay stages can be compensated using optical amplifiers. In this way, the implementation of an optical TDMA coder capable of addressing 1000 1-Gb/s channels with subnanosecond reconfiguration time may be feasible.

C. Integrated Modulator/Transmitter

The main reason a mode-locked laser with an external modulator is used in Figure 3 is that we can modulate arbitrarily short optical pulses ($\tau < 1$ ps) with a modulator that need only operate at the bit rate $1/T$, which in turn provides extremely high throughput $1/\tau$ in the TDMA network. Note that to increase the pulse energy arriving at each modulator, a synchronous array of N mode-locked lasers may be used instead of a single laser.

The use of an off-chip optical source and an on-chip modulator also has other advantages compared with an on-chip laser diode from the perspective of circuit integration. These advantages include lower electrical drive power, electrical drive power that is independent of output optical power, greater ease of hybrid integration on silicon substrates, and occupying far

TIME-DIVISION OPTICAL MICROAREA NETWORKS

less real estate on the chip. Clearly, the electrical drive power required should be compatible with the other devices on the chip, and the ease of hybrid integration will influence the compatibility of these devices with existing silicon integrated circuit technology. For this reason, integrated modulators, particularly of the multiple quantum well type, have received considerable attention in recent years. A notable example is the demonstration of a GaAs–AlGaAs multiple quantum well (MQW) modulator grown on a silicon substrate [52].

Experimental measurements were made of the modulation depth, excess transmission loss, and required electrical drive power, to evaluate the performance of an MQW modulator in the system shown in Figure 3. The experimental setup is shown in Figure 7. The modulator consists of a p-i-n diode in which the intrinsic layer is an MQW composed of 60 140 Å thick GaAs quantum wells separated by 160 Å thick $Al_{0.3}Ga_{0.7}As$ barriers [53,54]. When a reverse voltage is applied to the diode, the band edge of the MQW shifts toward lower energy, and the exciton absorption peak flattens. These effects change the optical absorption. If the wavelength of the optical signal

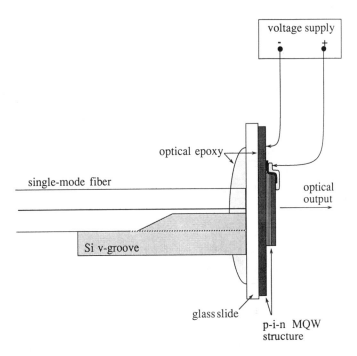

Figure 7 Experimental setup to evaluate performance of MQW modulator.

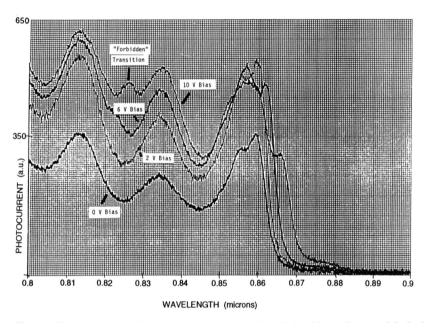

Figure 8 Measured photocurrent versus wavelength for bias voltages of 0, 2, 6, and 10 volts.

lies just above the band edge, a small change in applied reverse voltage can result in a large change in transmitted light.

In Figure 8, the photocurrent is shown as a function of wavelength for bias voltages of 0, 2, 6, and 10 volts. As the bias voltage is increased, the absorption edge shifts to longer wavelengths, as expected. The $n = 1$ heavy hole and $n = 1$ light hole exciton peaks are both clearly visible near the absorption edge. Periodic variations of the photocurrent with wavelength are due to residual Fabry–Perot interference effects. The additional peak near $\lambda = 0.825$ μm for large bias voltage is evidence of the formerly forbidden $n = 2$ heavy hole to $n = 1$ electron transition.

The normalized photocurrent is shown as a function of applied voltage in Figure 9 for $\lambda = 0.8597$ μm and for $\lambda = 0.8583$ μm. If the modulation depth β is defined in terms of the transmitted optical power P as

$$\beta = \frac{P_{max} - P_{min}}{P_{max}} \qquad (3)$$

then the maximum value of the modulation depth is $\beta = 0.5$, obtained for $\lambda = 0.8597$ μm and a bias voltage swing from $V_{min} = 0$ volts to $V_{max} = 5$ volts.

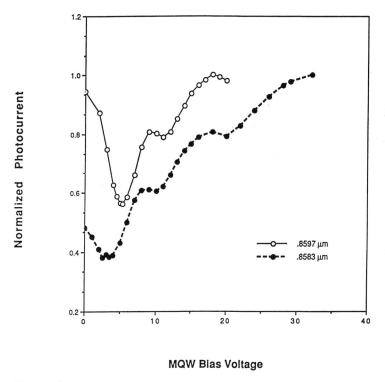

Figure 9 Normalized photocurrent versus applied voltage for $\lambda = 0.8597$ μm (solid line) and $\lambda = 0.8583$ μm (dashed line).

The sharp absorption edge in MQW devices is essential in achieving high modulation depths with low bias voltages, but also accounts for the extreme wavelength sensitivity of MQW modulators. For example, as seen in Figure 9, if the input light is detuned from the optimum wavelength by only 1.4 nm (i.e., $\lambda = 0.8583$ μm), the modulation depth is reduced substantially. The non-unity depth of the modulator results in a power penalty, which in turn can degrade the performance of the system, as will be discussed below.

The excess transmission loss of the modulator was measured to be 2.2 dB, which is caused primarily by Fresnel reflections due to index of refraction mismatch at the glass–GaAs and GaAs–air interfaces [55]. If a single-mode fiber were used rather than a power meter to collect the light exiting the modulator, the excess loss might exceed 2.2 dB. In the analysis below, we will assume that the total excess loss in this case is about 3 dB.

The average electrical drive power of the modulator is given by

$$\langle P_{\text{electrical drive}} \rangle = \frac{\omega C}{2} (V_{\text{min}}^2 - V_{\text{max}}^2) \tag{4}$$

where ω is the modulation frequency, C is the capacitance of the modulator, and the input electrical data stream is comprised of equiprobable 0 and 1 bits. The capacitance of the modulator was measured to be 20 pF when no light was applied and would be slightly higher with reverse bias applied and light absorbed [55]. The device tested had a large surface area (0.2 mm²), however. Since the area of the device needs to be only slightly larger than the 16 μm^2 of a single-mode fiber core (say, 20 μm^2), the capacitance could be scaled down to approximately $C = 2$ fF. The average electrical drive power of a 20 μm^2 area device is shown in Figure 10 for modulation frequencies $\omega = 5$ and 10 GHz. Note that the electrical drive power is independent of the average optical output power of the modulator,

$$\langle P_{\text{out}} \rangle = \frac{P_{\text{max}} + P_{\text{min}}}{2} \tag{5}$$

The electrical drive power of a laser diode, on the other hand, is directly proportional to the optical power output, where the constant of proportion-

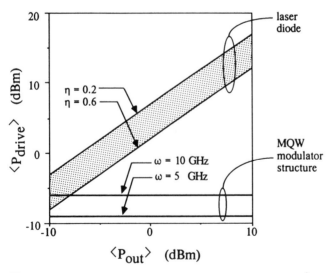

Figure 10 The average electrical drive power of a 20 μm^2 area MQW modulator for modulation frequencies $\omega = 5$ and 10 GHz. The electrical drive power of a laser diode is also plotted as a function of power output, for external quantum efficiencies ranging from $\eta = 0.2$ and 0.6 (shaded region).

ality is defined as the external quantum efficiency η. The electrical drive power of a laser diode is also plotted in Figure 10 as a function of power output, for external quantum efficiencies ranging from $\eta = 0.2$ to 0.6 (shaded region). It is seen in Figure 10 that for 0 dBm average optical power output, the MQW modulator requires about 10 dB less electrical drive power than the laser diode, and that this difference increases with optical power output.

D. Optical Correlation

In the correlation receiver, a delayed, attenuated version of the framing pulse is correlated with the TDMA frame. This process is equivalent to the logical AND of the framing pulse with the appropriate slot in the TDMA frame. This correlation function has been demonstrated with 2-ps optical pulses using a two-gap photoconductive AND gate shown in Figure 11a. In brief, the dual-gap photoconductive device consists of a 50-Ω microstripline circuit on an insulating substrate connecting two Fe-doped InGaAs photoconductive switches. When light impinges on a photoconductive gap in the electrically biased stripline, the resistance of the gap changes from high to low, allowing the current to propagate down the stripline. When two optical pulses (each of 2 ps width) simultaneously excite the two photoconductive gaps (points A and B in Figure 11b), an electrical pulse appears at the stripline output. On the other hand, if only one gap conducts (the two incident optical pulses do not coincide in time), a spurious signal of low amplitude is observed at the output. This spurious signal corresponds to dark current through either gap A or B, depending on whether an optical pulse is incident on gap A or B.

A more complete description of the photoconductive AND gate and its operation is now presented with the aid of Figure 11a. The device consists of a 1.4 μm thick layer of Fe-doped $In_{0.53}Ga_{0.47}As$ and a 0.5-μm Fe-doped InP buffer layer grown on a semiinsulating Fe-doped InP substrate by metal organic vapor-phase epitaxy (MOVPE). The electrical contacts consist of Au:Ge/Ni/Au conducting electrodes. Optically sensitive gaps in the electrode were produced by mesa etching and resulted in 5×5 μm^2 photoconductive gaps. The photoconductive gaps were then ion-damaged with Be at 750 keV with a bombardment dose of 6×10^{10} cm^{-2}. The ion bombardment reduced the free carrier lifetime from greater than 1 ns to less then 150 ps. The electron mobility was found to be 6000 $cm^2/V \cdot s$ by sheet resistance and Hall effect measurements [56]. The dark resistance of the device was measured to be 19 kΩ.

The operating speed of the gate was determined by using a time-resolved correlation technique. The laser source for this experiment was a

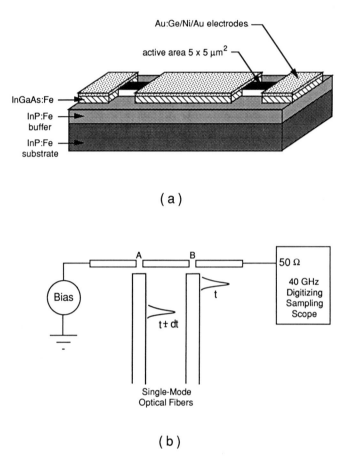

Figure 11 (a) Two-gap photoconductive AND gate consisting of a 50-Ω microstripline circuit on an insulating substrate connecting two Fe-doped InGaAs photoconductive switches. (b) When light impinges on a photoconductive gap in the electrically biased stripline, the resistance of the gap changes from high to low, allowing the current to propagate down the stripline. When two optical pulses simultaneously excite the two photoconductive gaps (points A and B), an electrical pulse appears at the stripline output. On the other hand, if only one gap conducts (the two incident optical pulses do not coincide in time), a spurious signal of low amplitude is observed at the output.

Quantronix Model 416 1.319-micron, mode-locked, Nd : YAG laser with 120-ps pulsewidth and 100-MHz repetition rate. The mode-locked pulses were pulse-compressed using a fiber-grating technique to obtain 2-ps pulses as measured with an optical autocorrelator. The output of the pulse compression system was divided into two equal intensity pulse trains, which initially traverse different but equal length paths. A relative delay was introduced between the two pulse trains by changing the length of one path with a stepping-motor-driven stage. Then, each pulse train was coupled into a short, equal length single-mode optical fiber. The optical coupling and electrical biasing arrangement of the photoconductive device for the correlation measurement is shown in Figure 11b. Each fiber is butt-coupled to one of the photoconductive gaps. The coupling efficiency between the 10-μm mode field diameter of the optical fiber and the 5 × 5 μm^2 gap, assuming a Gaussian transverse mode profile, was calculated to be 46.6%. The optical coupling arrangement delivered 1 pJ optical pulses to both of the photoconductive gaps. The energy incident on each gap was then 466 fJ. A 2.0 V dc bias was applied to the device. The response of the device was monitored by a Tektronix Model 11801 digitizing, sampling oscilloscope and a Tektronix SD-30 40-GHz sampling head with 50 Ω input impedance.

Figure 12 shows the measured impulse response of each of the photoconductive gaps. The trace denoted by the thin line is the impulse response of gap A. The trace denoted by the thick line is the impulse response of gap B. Each of the traces has the 8 mV dc potential due to the gate dark current removed. As can be seen from this figure, each gap has a similar impulse response. The measured rise time and fall time of the impulse response for each gap was 17 and 21 ps, respectively, and the FWHM was measured to be 24 ps. The fast impulse response of the photoconductive gaps can be attributed to the reduced carrier lifetime in this region resultant from ion-bombardment processing. Indeed, the measured fall time of the response of both gaps suggests that the carrier lifetime is less than 20 ps. It should be noted, however, that the measured impulse responses are approaching the measurement limitations of the oscilloscope imposed by the specified 8.8 ps time constant of the sampling head. A very notable feature apparent in the impulse response traces is the existence of slowly decaying, low amplitude tails that persist after the initial fast recovery of the photoconductance. These low amplitude tails have been observed in Si and GaAs photoconductor work performed by others [57,58]. For InGaAs-on-InP systems, the tails have been attributed to surface states or defects resident at the ternary–binary interface that trap and reemit carriers [59]. From the FWHM of the impulse response data, it would appear as though the device could operate as a logical AND gate at speeds approaching 40 Gbps. As

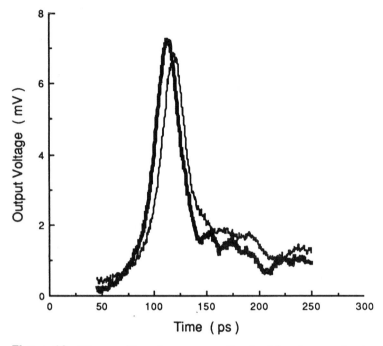

Figure 12 Measured impulse response of each of the photoconductive gaps. The trace denoted by the thin line is the impulse response of gap A. The trace denoted by the thick line is the impulse response of gap B. Each of the traces has the 8 mV dc potential due to the gate dark current removed.

will be seen, however, the slowly decaying tails severely limit the operating speed of the device.

The measured time-resolved correlation data is shown in Figure 13. The vertical axis corresponds to the peak output voltage of gap B at the time t as shown in Figure 11b. The horizontal axis is the time difference dt between the time of arrival of optical pulses at the gaps. Negative horizontal values indicate the optical pulse arriving at gap B is leading the optical pulse arriving at gap A, while positive horizontal values indicate the optical pulse arriving at gap B is lagging the optical pulse arriving at gap A. As in the impulse response measurements, the exciting pulses were 1-pJ, 2-ps optical pulses, and the correlation trace has the 8 mV dc potential from the gate dark current removed. For positive values of dt, the device has a slowly decreasing response. In this region of the curve, the pulse generated at gap B is sampling the tail of the signal generated at gap A. For negative values of dt, the device response is relatively flat. This result is due to the causality

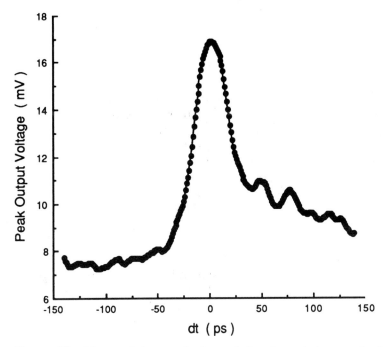

Figure 13 Measured time-resolved correlation data of photoconductive AND gate. The vertical axis corresponds to the peak output voltage of gap B at the time t. The horizontal axis is the time difference dt between the time of arrival of optical pulses at the gaps.

of the measurement system; i.e., the sampling pulse arrives at gap B before the signal pulse arrives at gap A. In the region around $dt = 0$, the response resembles the cross-correlation of the individual gap impulse responses [60]. Sampling is not a valid description of the operation of the device in this region, since the temporal width of the sampling and signal pulses are similar. If the device is performing a true cross-correlation operation (sampling is a special case of cross-correlation), then the response of the device in the region where dt is negative should be zero because the cross-correlation operations $S(T)$ involves a multiplication,

$$S(T) \propto \int_{-\infty}^{+\infty} f(t)h(t - T)\, dt \tag{6}$$

where $f(t)$ is the impulse response of gap A and $h(t)$ is the impulse response of gap B. In the case of sampling, $h(t)$ can be approximated by a delta function $\delta(t)$. In this case, the correlation operation becomes,

$$S(T) \propto \int_{-\infty}^{+\infty} f(t)\, \delta(t-T)\, dt \tag{7}$$

$$S(T) \propto f(T) \tag{8}$$

From Figure 13, however, it is clear that the response of the device is nonzero for negative values of dt. The residual 7.5-mV peak signal is a result of the nonzero dark conductivity of the InGaAs photoconductive gaps.

From a logical AND operation perspective, the nonzero dark conductivity of the gaps reduces the contrast ratio of the device. In this case, contrast ratio is defined as the ratio of the peak output voltage when the arrival of optical pulses on gaps A and B is coincident (i.e., $dt = 0$) to the peak output voltage when the arrival of optical pulses on gaps A and B differs by a minimum allowable pulse separation t (i.e., $dt \geq t$). The minimum allowable pulse separation is dictated by the recovery of the tail associated with the impulse response of the gaps. From Figure 13, it can be seen that the device is fully recovered for values of dt approximately equal to 200 ps. In this case, the contrast ratio is estimated from Figure 13 and found to be 2:1. Thus, the time-resolved measurement has shown the InGaAs AND gate to be capable of operating at 5 GHz (= 1/200 ps) with a contrast ratio of 2:1 at 466 fJ pulse energies incident on the photoconductive gap.

IV. OPTICAL TDMA μAN DEMONSTRATION AT 5 Gb/s

A. Experimental Setup and Results

Using the components previously described, an experimental 5-Gb/s optical TDMA network was built. Rather than adding the framing pulse directly to the TDMA frame, as described previously and shown in Figure 3, in the experimental configuration of the optical TDMA network we have chosen to deliver the framing pulse directly to the variable-integer-delay line coder, bypassing the star and splitter. This approach was necessary due to power constraints with the photoconductive AND gate. A schematic diagram of the optical TDMA network developed in this work is illustrated in Figure 14. The primary pulse source for this network is obtained from a Quantronix Model 4217 Nd : YLF, 1.313-μm wavelength, mode-locked laser operating at a 100-MHz repetition rate with 65-ps FWHM pulses. Thus, the system frame period T is 10 ns, which is equivalent to a 100 Mb/s baseband data rate. The user time slot τ was set at 200 ps, corresponding to a 5 Gb/s multiplexed data rate with N equal to 50 user channels. The optical pulses are passively distributed to the network by a -3-dB 1×2 splitter. This power splitting of the pulse performs the task of creating a framing

TIME-DIVISION OPTICAL MICROAREA NETWORKS 145

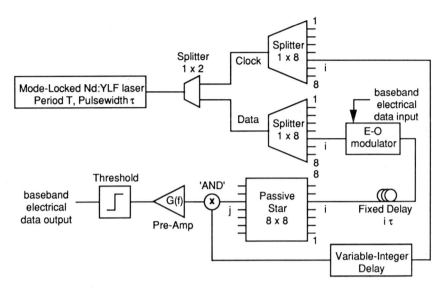

Figure 14 Schematic diagram of optical TDMA microarea network demonstration.

pulse separate from the data pulses, which is sent directly to the variable-integer-delay line coders of the receivers.

The clock and data pulses are distributed to the network by two Interfuse 1 × 8 passive spitters. The typical measured power loss between the input port and any output port for these 1 × 8 splitters was −8.7 dB. Seven of these output ports were used to transmit data to the network comprising 14 percent total network usage. Five of these user channels were fixed to transmit all "ones" data, while the other two channels had full modulation capability. Lithium niobate electrooptic modulators were used to transduce the 100-Mb/s baseband electrical signals onto the optical channels. Polarization sensitive Crystal Tech electrooptic modulators with electrical modulation bandwidths of 3 GHz, drive voltages of 8–10 V, and crosstalk ratios of approximately 31 dB were used in the setup. The measured insertion losses of these modulators were −7.3 and −6.7 dB. These losses are primarily due to waveguide-to-fiber coupling losses. The data channels were then transmitted through fixed optical delays corresponding to integer multiples of the 200 ps user time slot and combined in an Interfuse 8 × 8 passive star optical coupler. The measured insertion loss between any input port and any output port of the passive star-coupler was −9.4 ± 0.3 dB. Now, all of the output ports contain the multiplexed optical data composed of the user channels transmitting on the network. One of these output ports is con-

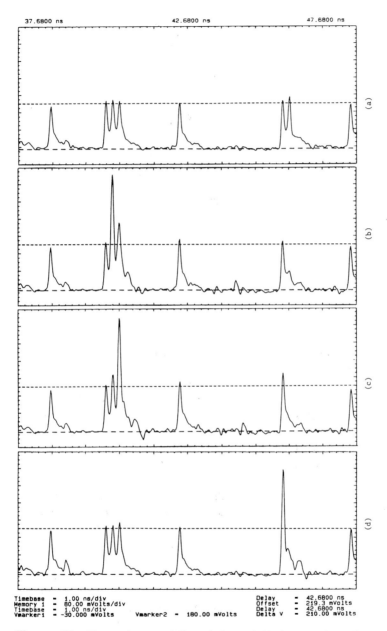

Figure 15 The tuning capability of the variable-integer-delay line coder. The series of traces shown here are the output from the photoconductive AND gate amplified by Anritsu A3H1001 10-GHz wide band amplifiers and measured with a

TIME-DIVISION OPTICAL MICROAREA NETWORKS 147

nected to each receiver, consisting of a photoconductive AND gate detector, a preamplifier, and the baseband recovery electronics.

The star-bypassed framing pulse is sent to a variable-integer-delay line coder capable of coding four user addresses. This variable-delay line coder consists of two $LiNbO_3$ electrooptic directional couplers, discrete lengths of single-mode optical fiber, and a 2 × 1 passive coupler. The total insertion loss of this variable-delay line coder was measured as -14.7 dB. Each channel receiver has a variable-delay line coder that delivers the framing pulse to an Fe-doped InGaAs photoconductive AND gate for correlation with a time slot from the multiplexed data stream emanating from the output of the passive star. The tuning capability of the variable-integer-delay line coder is shown in Figure 15. The series of traces shown here are the output from the photoconductive AND gate amplified by Anritsu A3H1001 10-GHz wide band amplifiers and measured with a 20-GHz Hewlett-Packard Model 54120A digitizing oscilloscope. Figure 15a shows the framing pulse tuned to channel 26 in the 10-ns frame of channels. Channel 26 is not transmitting data; thus, no correlation is performed. The detected pulses shown here are due to the dark current of the photodetector and are an artifact of the nonideal behavior of the photoconductive AND gate. Figure 15b shows the framing pulse tuned to channel 1, where the framing pulse is now correlated with the data transmitted in that channel. Figure 15c shows the framing pulse tuned to channel 2, which is 200 ps from channel 1, and Figure 15d shows the framing pulse tuned to channel 25, which is 5 ns from channel 1.

The electrical output from the photoconductive AND gate is further investigated with the aid of Figure 16. This figure shows the oscilloscope trace of an eye diagram resultant from the correlation of the framing pulse with a psuedorandom data pattern transmitted in user channel 1. The three consecutive pulses correspond to channels 50, 1, and 2 and are displayed on a 100 ps/div time scale. As can be seen from this figure, the contrast ratio, or the ratio of the peak signal received when the framing pulse is correlated with a transmitted "one" to the peak signal received when the framing

---

20-GHz Hewlett-Packard Model 54120A digitizing oscilloscope. (a) The framing pulse is tuned to channel 26 in the 10-ns frame. Channel 26 is not transmitting data; thus, no correlation is performed. The detected pulses shown here are due to the dark current of the photodetector and are an artifact of the nonideal behavior of the photoconductive AND gate. (b) The framing pulse is tuned to channel 1, where the framing pulse is now correlated with the data transmitted in that channel. (c) The framing pulse is tuned to channel 2, which is 200 ps from channel 1. (d) The framing pulse is tuned to channel 25, which is 5 ns from channel 1.

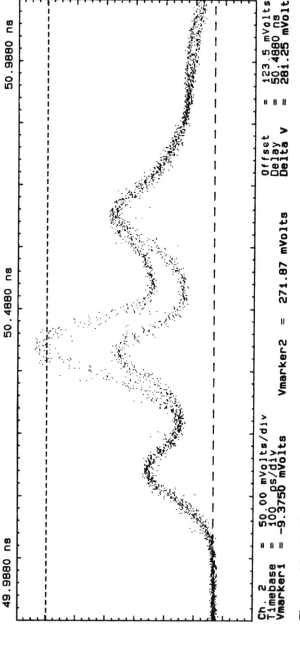

Figure 16 Oscilloscope trace of an eye diagram resultant from the correlation of the framing pulse with a psuedorandom data pattern transmitted in user channel 1. The three consecutive pulses correspond to channels 50, 1, and 2 and are displayed on a 100 ps/div time scale.

pulse is correlated with a transmitted "zero," is approximately 2:1. The energy per optical pulse incident on the photoconductive device was calculated to be approximately 1.4 pJ using the measured average transmitted power per channel and the component insertion loss data. Including the estimated optical fiber to photoconductive gap coupling efficiency of 46.6%, the energy per optical pulse incident on a photoconductive gap is 652 fJ. An increase in the contrast ratio can be achieved by increasing the energy of the optical pulses incident on the photoconductive gaps. This increase can be realized by reducing the component insertion losses in the network, increasing the optical fiber to photoconductive gap coupling efficiency, or increasing the energy of the transmitted pulse.

The next task is to threshold detect the correlated data from the photoconductive detector and recover the baseband 100-Mb/s electronic signal. As will be seen, given the low-noise characteristics of the TDMA network and the photoconductive detector, this contrast ratio is found to be sufficient to insure a 10^{-9} bit error rate (BER) transmission system.

B. Noise and BER Analysis

The BER performance of the photoconductive AND gate is dependent not only on the contrast ratio of the device, but also on the noise characteristics of the photoconductor itself and the surrounding electronic circuitry. The noise current generator equivalent circuit model used for the AND gate is shown in Figure 17. To simplify the analysis, only one of the 10-GHz preamplifiers necessary for the receiver circuitry has been included in the noise and BER analysis. The noise generated in the photoconductive gate consists predominantly of generation–recombination and thermal noise, modeled as noise current sources i_{Ngr} and i_{Nt}, respectively. The mean-squared generation–recombination noise term $\langle i_{Ngr}^2 \rangle$ is similar in form to shot noise. For photoconductors, however, the photoconductive gain τ_0/τ_d must be included in the noise estimation, where τ_0 is the hole lifetime and τ_d is the electron transit time across the photoconductive gap. The addition of the photoconductive gain in the generation–recombination noise equation is the origin of a term that decreases as $1/\nu$, where ν is the frequency [61]. The amplifier contributes its own noise to the system, as shown by the noise voltage source at the output of the amplifier in Fig. 17. The amplifier mean-squared noise voltage can be written as

$$\langle v_{Na}^2 \rangle = 4k(F-1)\,290\,\Delta\nu\,R_i \tag{9}$$

The mean-squared noise currents for the gate are estimated from

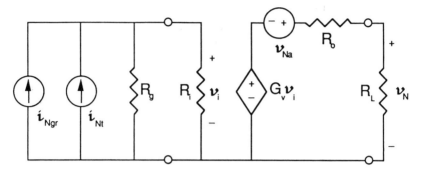

Figure 17 The noise current generator equivalent circuit model used for the photoconductive AND gate.

$$\langle i_{Ngr^2} \rangle = 4e\,\Delta\nu \left(\frac{\tau_o}{\tau_d}\right) \left[\frac{\langle i_s \rangle + i_d}{1 + 4\pi^2 \tau_o^2 \nu^2}\right] \quad (10)$$

$$\langle i_{Nt^2} \rangle = \frac{4kT\,\Delta\nu}{R_g} \quad (11)$$

The terms used in Eqs. (9), (10), and (11) are defined as

$\langle i_s \rangle$ = mean signal current in one bit period (312.8 µA),
T = room temperature (290 K)
i_d = gate dark current (160 µA)
$\Delta\nu$ = circuit bandwidth (10 GHz)
ν = reference frequency (0 Hz)
e = electron charge (1.602 × 10^{-19} C)
τ_o = hole lifetime (< 20 ps)
τ_d = electron transit time (42 ps)
k = Boltzmann's constant (1.381 × 10^{-23} J/K)
F = amplifier noise figure (5)
R_g = gate dark resistance (19 kΩ)
R_i = amplifier input impedance (50 Ω)

The mean-squared noise voltage at the output of the amplifier is given by

$$\langle v_{N^2} \rangle = \left(\frac{1}{2}\right)^2 \left[\langle v_{Na^2} \rangle + \left[\frac{G_v R_g R_i}{R_g + R_i}\right]^2 (\langle i_{Nt^2} \rangle + \langle i_{Ngr^2} \rangle)\right] \quad (12)$$

where G_v is the voltage gain of the amplifier, measured to be approximately 5. The calculated noise power for the amplifier, generation–recombination, and thermal noise sources at the output of the amplifier are -67.9, -67.5, and -85.8 dBm, respectively. From these values, it is clear that the AND gate circuit is dominated by amplifier and generation–recombination noise. For the case of coincident clock and signal pulses, substitution of the numeric values shown above into Eq. (12) result in a calculated noise power of -64.7 dBm over the 10-GHz bandwidth of the AND gate circuit. The calculated signal power for this case, using experimental data similar to the data shown in Figure 16 and including a voltage gain of 5, is -8 dBm. The resultant theoretical signal-to-noise ratio (SNR) is 56.7 dB. The low-noise characteristics of the gate become even more evident when the calculated noise power is normalized to a 1-Hz frequency band. This results in a noise power spectral density of -104.7 dBm/Hz.

The low-noise characteristics of the gate were verified using a Hewlett-Packard 70000 Series, 22-GHz spectrum analyzer, with a noise power spectral density of -147 dBm/Hz (measured with a 50-Ω load). The measured frequency spectrum of the gate output over a 12-GHz band is shown in Figure 18, and the noise power of the output spectrum is seen to be approximately -65 dBm. This measured value is in good agreement with the

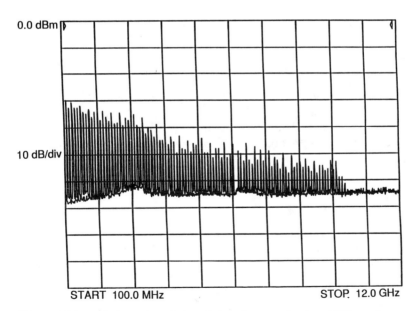

Figure 18 Noise characteristics of the photoconductive AND gate over a 12-GHz band, measured using a Hewlett-Packard 70000 Series, 22-GHz spectrum analyzer, with a noise power spectral density of -147 dBm/Hz (measured with a 50-Ω load).

calculated noise power. Clearly, the high SNR of the photoconductive AND gate should lead to satisfactory BER performance despite the low contrast ratio of the device.

Indeed, the BER of the photoconductive AND gate operating at 5 Gb/s can be estimated from the following expression, assuming Gaussian statistics and equiprobable 1 and 0 events:

$$P_e = \frac{1}{2\sqrt{2\pi}\,\sigma_1} \int_{-\infty}^{v_T} \exp\left[-\frac{(v-\langle v_1\rangle)^2}{2\sigma_1^2}\right] dv$$
$$+ \frac{1}{2\sqrt{2\pi}\,\sigma_0} \int_{v_T}^{+\infty} \exp\left[-\frac{(v-\langle v_0\rangle)^2}{2\sigma_0^2}\right] dv \qquad (13)$$

After some algebra, Eq. (13) can be reduced to

$$P_e = \frac{1}{4}\left[\operatorname{erfc}\left(\frac{\langle v_1\rangle - v_T}{\sqrt{2}\,\sigma_1}\right) + \operatorname{erfc}\left(\frac{v_T - \langle v_0\rangle}{\sqrt{2}\,\sigma_0}\right)\right] \qquad (14)$$

The terms in Eq. (14) are defined as

v_T = threshold voltage
$\langle v_1\rangle$ = mean signal voltage in one bit period when a 1 is received
$\langle v_0\rangle$ = mean signal voltage in one bit period when a 0 is received
$\sigma_1 = \sqrt{\langle v_{N^2}\rangle}$, rms noise voltage when a 1 is received
$\sigma_0 = \sqrt{\langle v_{N^2}\rangle}$, rms noise voltage when a 0 is received

The rms noise voltages, σ_1 and σ_0, have values of 130.8 and 118.9 μV, respectively. The mean signal voltages $\langle v_1\rangle$ and $\langle v_0\rangle$ are 78.2 and 39.1 mV, respectively, as evaluated from recorded data. The threshold voltage for the BER calculation was set at the mean of $\langle v_1\rangle$ and $\langle v_0\rangle$, i.e., $v_T = 58.66$ mV. The small rms noise voltages with respect to the mean signal voltages result in a calculated BER of less than 10^{-15}.

C. Power Budget Analysis

The maximum size of the μAN is limited by either the duty cycle of the mode-locked laser or the optical power budget. The mode-locked pulse-compressed laser described previously would permit $N = 5000$, $\tau = 2$-ps time slots in a $T = 10$-ns TDMA frame. The question then arises as to whether the transmitter can deliver sufficient energy to the receiver to recover the data with less than 10^{-9} bit error rate, taking into account the splitting and excess losses in the system.

If the energy per pulse from the optical source is E_L, and the total optical loss in the system is L (dB), then the energy E_d reaching the detector is

$$E_d = E_L \, 10^{-L/10} \quad [J] \tag{15}$$

where

$$L = L_{\text{mod}} + L_{\text{cod}} + 3 + 10 \log\left(\frac{N}{\beta}\right) \quad [\text{dB}] \tag{16}$$

Here, L corresponds to the case where an array of N synchronous mode-locked lasers are used, and β is the modulation depth. The last term represents the splitting loss in the star-coupler. If instead a single mode-locked laser is used, it would be followed by a $1 \times N$ splitter (as shown in Fig. 3), and N in the last term would be replaced by N^2. The loss of 3 dB in this expression corresponds to the splitting loss at the input to the correlator. The excess loss in the coder, L_{cod}, depends on the number of stages, and it can be quite large. As an example, we will assume that the entire coder is integrated on a single substrate, so that the insertion loss is primarily due to the fiber-to-waveguide coupling, and $L_{\text{cod}} = 14$ dB. In this example, the excess loss in the modulator L_{mod} is taken to be 3 dB, and the modulation depth is assumed to be unity. Finally, the energy per pulse produced by each laser is taken to be $E_L = 20$ nJ. With these assumptions, the energy reaching the detector is plotted as a function of N in Figure 19 for the case of an array of N synchronous mode-locked lasers (dashed line) and the case of a single mode-locked laser (solid line). In order for the receiver to recover the data with an acceptable bit error rate (say, 10^{-9}), the sensitivity of the detector must be less than the received energy. Sensitivity is defined as the minimum amount of energy required at the receiver input to yield a given acceptable bit error rate. The sensitivities of several detectors at 1.3 μm are indicated in Figure 19, including the photoconductive AND gate (466 fJ), a pn diode (8 fJ), a pin diode (0.32 fJ), an avalanche photodiode (15.8 aJ), and the quantum limit (3.2 aJ). With the photoconductive AND gate, which has rather poor sensitivity, the number of nodes that can be accommodated in the network is between about $N = 15$ and $N = 1000$, depending on whether a single laser or an array of lasers is used. For more sensitive detectors, the allowed number of nodes on the network increases rapidly.

V. CONCLUSION

Optical micro-area networks (μANs) can provide flexible communications among VLSI processors and eliminate electrical I/O bottlenecks. Shared-medium multiple access protocols avoid the access delays associated with statistical multiple access protocols (which are unacceptable in multi-processor applications) and increase the throughput of high data-rate

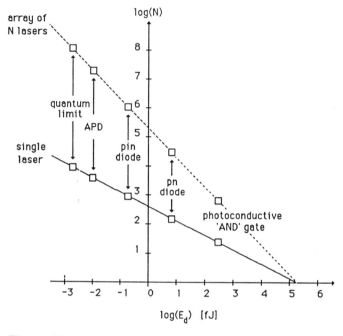

Figure 19 Dimension N of microarea network versus the energy reaching the detector for the case of an array of N synchronous mode-locked lasers (dashed line) and the case of a single mode-locked laser (solid line). The sensitivities of several detectors at 1.3 µm are indicated, including the photoconductive AND gate (466 fJ), a pn diode (8 fJ), a pin diode (0.32 fJ), an avalanche photodiode (15.8 aJ), and the quantum limit (3.2 aJ).

µANs, at the expense of wasting the fiber's bandwidth. Time-division multiple access (TDMA) may be more practical to implement in a µAN than other shared-medium multiple access protocols such as frequency division or code division.

Since the total throughput of TDMA is given by $1/\tau$, the throughput can be increased by making τ small. Accomplishing this goal requires avoiding the use of low-bandwidth electronics in the portion of the µAN that directly processes these short pulses. Instead, optical processing can be used in those portions of the network. The architecture of a TDMA µAN that uses optical multiple access processing and is self-clocking was described in detail. Experimental demonstrations of key subsystems for optically generating, modulating, synchronizing, delaying, and correlating short optical pulses were presented.

The feasibility of a variable-integer-delay line, which provides rapid tuning, wide tuning range, and high precision, was investigated and demonstrated for $N = 64$ 100-Mb/s channels ($T = 10$ ns and $\tau = 156.25$ ps). A delay-line capable of addressing $N = 1000$ 1-Gb/s channels with subnanosecond reconfiguration time appears to be feasible.

The main reason a mode-locked laser with an external modulator is used in the TDMA μAN is that arbitrarily short pulses ($\tau < 1$ ps) can be modulated with a modulator that need only operate at the bit rate $1/T$, which in turn provides extremely high total throughput ($1/\tau > 1$ THz). The use of an off-chip optical source and an on-chip modulator also has circuit-integration advantages over a laser diode, including lower electrical drive power, electrical drive power that is independent of output optical power, greater ease of hybrid integration on silicon substrates, and less required real estate on the chip. Experimental measurements of the modulation depth, excess transmission loss, and required electrical drive power of an MQW modulator were presented and related to system performance.

The optical correlation receiver was demonstrated using 2-ps optical pulses and a two-gap photoconductive AND gate. The sensitivity, rise time, and fall time of the device were measured. The demonstration of a 5-Gb/s optical TDMA μAN was presented, and its performance was analyzed.

The maximum size of the TDMA μAN is limited by either the duty cycle of the mode-locked laser or the optical power budget. The mode-locked pulse-compressed laser used would permit $N = 5000$, $\tau = 2$-ps time slots in a $T = 10$-ns TDMA frame (i.e., 5000 100-Mb/s users). Given the 20-nJ pulse energy produced by the laser, the 250-fJ sensitivity of the photoconductive AND gate, the splitting loss in the star-coupler, the excess loss (3 dB) and modulation depth (assumed to be unity) of the modulator, and the excess loss of the variable optical delay (assumed to be 14 dB), the power budget of the TDMA μAN was calculated. From power budget considerations, on the order of $N = 1000$ nodes in the TDMA μAN is possible.

In practice, as N increases toward 1000 nodes or the baseband data rates approach Gb/s speeds, the performance of an optical TDMA network places strict requirements on very precise, low-loss components. In reality, much advancement in selected areas of photonics and microelectronics needs to be achieved before large optical networks of this or any type can be implemented in the field. For example, as evidenced from the experimental network previously presented, very high insertion losses were encountered in incorporating the lithium niobate modulators into the experimental system. Indeed, fiber-to-waveguide and waveguide-to-fiber coupling efficiency improvements are extremely important in attaining the low insertion loss requirements of L_{mod} and L_{cod} in Eq. (16).

Coupling efficiencies can be improved by using tapered lens, hyperbolic, or D-shaped optical fibers or thin-film layered structures [62]. These techniques better match the electromagnetic field patterns between the optical fiber and waveguide structures. A fiberless waveguide media for μANs, however, such as integrated photonic components on a substrate or board, would eliminate the fiber coupling losses completely. To do so, the integrated photonic materials must have the low propagation loss of optical fiber as well as fabrication (as for a substrate) or packaging (as for a hybrid board) compatibility with the interconnect's active components. Lithographically patterned silica [63] and polymer [64] materials show great promise in meeting these concerns as well as having the advantage of being compatible with current microfabrication technology. Also, as N grows large in a network, micron level tolerances are required to precisely set the passive delays for the address signatures and coding of the channels in the TDMA environment. Microfabrication techniques easily meet these tolerances and can also be processed over wide areas. In fact, these materials have already been used in the successful development of passive components such as optical star-couplers [65], which reduce the physical size of the interconnect.

Apart from the research in passive interconnect materials, further development of the active components such as transmitters, modulators, switches, and receivers is necessary in order to reduce the size and power consumption of the current experimental testbed presented in Section IV.A. The large Quantronix laser source could be replaced with a compact, 5-GHz mode-locked, optically pumped Nd : YLF laser [66] or a 40-GHz semiconductor colliding-pulse mode-locked laser [67]. These types of lasers have the capability of providing very high baseband signals for the network, but still rely on high rf-bandwidth electronics to drive the off-chip modulators. For a distributed array of N laser sources, external cavity semiconductor traveling-wave amplifiers [68] could prove useful as a compact, low-power transmitter. These sources can easily achieve mode-locked repetition rates on the order of hundreds of megahertz to a few gigahertz and achieve the narrow pulses required for an interconnect with large N.

The most critical component as N grows large and the multiplexed data rate increases is the high-speed detection, correlation, and thresholding of the optoelectronic receiver. As presented, the current photoconductive AND gate has operating speeds of at least 5 Gb/s. The nonideal behavior of operation resultant from the gap leakage currents and resultant low con-

TIME-DIVISION OPTICAL MICROAREA NETWORKS

trast ratio, however, places stringent requirements on the electronic thresholding device. This thresholding device must have gigahertz bandwidth in order to follow the fast rising edge of the incoming signal (see Fig. 16), and have the sensitivity to trigger a pulse when thresholding in a selected time slot, while not integrating over successive time slots. This behavior is not readily available with current commercial electronics; thus, improvements in the photonic device's correlation operation (i.e., a higher contrast ratio and thus more ideal AND operation) is needed. Indeed, it is the intent of the TDMA architecture to reduce the need for high-bandwidth electronics (i.e., using electronics that need operate at only the baseband channel rates) and photonically process the high-bandwidth, multiplexed data stream.

One method to photonically process the multiplexed data stream is to use an all-optical logic device such as a multiple quantum well (MQW) etalon. Such a device would be designed to perform the thresholding function based on input optical intensity. As illustrated in Figure 20, incoming multiplexed data pulses have optical intensities below the threshold level and do not affect the resonant characteristics of the etalon. These optical pulses are blocked by the device. When the framing and data pulses are coincident, however, a larger optical intensity is incident on the device, which changes the resonant characteristics of the etalon. The device becomes more transparent and allows the coincident optical data pulse to pass through, forming the demultiplexed data stream. The demultiplexed data would subsequently be photodetected and converted to the baseband electrical signal.

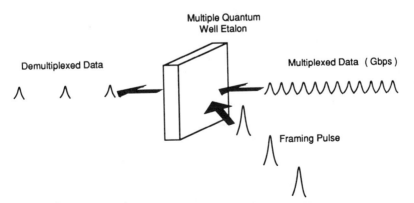

Figure 20 MQW etalon as an all-optical logic device to perform the thresholding function based on input optical intensity.

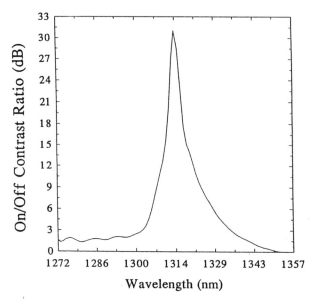

Figure 21 On/off constrast ratio (in dB) versus wavelength for a GaAlInAs/AlInAs multiple quantum well asymmetric Fabry–Perot reflector etalon, which takes advantage of the large absorptive and refractive nonlinearities associated with pump beam induced saturation of the heavy hole exciton resonance. The contrast ratio exceeds 30 dB for an optical pump intensity of 30 kW/cm^2.

These devices have the ability to achieve very large contrast ratios. Recently, we have designed and fabricated a high contrast, low intensity of GaAlInAs/AlInAs multiple quantum well asymmetric Fabry–Perot reflector etalon for operation at 1.3-μm wavelengths [69]. The reflection etalon takes advantage of the large absorptive and refractive nonlinearities associated with pump beam induced saturation of the heavy hole exciton resonance. As shown in Figure 21, these effects resulted in an all-optical modulator that achieved a contrast ratio exceeding 30 dB for an optical pump intensity of 30 kW/cm^2. Now, the demultiplexing burden is on the speed of the optical etalon device. These devices also need high bandwidths in order to follow the rising edge of the incoming signals and recover before the next time slot arrives. Etalon devices typically have turn-on times of less than 1 ps; however, the recovery time of these devices can be on the order of nanoseconds. Recently, we measured an MQW etalon with a recovery time of approximately 125 ps [61]. This recovery time is on the order of what is necessary for an all-optical logical AND gate to work in gigabit–picosecond optical TDMA networks.

REFERENCES

1. R. W. Keyes, Physical limits on digital electronics, *Proc. IEEE*, *63*:740 (1975).
2. J. W. Goodman, F. J. Leonberger, S. Y. Kung, and R. A. Athale, Optical interconnections for VLSI systems, *Proc. IEEE*, *72*:850 (1984).
3. P. R. Prucnal, VLSI fiber optic LAN, *Proc. International Optical Computing Conf.*, *700*:230 (1986).
4. P. R. Prucnal, VLSI optical interconnection networks, *SPIE Proc. Integration and Packaging of Optoelectronic Devics*, *703*:106 (1986).
5. P. R. Prucnal, Optical interconnections for VLSI local area networks, *IEEE Electrotechnol. Rev.*, 97 (1986).
6. S. E. Elby and P. R. Prucnal, Fiber-optic interconnect devices for VLSI micro-area networks, SPIE *Proc. Soc. Photoopt. Inst. Eng.*, *994*:77 (1988).
7. P. R. Prucnal and P. A. Perrier, Optical interconnect switch, *SPIE Proc. Soc. Photoopt. Inst. Eng.*, in press.
8. P. R. Prucnal, An optical interconnect switch, *Proc. Materials Res. Soc.*, MRS, Pittsburgh, PA, 1989, p. 354.
9. W. Stallings, *Local Networks, An Introduction*, Macmillan, New York, 1984.
10. R. A. Linke, "Frequency division multiplexed optical networks, *IEEE Network Magazine*, *3*:13 (1989).
11. I. P. Kaminow, Non-coherent photonic frequency-multiplexed access networks, *IEEE Network Magazine*, 3:4 (1989).
12. P. A. Perrier and P. R. Prucnal, Wavelength-division integration of services in fiber-optic networks, *Int. J. Analog and Digital Cabled Systems*, *1*:149 (1988).
13. P. R. Prucnal, M. A. Santoro, and T. R. Fan, Spread spectrum fiber optic local area network using optical processing, *IEEE J. Lightwave Technol.*, *LT-4*:547 (1986).
14. P. R. Prucnal, M. A. Santoro, and S. K. Sehgal, Ultrafast all-optical synchronous multiple access fiber networks, *J. Sel. Areas Comm.*, *SAC-4*:1484 (1986).
15. M. A. Santoro and P. R. Prucnal, Asynchronous fiber optic LAN using CDMA and optical correlation *Proc. IEEE*, *75*:1336 (1987).
16. W. C. Kwong, P. A. Perrier, and P. R. Prucnal, Performance comparison of code-division multiple access and synchronous spread spectrum multiple access, *IEEE Trans. Comm.*, *COM-39*:1625 (1991).
17. G. R. Cooper and R. W. Nettleton, A spread spectrum technique for high capacity mobile communications, *IEEE Trans. Vehicular Technol.*, *27*:264 (1978).
18. H. J. Kochevar, Spread spectrum multiple access communication experiment through a satellite, *IEEE Trans. Comm.*, *COM-27*:853 (1979).
19. R. L. Pickholtz, D. L. Schilling, and L. B. Milstein, Theory of spread-spectrum communications: A tutorial, *IEEE Trans. Comm.*, *COM-30*:855 (1982).
20. M. B. Pursley, Spread spectrum multiple-access communications in multi-user communication systems (G. Longo, ed.), Spring-Verlag, New York 1981.
21. L. Kleinrock and F. A. Tobagi, Packet switching in radio channels: Part I—

Carrier sense multiple-access modes and their throughput-delay characteristics, *IEEE Trans. Comm.*, *COM-23*:1400 (1975).
22. J. Y. Hui, Throughput analysis for code division multiple accessing for spread spectrum channel, *J. Sel. Areas Comm.*, *SAC-2*:482 (1984).
23. J. Y. Hui, Pattern code modulation and optical decoding: A novel code division multiplexing technique for multifiber network, *J. Sel. Areas Comm.*, *SAC-3*:916 (1985).
24. W. C. Kwong and P. R. Prucnal, Synchronous CDMA demonstration for fiber-optic networks with optical processing, *Electron. Lett.*, *26*:1990 (1990).
25. J. A. Salehi, Code division multiple-access techniques in optical fiber networks—Part I: Fundamental principles, *IEEE Trans. Comm.*, *COM-37*:824 (1989).
26. J. A. Salehi and C. A. Brackett, Code division multiple-access techniques in optical fiber networks—Part II: Systems Performance Analysis, *IEEE Trans. Comm.*, *COM-37*:834 (1989).
27. J. A. Salehi, Emerging optical code-division multiple access communications systems, *IEEE Network*, *1*:31 (1989).
28. G. J. Foschini and G. Vannucci, Using spread-spectrum in a high capacity fiber-optic local area network, *J. Lightwave Technol.*, *LT-6*:370 (1988).
29. G. Vannucci, Combining frequency-division and code-division multiplexing in a high-capacity optical network, *IEEE Network*, *3*:21 (1989).
30. G. Vannucci and S. Yang, Experimental spreading and despreading of the optical spectrum, *IEEE Trans. Comm.*, *COM-37*:770 (1989).
31. D. D. Sampson and D. A. Jackson, Coherent optical fiber communications system using all-optical correlation processing, *Op. Lett.*, *15*:585 (1990).
32. D. D. Sampson and D. A. Jackson, Spread-spectrum optical fibre network based on pulsed coherent correlation, *Electron. Lett.*, *26*:1550 (1990).
33. M. E. Marhic and Y. L. Change, Pulsed coding and coherent decoding in fibre-optic ladder networks, *Electron. Lett.*, *25*:1535 (1989).
34. S. Matsunaga and R. Gagliardi, Digital signaling with code-division multiple access in optical fiber communications, Univ. Southern Calif., Tech. Rep. CSI-88-02-03, 1988.
35. R. E. Blahut, *Theory and Practice of Error Control Codes*, Addison-Wesley Reading, MA, 1984.
36. A. A. Shaar and P. A. Davis, Prime sequence: quasi-optimal sequences for OR channel code division multiplexing, *Electron. Lett.*, *19*:888 (1983).
37. Z. Kostic, S. V. Maric, and E. L. Titlebaum, A new family of algebraically designed optical orthogonal codes, Proc. 28th Annual Alerton Conf. Commun., Control, and Computing, Univ. of Illinois, Urbana-Champaign, IL, 1990.
38. S. Tamura, S. Nakano, and K. Okazaki, Optical code-multiplex transmission by gold sequences, *J. Lightwave Technol.*, *LT-3*:121 (1985).
39. F. R. K. Chung, J. A. Salehi, and V. K. Wei, Optical orthogonal codes: Design, analysis, and applications, *IEEE Trans. Inform. Theory*, *IT-35*:595 (1989).
40. R. Petrovic and S. Holmes, Orthogonal codes for CDMA optical fibre LANs with variable bit interval, *Electron. Lett.*, *26*:662 (1990).

41. R. I. MacDonald, Fully orthogonal optical-code multiplex for broadcasting, *Opt. Lett.*, *13*:539 (1988).
42. A. S. Holmes and R. R. A. Syms, All-optical CDMA using "Qusai-prime" code, *J. Lightwave Technol.*, *LT-10*:279 (1992).
43. Y.L. Chang and M. E. Marhic, 2^n codes for optical CDMA and associated networks, Proc. IEEE/LEOS Summer Topical Meetings, Monterey, CA, 1990, pp. 23–24.
44. H. Chung and P. V. Kumar, Optical orthogonal codes—new bounds and an optimal construction, *IEEE Trans. Inform. Theory*, *IT-36*:866 (1990).
45. S. E. Miller and I. P. Kaminow (eds.), *Optical Fiber Telecommunications* II, Academic Press, California, 1988.
46. K. P. Jackson, S. A. Newton, B. Moslehi, M. Tur, C. C. Cutler, J. W. Goodman, and H. J. Shaw, Optical fiber delay-line signal processing, *IEEE Trans. Microwave Theory and Techniques*, *MTT-33*:193 (1985).
47. B. Moslehi, J. W. Goodman, M. Tur, and H. J. Shaw, Fiber-optic lattice signal processing, *Proc. IEEE*, *72*:909 (1984).
48. P. R. Prucnal, M. A. Santoro, S. K. Sehgal, and I. P. Kaminow, TDMA fiber optic network with optical processing, *Electron. Lett.*, *22*:1218 (1986).
49. P. R. Prucnal, D. J. Blumenthal, and M. A. Santoro, A 12.5 Gbps fiber-optic network using all-optical processing, *Electron. Lett.*, *23*:629 (1987).
50. R. A. Thompson, Optimizing photonic variable-integer-delay circuits, Topical Meeting on Photonic Switching, Incline Village, NV, 1987, Technical Digest, vol. 13, paper FD4.
51. P. R. Prucnal, M. F. Krol, and J. L. Stacy, Demonstration of a rapidly tunable optical time-division multiple-access coder, *IEEE Photonics Technol. Lett.*, *3*:170 (1991).
52. K. W. Goossenn et al. GaAs–AlGaAs multiquantum well refection modulators grown on GaAs and silicon substrates, *IEEE Photonics Technol. Lett.*, *1*:304 (1989).
53. W. D. Goodhue et al. Quantum well charge-coupled device-addressed MQW spatial light modulators *J. Vac. Sci. Tech.*, *B4*:769 (1986).
54. H. S. Cho and P. R. Prucnal, Effect of parameter variations on the performance of GaAs/AlGaAs multiple quantum well electroabsorption modulators, *IEEE J. Quantum Electron.*, 25 (1989).
55. P. R. Prucnal, S. D. Elby, and K. B. Nichols, Optical transmitter for fiberoptic interconnects," *Opt. Eng.*, *29*:1136 (1991).
56. E. Desurvire, B. Tell, I. P. Kaminow, K. F. Brown-Goebeler, C. A. Burrus, B. I. Miller, and U. Koren, 1 GHz GaInAs:Fe photoconductive optical AND gate with ~100 fJ switching energy for time-division access networks, *Electron. Lett.*, *25*:105 (1989).
57. C. H. Lee, *Picosecond Optical Devices*, Academic Press, Orlando, FL, 1984, pp. 73–117.
58. M. B. Johnson, T. C. McGill, and N. G. Paulter, Carrier lifetimes in ion-damaged GaAs, *Appl. Phys. Lett.*, *54*:2424 (1989).
59. P. M. Downey, R. J. Martin, R. E. Nahory, and O. G. Lorimor, High speed, ion bombarded InGaAs photoconductors, *Appl. Phys. Lett.*, *46*:396 (1985).

60. A. Yariv, *Optical Electronics*, 4th ed., Saunders College Publishing, Philadelphia, PA, 1991, pp. 411–418.
61. C. C. Hsu, B. P. McGinnis, J. P. Sokoloff, G. Khitrova, H. M. Gibbs, N. Peyghambarian, S. T. Johns, and M. F. Krol, Room-temperature optical nonlinearities of GaInAs/AlInAs and GaAlInAs/AlInAs multiple quantum wells and integrated-mirror etalons at 1.3 µm, submitted for publication.
62. Y. Cai, T. Mizumoto, E. Ikegami, and Y. Naito, An effective method for coupling single-mode fiber to thin-film waveguide, *J. Lightwave Technol.*, 9:577 (1991).
63. A. Himeno, M. Kobayashi, and H.Terui, High-silica single-mode optical reflection bending and intersecting waveguides, *Electron. Lett.*, 21:1020 (1985).
64. R. R. Krchnavek, G. R. Lalk, and R. Denton, *Materials for Optical Information Processing* (C. Warde, J. Stamatoff, and W. Wang, eds.), Materials Research Society, Symposia Proceedings, Vol. 228, Pittsburgh, PA, 1992, pp. 95–100.
65. C. Dragone, C. A. Edwards, and R. C. Kistler, Integrated optics $N \times N$ multiplexer on silicon, *IEEE Photonics Technol. Lett.*, 3:896 (1991).
66. P. A. Schulz and S. R. Henion, 5-GHz mode locking of a Nd:YLF laser, *Op. Lett.*, 16:1502 (1991).
67. Y. K. Chen, M. C. Wu, T. Tanbun-Ek, R. A. Logan, and M. A. Chin, Subpicosecond monolithic colliding-pulse mode-locked multiple quantum well lasers, *Appl. Phys. Lett.*, 58:1253 (1991).
68. P. J. Delfyett, C.-H. Lee, G. A. Alphonse, and J. C. Connolly, High peak power picosecond pulse generation from AlGaAs external cavity mode-locked semiconductor laser and traveling-wave amplifier, *Appl. Phys. Lett.*, 57:971 (1990).
69. M. F. Krol, T. Otsuki, R. K. Boncek, B. P. McGinnis, G. Khotrova, H. M. Gibbs, and N. Peyghambarian, unpublished work.

4
Ultrafast All-Optical Switching Devices

Mohammed N. Islam
University of Michigan, Ann Arbor, Michigan

I. INTRODUCTION

This chapter focuses on all-optical switching devices that can be used for ultrafast serial processing. Ultrafast devices are routing and logical switches that have the potential of operating at speeds greater than 50 Gb/s, which is beyond the limits where electronics might be expected to operate. Devices that are based on "all-optical" interactions rely on virtual transitions in the material, i.e., the interaction is through deformation of wavefunctions, which is nonresonant and can be almost instantaneous. Applications in which serial devices will be important include high-performance front and back ends of telecommunications systems as well as fiber local area networks. The interest in such ultrafast switches stems from their ability to answer two questions. First, how can processing beyond electronic speeds be accomplished? Second, how can the bandwidth-rich environment provided by optical fibers be further utilized? For instance, in the low-loss window between 1.3 and 1.6 μm there is about 40 THz of bandwidth, and to exploit this advantage for time-division-multiplexed systems requires ultrafast switches.

The discussion in this chapter is restricted to nonlinear guided-wave devices in glass or optical fibers. Although the nonlinearity in fused silica fibers is weak, the very low loss and excellent guiding properties of optical fibers mean that long interaction lengths can be employed. In addition, fused silica has for all practical purposes an instantaneous response time since it is generally used far off resonance. Fibers also have a mature fabrication technology, and the equations governing their behavior are well understood. Therefore, although fiber devices may have a long latency (delay from the input to the output), they at least permit exploration of various switch architectures. The long latency restricts the use of these devices to pipelined, feed-forward applications.

In addition to using fibers, many of the interesting devices described in this chapter use solitons. Solitons can be defined as pulses that propagate nearly distortion-free for long distances in fibers. They occur in the anomalous group-velocity dispersion regime of fibers (e.g., wavelengths longer than 1.3 μm), and solitons represent a balance between the nonlinearity and the dispersion in the fiber. The key feature of solitons that is used for switching applications is that they act in many ways like fundamental data bits and the entire pulse switches as a unit [1]. i.e., because the two opposing forces of nonlinearity and dispersion act on a soliton, it is internally balanced, has a restoring force, and, therefore, acts as a stable, robust pulse. For terabit-rate switches that use picosecond or femtosecond optical pulses in fibers, solitons are the preferred and natural data bits.

A. Routing Versus Logical Devices

Ultrafast devices can be divided into two general categories that are illustrated in Figure 1. The first is a routing switch in which the input is connected to one of several output ports, and the routing is based on either the intensity of the signals or an externally supplied control beam. If only one output port is employed, then the routing switch works like an on–off switch. Also, if the routing is based on the intensity of the input, then the device may be used as a limiter or a saturable absorber. Routing switches are "physical" switches, since photons are physically moved from one port to another. The other category is a logic gate (Fig. 1b) in which a Boolean operation is performed based on the values of the input signals. The logical approach can be powerful, because it allows intelligence to be distributed throughout the system (in the sense that one data stream can control another), and this is one reason why modern electronic systems operate based on digital logic.

Routing and logic switches differ fundamentally in the manner of the control. In routing switches, the control is typically in a different physical format than the data, and the control network may be external to the switching fabric. In a logic gate, on the other hand, the control is in the same physical format as the data, and, therefore, the control can be distributed throughout the switching fabric. Another difference between the two

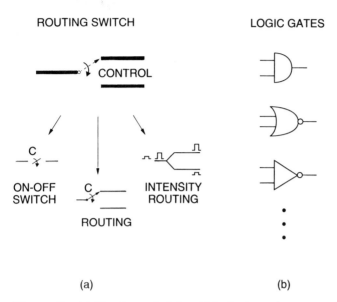

Figure 1 (a) Routing switch in which the input is connected to one of several output ports. (b) Logic gate in which a Boolean operation is performed based on the values of the input signals.

devices of Figure 1 is the representation of the decision. A routing switch represents its decision by the position or location of the data, while the output of a logic gate has a "0" or "1" logic level. Since routing switches route the same photons from the input to the output, the signals may degrade because of loss, dispersion, or crosstalk. In digital logic gates, the signal level and timing is regenerated at the output of each gate by replacing the input photons with new photons from a local power supply. The penalty for high-speed, digital logic based systems is that the switching energy and power supply requirements are major constraints.

B. Challenges for All-Optical Switching Devices

There are many challenges in implementing devices that rely on all-optical interactions, and listed here are some of the key issues. First, how can we make three terminal devices whose behavior is independent of the phase between signals? In general, the two input signals to a device may originate at different points in the system and their relative phase will be arbitrary. All-optical interactions are coherent processes in which the input signal phases are preserved. Phase insensitivity restricts the useful nonlinear processes to those that depend on the intensity but not the electric fields of the inputs. Second, how do we make cascadable gates where the output looks like the input? Except in a few trivial situations, to complete a useful operation requires several decisions and corresponding levels of logic, and so we must have devices that can drive similar devices. Third, how do we achieve small signal gain where a smaller signal controls a larger pulse? Devices, in general, need gain to fan-out to several devices and to compensate for system losses in connecting to the next gate. The splitting and coupling losses between stages can be partially compensated by introducing amplifiers, but at the cost of increased system-hardware complexity and additional spontaneous noise. Fourth, how can we lower the switching energy without a large increase in the device size? At this point, the fundamental problem of potential terabit systems appears to be power supply limitations. For example, the average power required for a gate is the switching energy times the bit rate, which means that to switch a picojoule energy device at a terabit requires a laser with a watt of average power. The lasers must provide this large power at high repetition rates, in short pulses, and at wavelengths compatible with the remainder of the system. Finally, how can we handle the timing constraints and synchronization required for terabit systems? As the bit rate increases, the bit period decreases and the tolerance to timing jitter and clock skew decreases. In a synchronous system all parts of the system must be phase and frequency locked to a master clock, and timing jitter must be minimized at each stage to avoid accumulation of errors. This chapter shall describe devices that address each of these issues.

A summary of this chapter follows. In Section II routing devices are described such as Kerr gates, nonlinear directional couplers, and nonlinear optical loop mirrors. Section III concentrates on digital soliton logical devices, and the discussion includes soliton dragging logic gates, soliton trapping logic gates, and a cascade of these gates. Finally, technological challenges for the future and potential applications for all-optical devices are described in Section IV. In this brief review only a few examples are chosen to illustrate various approaches and to raise the key issues relevant to all-optical switching. A more comprehensive review of ultrafast fiber switching devices and systems is given in [2].

II. ROUTING DEVICES

A. Kerr Gates

A Kerr modulator uses the change in polarization state that is due to the intensity-dependent refractive index n_2. The typical configuration for a Kerr modulator is shown in Figure 2, where a weak signal at frequency ω_2 is gated by a strong pump at ω_1. The strong pump is polarized along one

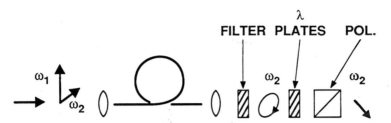

Figure 2 Schematic of a fiber Kerr modulator. A strong optical pulse at ω_1 changes the polarization state of a signal at ω_2 by the intensity-dependent nonlinear index [20].

axis of a polarization maintaining fiber, while the weak signal is polarized at 45° to the axis. The frequency filter at the fiber output removes the pump at ω_1. The wave plates are adjusted so that the polarizer blocks the weak signal in the absence of the pump, and the pump increases the probe transmission through the optically induced birefringence. The power transmission through the polarizer is proportional to $\sin^2(\Delta\phi/2)$, where

$$\Delta\phi = \frac{2\pi L}{\lambda}(\delta n_{\parallel} - \delta n_{\perp}); \delta n_{\parallel} - \delta n_{\perp} = n_2 I_p \qquad (1)$$

and I_p is the pump intensity and L is the length of the fiber.

Morioka et al. [3], have used the Kerr effect to demultiplex a train of 30-ps optical pulses from a gain-switched laser diode using control pulses from a mode-locked Nd:YAG laser. A schematic diagram of their Kerr demultiplexer setup is shown in Figure 3. Two problems of using long lengths of high birefringent fibers are (a) polarization dispersion limits the effective interaction length for short pulses; and (b) the temperature-dependent birefringence causes polarization fluctuations of the signal pulse. Both of these limitations can be circumvented by canceling the overall birefringence by splicing together two equal length fibers with their axes crossed at right angles. For example, in Figure 3 the Kerr medium consists of two 10 m lengths of polarization maintaining fiber spliced together with crossed axes. Synchronizing the 82-MHz repetition rate of the pump laser at 1.06 μm to every 24th pulse of the 1.3-μm probe laser diode ($\sim$ 2 GHz repetition rate) causes the pump to remove every 24th pulse. Figure 4 shows the demultiplexed probe pulses that are observed with a PIN photodiode. The cross-coupling from (b) to (c) in Figure 4 was measured using a streak camera and found to be less than $\sim$ 20 dB. Although the birefringence is compensated for in this experiment, the group velocity difference between the two optical pulses still limits the walk-off length. One solution is to choose the

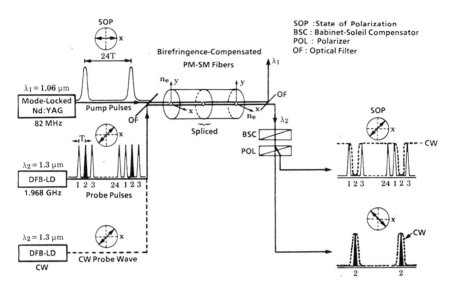

Figure 3 Experimental setup for the all-optical demultiplexer based on the optical Kerr effect in fibers [3].

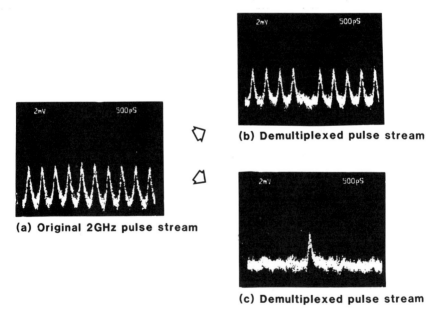

Figure 4 Result of all-optical demultiplexing using the Kerr effect in fibers: (a) original 2-GHz pulse stream; (b) demultiplexed pulse stream in one arm; (c) demultiplexed pulse stream in the other arm [3].

ULTRAFAST ALL-OPTICAL SWITCHING DEVICES

two wavelengths on opposite sides of the zero-dispersion wavelength of the fiber, so that the two group velocities are equal.

B. Nonlinear Directional Couplers

A nonlinear directional coupler (NLDC) exhibits sharper switching characteristics than a Kerr gate, although it is typically used as a single input, intensity-dependent routing switch. The NLDC is an example of a two-coupled-mode system, where the intensity-dependent change in index blocks the normal coupling between guides to cause switching. Jensen [4] first proprosed and gave a theoretical treatment of the NLDC. Experimental implementations of the NLDC (Fig. 5) include dual-core fibers [5],

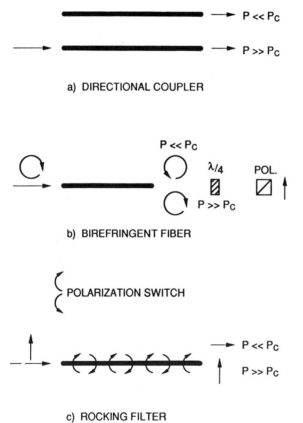

Figure 5 Three analogous implementations of a nonlinear directional coupler: (a) dual-mode fiber; (b) polarization instability in a birefringent fiber; and (c) a periodic rocking filter in fiber [20].

polarization switching in birefringent fibers [6], and polarization switching in periodic fiber filters [7]. All three experiments are described by analogous equations and show the pulse break-up problem inherent to switching with an instantaneous nonlinearity. The dual-core fiber is perhaps easier to understand and raises the major issues.

Consider a dual-mode coupler in which light is coupled into only one of the two waveguides (waveguide 1). There is a coherent interaction of the two optical waveguides in close proximity, and these waveguides periodically exchange power because their evanescent fields overlap. Jensen [4] derived the relation for the fraction of power in waveguide 1 as a function of distance along the guide, z, to be of the form

$$\frac{P_1}{P} = \frac{1}{2}\left[1 + cn\left[\frac{\pi z}{2L_c} \mid \left(\frac{P}{P_c}\right)^2\right]\right] \tag{2}$$

where $cn(\phi \mid m)$ is the Jacobi elliptical function and L_c is the coupling length. The coupling length is a function of the geometry and separation between the two waveguides. The critical power P_c is given by

$$P_c = \frac{\lambda A_{\text{eff}}}{n_2 L_c} \tag{3}$$

where A_{eff} is the effective area of the waveguide and P_c corresponds to the power needed for a 2π nonlinear phase shift in a coupling length. Note that the critical power is inversely proportional to the coupling length. In the dual-mode coupler, at low intensities the light couples back and forth periodically between the guides with a periodicity of $2L_c$. As the power is increased toward P_c, the nonlinear index detunes the two waveguides, thereby disturbing the periodicity of the energy transfer. Figure 6 shows the relative output from waveguide 1 as a function of input power for a CW input, and we find a sharp slope and peak transmission as large as one. As the overall length of the coupler increases, the transfer function becomes increasingly sharp around P_c. For example, we include in Figure 6 with the dotted curve the characteristics for a $2L_c$ long device, which is much sharper than the behavior for one coupling length shown by the solid curve.

Whereas the curves of Figure 6 are plotted for CW waves, most experiments with ultrashort pulses use Gaussian or squared hyperbolic secant pulses. When the bell-shaped pulses are switched in an NLDC, the weaker wings behave differently from the center peak and the pulse shape breaks up at the output. The result is a reduction in the contrast ratio when the output is averaged over the entire pulse. For example, we show in Figure 7 the calculated switching response for pulses with squared hyperbolic secant intensity profiles and a one-coupler-length device, and the degradation of

ULTRAFAST ALL-OPTICAL SWITCHING DEVICES

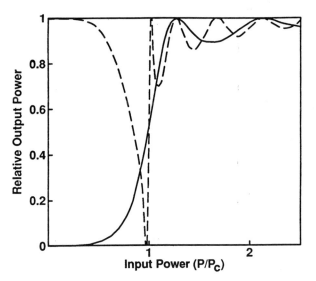

Figure 6 Calculated transmission function for a nonlinear directional coupler at the output of the guide that is excited at the input (guide 1). The input power is normalized to the critical power P_c. The length of the couplers are one coupling length L_c (solid line) and $2L_c$ (dashed line) [4].

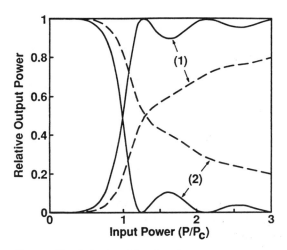

Figure 7 Calculated fractional output power emerging from waveguides (1) and (2) as a function of input power for a coupler of length L_c. Solid curves: constant-intensity input signal; dashed–dotted curves: coupler response integrated over a $\text{sech}^2(t)$ pulse intensity profile [5].

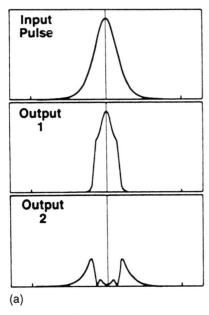

Figure 8 Calculated reshaping and break up of sech^2 (t) pulses due to the nonlinear transmission in Fig. 2.3.2: (a) coupler length L_c, peak power of $2P_c$ [8]; (b) coupler length $2L_c$ and peak powers of $0.75P_c$, $0.825P_c$, $0.975P_c$, $1.1P_c$, $2P_c$ and $3P_c$ [6].

the switching performance is evident. Furthermore, the NLDC also turns out not to be cascadable because the output pulse shape is different from the input. Figure 8 shows various temporal profiles calculated at the output of NLDCs that are (a) one- and (b) two-coupling lengths long. Potential solutions of this problem will be addressed at the end of this section.

1. Two-Core Fiber Directional Coupler

Friberg et al. [5] demonstrated a two-core fiber directional coupler using 100-fs optical pulses. Their coupler consists of a 5 mm length of dual-core fiber, which contains two 2.8-μm diameter germanium-doped cores (core–cladding index difference of 0.003) with 8.4 μm separation between core centers. Each fiber core is single mode for wavelengths longer than 500 nm, and the coupling length is determined by a white-light measurement technique to be approximately 4.7 mm at 620 nm. The pulses for the experiment are derived from a colliding-pulse mode-locked dye laser and a copper-vapor-laser pumped dye amplifier system (Fig. 9). The laser produced 100-fs pulses at a wavelength of 620 nm, which were amplified at an

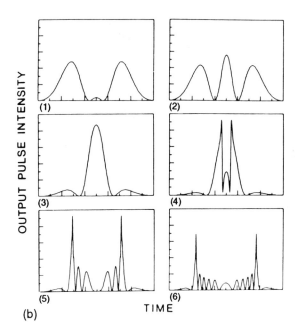

(b)

8.6-KHz repetition rate to 100 nJ. Amplified pulses were focused into one fiber core, and the other input core was carefully blocked by the edge of a razor blade. The output from each core was focused onto a separate power meter, and the average power emerging from each core was measured as a function of the input power.

The fraction of the output power emerging from each of the two waveguides is illustrated in Figure 10. Since the power meters respond slowly compared with the pulse duration, the response is integrated over the entire pulse. Reasonable agreement is found with the expected response for $sech^2$ intensity profiles that was predicted in Figure 7. From the data, Friberg et al. estimate the critical power to be approximately $P_c = 32$ kW, which is relatively high because of the short length of the coupler.

In addition, the reshaping of pulses as seen in Figure 8a is also confirmed experimentally. For example, Figure 11 illustrates the autocorrelation measurements obtained for 200-fs pulses at a power of $\cong P_c$ at the output of the nonlinear coupler for (a) the same guide as the input (waveguide 1) and (b) the other guide. While the pulses from waveguide 1 are similar to those at the input, the pulses from the other guide are strongly reshaped. The triply peaked autocorrelation trace corresponds to a doubly peaked intensity profile, and the 340-fs peak separation is consistent with the duration of the

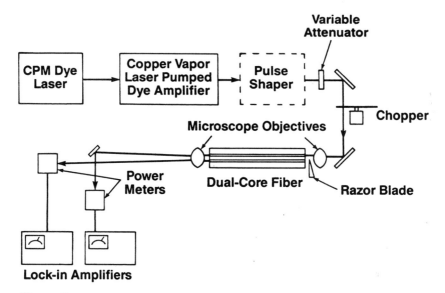

Figure 9 Schematic of experimental apparatus for testing femtosecond switching in dual-core fibers [8].

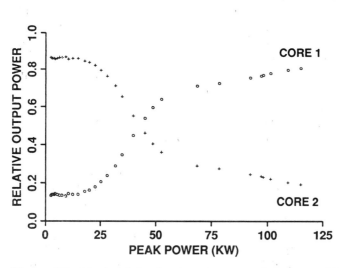

Figure 10 Measured fractional output power from waveguides (1) and (2) for the 5-mm, dual-core fiber nonlinear coupler. These data are the response for 100-fs input pulses [5].

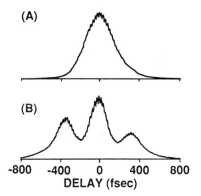

Figure 11 Autocorrelation traces for output pulses from the coupler. The input pulses were 200 fs in duration. (a) Guide (1), peak power $P \cong 2P_c$; (b) guide (2), peak power $P \cong 2P_c$ [5].

input pulse. Furthermore, the individual peaks are significantly narrower than the input. These data show that for input powers greater than P_c the nonlinear action selects the central part of the input pulse and directs it to the other output port.

The pulse break-up problem exemplified by the NLDC is a universal problem for all-optical interactions, since the switching is controlled by the instantaneous optical intensity. One solution for improving the contrast and cascadability for pulsed operations is to use square-shaped input pulses, in which case the ideal CW response should be obtainable. For example, Weiner et al. [8] have used pulse shaping techniques in the picosecond and femtosecond regime to generate ~ 540-fs square pulses. The switching characteristics using these square pulses are shown in Figure 12, and we find improved extinction ratio and lower switching power as compared with Gaussian pulses. One drawback of this technique is that pulses of finite bandwidth will still have finite rise times, and the square edges will be adversely affected by group-velocity dispersion in a fiber or glass waveguide.

A more attractive solution is to use fundamental soliton pulses that have a uniform phase shift, so that entire pulses switch as a unit. Now the device must be in the anomalous group-velocity dispersion regime, but the problems of dispersion and self-phase modulation are kept in balance. For example, Trillo et al. [9] have predicted that solitons can improve the NLDC performance when the soliton period is on the order of the coupling length. Because of the typically short lengths and high switching energies of pulses used with NLDCs, however, experiments with solitons in NLDCs

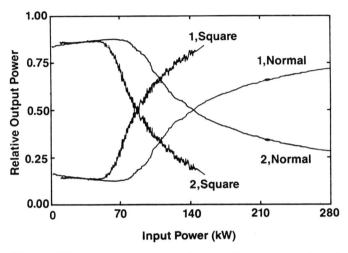

Figure 12 Improved switching in a nonlinear coupler by using square optical pulses. Plots show the relative output power from the two cores (marked 1 and 2) for 100-fs normal bell-shaped input pulses (Normal) and for 540-fs square pulses (Square) [8].

have not been reported. In fact, although longer coupling lengths would reduce the switching energy, obtaining long L_c is impractical both because the tolerances for fabrication are then too tight and because the device becomes extremely sensitive to external perturbations such as bends. Solitons have been successfully applied to fiber loop mirrors to avoid the pulse reshaping problem.

C. Nonlinear Fiber Loop Mirrors

Another example of a dual-mode device is a nonlinear Mach–Zehnder interferometer, where the phase difference between the two channels varies with optical intensity. Whereas an NLDC requires a 2π phase shift to switch, an interferometer requires only a π phase shift difference between the two arms to change from constructive to destructive interference. The response of an interferometer follows a squared sinusoid function of the nonlinear phase, however, which is less sharp than an NLDC.

An implementation of the interferometer in fibers that has received much attention is based on a nonlinear Sagnac interferometer or a nonlinear optical loop mirror (NOLM). As depicted in Figure 13, the NOLM consists of a four-port directional coupler in which two ports on one side are connected by a loop of fiber. The two arms of the interferometer

ULTRAFAST ALL-OPTICAL SWITCHING DEVICES

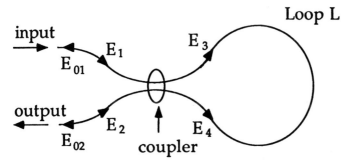

Figure 13 Schematic of the configuration for a nonlinear optical loop mirror [1].

correspond to the two counterpropagating directions around the loop, and this configuration is very stable since both arms involve exactly the same optical path. When the coupler divides the input equally, the NOLM acts as a perfect mirror. For other than the 50:50 splitting ratio, the nonlinear phase shift is different in the two directions, and the NOLM acts as an intensity-dependent mirror. Doran and Wood [1] first proposed the NOLM to demonstrate that solitons permit complete pulse switching and that a soliton may act as a fundamental data bit in ultrafast optical processing.

1. Solitons in Optical Loop Mirrors

Experiments confirming the complete switching properties of solitons have been performed by Blow et al. [10] and my colleagues and myself [11]. As an example, we observe complete switching of 310-fs soliton pulses at an energy of ~ 55 pJ with 90% peak transmission in an NOLM. A passively mode-locked NaCl color-center laser [12] provides τ = 310-fs Gaussian pulses at λ = 1.692 μm. In the experimental apparatus of Figure 14, a variable attenuator is used to vary the input power, and an isolator prevents feedback into the laser. A 1-cm thick uncoated quartz beam splitter picks off a fraction of the input and reflected light, and apertures are used to block multiple reflections. The NOLM is made of 25 m of single-mode, polarization maintaining, dispersion-shifted fiber. The coupler is made of quartz blocks in which the polarization maintaining fiber is glued and polished. Thirty-two percent of the light is coupled over, and the loss across the coupler is ~ 8%. The loop is closed at a ~ 90% transmitting splice. Throughout the setup, the polarization extinction ratio is better than 15:1.

Figure 15 illustrates the nonlinear transmission (E_{out}) and reflection (E_{refl}) as a function of input power (E_{in}). The ratio of E_{out} to ($E_{out} + E_{refl}$) increases up to a peak and is then followed by a dip and an approximately flat region. After correcting for the various losses, the peak occurs at E_{in}/E_1 ~ 1.66 and

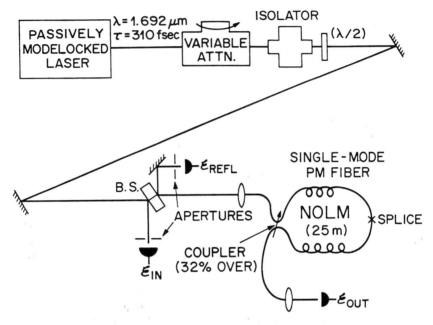

Figure 14 Experimental configuration for the fiber nonlinear optical loop mirror (NOLM) ($\lambda/2$ = half-wave plate, B.S.= beam splitter).

the transmission at the peak is $\sim 90\%$. This peak corresponds to a pulse switching energy of $1.66 \times P_1 \times 310$ fs $\cong 55$ pJ. The peak and null are less pronounced than in the ideal case [1] because of soliton self-frequency shift effects [13,14]. Since the coupler unequally divides the light in the two directions, the intensity-dependent frequency shift is different in the two directions, and the interference is incomplete. At higher powers, the transmission is nearly flat since the two counter propagating pulses no longer meet at the coupler because their unequal frequency shifts separate them both temporally and spectrally. In the limit of no-pulse overlap, the expected transmission is $(0.32)^2 + (0.68)^2 = 56\%$. In the experiments by Blow et al. [10], 710-fs pulses (autocorrelation width of 1.1 ps) are used and a 58:42 coupler connects to the two ends of a 100-m fiber. They observe 93% peak transmission at 46 pJ of energy, but because of their broader pulses they do not find frequency shift effects.

At the peak of transmission, we see switching of the complete soliton waveform. In Figure 16, we show the autocorrelations of the transmitted pulses and include inserts of the intensity profile from computer simulations that include the Raman effect at the corresponding power levels. At

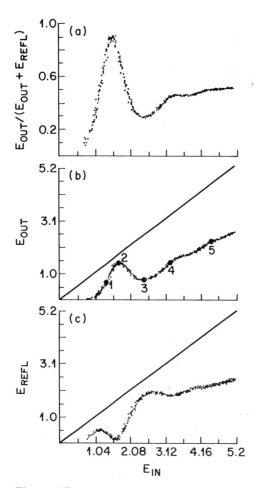

Figure 15 Measured transfer characteristics of the nonlinear optical loop mirror: (a) $E_{out}/(E_{out} + E_{refl})$; (b) E_{out}; and (c) E_{refl} versus E_{in}. The numbers in (b) correspond to the plots in Fig. 16. All energies are normalized to the fundamental soliton energy $E_1 = 33.2$ pJ.

the peak of the transmission (curve 2), the output is slightly broader than the input pulse but otherwise identical in shape. If the pulses were not solitons, then only the center of the pulse would switch, and the autocorrelation would be narrower. The solitons broaden slightly because, after dividing in the coupler, one arm has lower-than-fundamental soliton power. If the power is lowered further, than the transmitted pulses broaden more (curve 1).

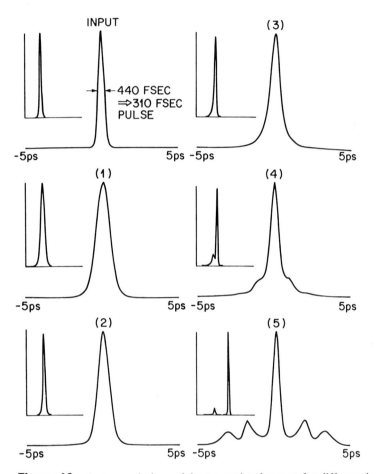

Figure 16 Autocorrelations of the transmitted output for different input powers. The numbers correspond to power levels shown in Fig. 15(b). The inserts are intensity curves versus time from numerical simulations including the Raman effect.

As confirmation of the influence of soliton self-frequency shift, we see pulse break up at higher powers. The frequency shift effect asymmetrizes the pulses, causes higher-order solitons to split, and, as already discussed, temporally separates the pulses from the two directions. The transmitted pulse starts to change shape around the dip in transmission (curve 3). At yet higher powers, the output splits into three pulses (corresponding to five peaks in the autocorrelation) that separate with increasing power. There is qualitative agreement between the calculations and the observed splitting.

ULTRAFAST ALL-OPTICAL SWITCHING DEVICES 181

Note that, because higher-order solitons no longer have uniform phase shift over the entire pulse [15], we do not expect complete switching at higher powers.

2. Loop Mirror as a Three-Terminal Device

Thus far, we have described the NOLM as a single-input device, which can act as a saturable absorber or a pulse shaper. By adding a control beam that is orthogonal either in frequency or polarization, the NOLM can also act as a three-terminal switch. The idea of using a Mach–Zehnder interferometer as an all-optical gate was first discussed by Lattes et al. [16], and their structure was an all-optical analogue of electrooptic devices. As shown in Figure 17, in an exclusive-OR (XOR) gate the control pulses propagate through both arms and interfere destructively at the output junction. When either signals A or B are injected orthogonally polarized to the control, then one arm is phase shifted by π so the two control pulses now interfere constructively at the output. When both pulses are incident, however, both arms receive equal phase shifts with no resulting output. A polarizer is used at the output of the interferometer to block both signal inputs. Lattes et al. first demonstrated the concept in a lithium niobate device, although the same basic idea can also be applied to a fiber loop mirror.

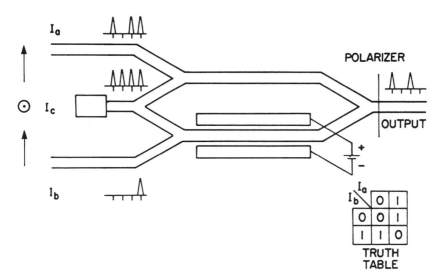

Figure 17 Schematic of a Mach–Zehnder interferometer used as an optical logic gate. A continuous stream c of pulses in modulated by the information carrying pulses incident in waveguides a and b [16].

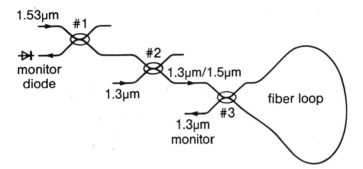

Figure 18 Experimental configuration for testing the two-wavelength operation of a nonlinear optical loop mirror [17].

The simplest technique for adding a control pulse is to use two separate wavelengths in the NOLM, which is analogous to the operation of the Kerr gates. In this mode of operation, a high-power signal at one wavelength switches a low-power signal at another wavelength, and the device behaves like an all-optical modulator. Blow et al. [17], demonstrated the two-wavelength operation of the NOLM in the configuration of Figure 18. The low-power signal (~ 5 mW) was obtained from a CW color-center laser (1.53 μm), and the pump or control was derived from a 1.3-μm mode-locked Nd : YAG laser (pulse width 130 ps). Coupler 1 has a 50:50 splitting ratio for 1.53 μm and provides a monitoring point for the backreflected signal. Coupler 2 combines the 1.3- and 1.53-μm signals, and the resultant signal is launched into the NOLM. The NOLM coupler (coupler 3) has a coupling ratio of 50:50 for the 1.53-μm signal and virtually 100:0 for the 1.3-μm control. Therefore, the cross-phase modulation induced phase shift is different in one direction than the other when the control pulse is added. The NOLM uses a 500-m polarization maintaining fiber, and the response of the device is monitored at coupler 1 with a photodiode that had a pulse response of 70 ps.

The experimental results for several pump powers are displayed in Figure 19. At an average (peak) power of 20 mW (2 W) at 1.3 μm (Fig. 19a) there is almost complete switching of the 1.53-μm probe signal. The flat-topped feature seen in the response results from group-velocity walk-off between the two beams, which leads to a uniform phase across the center of the pulse [17]. As the pump power is further increased (Figs. 19b and 19c), the behavior reflects the periodic intensity response of the NOLM. In Figure 19d, the central part of the pulse is fully reflected, and for this case Blow et al. use a 100-m NOLM and an input pump power of 55 mW.

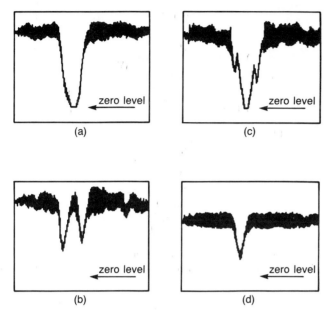

Figure 19 Reflected 1.53-μm signal with a loop length of 500 m for average pump input powers of (a) 20 mW, (b) 35 mW, and (c) 60 mW. In (d) the loop length is 100 m and the average pump input power is 55 mW[17].

3. Cross-Polarized Control and Signals in Loop Mirror

An alternate way of creating a three-terminal switch is to use a control pulse orthogonal to the signal pulse in a polarization maintaining fiber. Because of the birefringence or the velocity difference between the two axes, the control and signal pulses can be arranged to walk completely through each other. The two pulses interact through cross-phase modulation in the fiber and phase shift one another. Even if the phase shift in a single pass is less than π, the required π phase shift can be accumulated by using cross-splices (splices in which the slow and fast axes of the fiber are exchanged) to permit multiple passes.

Moores et al [18] first proposed a multiple slip-through scheme, and their configuration is shown if Figure 20. Their design involves using soliton pulses in which the control pulse slides through half of the signal pulse 11 times until the latter accumulates a π phase shift. The control pulse always leaves the device through the transmitted port, but the signal is transmitted only when the control is applied. If a polarizer is placed at the transmitted port along the signal axis, the device acts as an AND gate. In preliminary experiments Moores et al. demonstrated some enhancement of the output

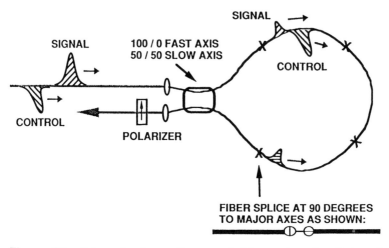

Figure 20 Schematic of a nonlinear optical loop mirror in which orthogonally polarized control and signal pulses interact repetitively because of cross-splices in the fiber [18].

with the addition of the control, but the contrast ratio was poor because of improper couplers. Nonetheless, a key feature of the configuration is that this switching mechanism allows for regenerative, synchronizing devices, since incoming pulses can be replaced by locally generated pulses.

Avramopoulos et al. [19] point out that another advantage of the slip-through interaction is that solitons or square pulses are not required to obtain complete switching, as we found for the NLDC. Since the nonlinear interaction is proportional to the integrated intensity that lies within a certain time window, which is fixed by the fiber length and the birefringence, the signal pulse may arrive at any time within the window and still have the same effect. Because of this integration effect, an arbitrarily shaped pulse can obtain a uniform phase shift and switch with good contrast ratio. Furthermore, the system can be insensitive to timing jitter in the signal stream if the pulse length is shorter than the window length.

A schematic of their slip-through device is shown in Figure 21, and in their experiments Avramopoulos et al. use polarization maintaining components throughout the 500-m loop. The control pulse is split and propagates through both directions in the NOLM, while the signal enters through a polarizing beam splitter, traverses the loop in only one direction, and then is removed by another polarizing beam splitter. By using separate polarizing beam splitters, the couplers can be of conventional design; therefore, they avoid the difficulties that Moores et al. [18] had in making couplers that have different

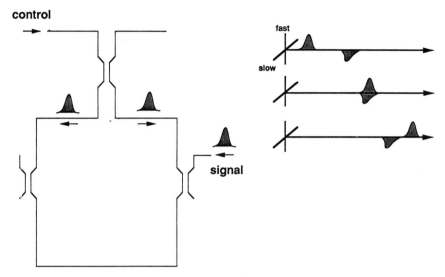

Figure 21 Schematic of nonlinear optical loop mirror in which orthogonally polarized control and signal pulses interact and the signal pulses enter the exit through separate polarizing beam splitters [19].

ratios for different polarizations. The signal and control pulses were derived from an electronic pulse generator that produced two independent but synchronized electrical pulses with adjustable delay. These electrical signals drive two laser diodes (one at 1.531 μm and the other at 1.534 μm) to produce $\sim$ 1 pJ optical pulses that were typically between 0.5 and 1 ns in width. The laser diode outputs were then amplified in erbium-doped fiber amplifier (EDFA), and to achieve large pulse energies ($>$ 1 nJ) the duty cycle of the lasers were kept low. To overcome the imperfect extinction of the signals in the polarizing beam splitters, a narrowband interference filter was used to distinguish the signals from the two diodes.

Figure 22 shows both the transmitted and reflected outputs of the device in its switched and unswitched states using $\sim$ 1-ns pulses at 2.5-MHz repetition rate. After using the interference filter in front of the detector, large contrasts in transmission ($\sim$ 40:1) and reflection ($\sim$ 10:1) were observed. In Figure 22c the residual reflected light is mostly amplified spontaneous emission from the EDFA, which is always reflected. The switching energy in this experiment was $\sim$ 4.4 nJ, which corresponds to a peak pulse power of 4.4 W. The maximum reflection was measured to be $\sim$ 41%, and the maximum transmission was $\sim$ 84%. Since a larger control pulse is required

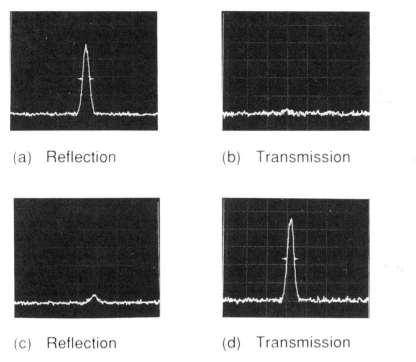

Figure 22 The transmitted and reflected outputs from the nonlinear optical loop mirror of Fig. 21 for signal and control pulse durations of 1 ns. Reflection (a) and transmission (b) at 10-ns relative delay. Reflection (c) and transmission (d) when the control and signal pulses overlap [19].

to switch a smaller signal pulse, the configuration in Figure 21 does not exhibit fan-out. Avramopoulos et al. quote a total gain for the device of ~ 840, however, because they consider the switch and amplifier together as the device. In addition, in Figure 23 we show the variation of reflectivity and transmission of the NOLM as a function of signal energy. As expected for the Mach–Zehnder interferometer, the response behaves like a squared sinusoid of the induced phase shift.

III. SOLITON LOGIC GATES

A. Soliton-Dragging Logic Gates

Soliton-dragging logic gates (SDLG) are three-terminal devices that restore both the logic level and timing, that have potential operating speeds

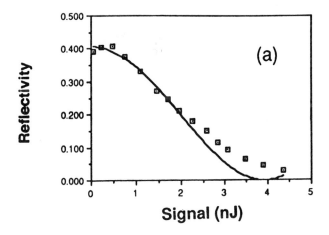

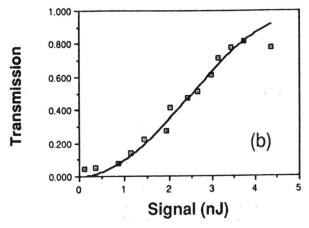

Figure 23 Switching curves showing (a) reflectivity and (b) transmission as a function of signal pulse energy for control and signal pulse durations of 1 ns [19].

of 0.2 Tb/s, and that have switching energies approaching a picojoule. In addition, SDLGs satisfy all requirements for a digital optical processor; i.e., SDLGs are cascadable, have fan-outs of between 6 and 30, and are Boolean complete. A low-energy, all-optical, NOR gate that is based on timing shifts from soliton dragging is demonstrated in moderately birefringent optical fibers [21]. Figure 24a shows a schematic of the NOR gate that consists of two lengths of fibers. The two fibers are connected through

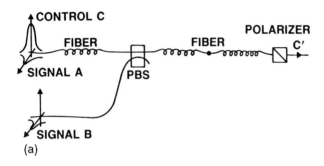

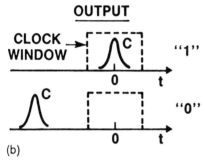

Figure 24 Schematic of a soliton dragging NOR gate with control or power supply C along one axis and signals A and B polarized orthogonally (PBS = polarizing beam splitter). (b) Example of time-shift keyed logic, where a Boolean "1" corresponds to a control pulse that arrives within the clock window.

a polarizing beam splitter, and the output is filtered by a polarizer or another polarizing beam splitter. The power supply or control pulse C provides gain, propagates along one principal axis in both fibers and corresponds to A NOR B at the output. For a cascadable gate, C should be approximately a fundamental soliton and should experience insignificant frequency shift from the self-Raman amplification effects [13]. The two signal pulses A and B are polarized orthogonal to C and are blocked by the polarizer at the output. The signals are timed so that A and C coincide at the input to the first fiber and B and C coincide (in the absence of A) at the input to the second fiber. Note that logic-level restoration is obtained by replacing the signal pulses by the control pulse at the output. Also, in a system consisting of such gates, the output from one gate (the control pulse) is flipped in polarization and serves as the input to the next gate, and at the output of the second gate it is removed from the system by the

ULTRAFAST ALL-OPTICAL SWITCHING DEVICES

polarizer. Consequently, the output of one gate passes through only one more level of logic before it is removed from the system.

The logic gate is designed for digital optical applications and operates based on time shifts from soliton dragging. In soliton dragging, two pulses that are coincident in time interact through cross-phase modulation [22,23] and, consequently, chirp in frequency and time shift by propagating in a dispersive delay line. The pulse along the slow axis speeds up, while the pulse along the fast axis slows down. We assume that a signal corresponds to a pulse with guard bands surrounding its time slot. Then, a "1" corresponds to a pulse that arrives within the clock window and a "0" either to no pulse or an improperly timed pulse (Fig. 24b). For the NOR gate, the fiber length is trimmed so that in the absence of any signal the control C arrives within the clock window and corresponds to a "1." When either or both signals are incident, they interact with the control pulse through soliton dragging and pull C out of the clock time window.

Figure 25 shows the experimental apparatus for testing a single NOR gate. We obtain $\tau \sim$ 500-fs pulses near 1.685 μm from a passively mode-locked NaCl color-center laser in which a 2 mm thick quartz birefringent plate deliberately limits the bandwidth and, thus, broadens the pulses [12]. The input stage separates the control C, signals A and B, and clock beams, and stepper-motor controlled delay stages are used to time properly signal

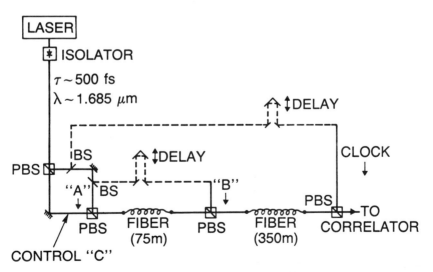

Figure 25 Experimental configuration for testing the all-optical soliton-dragging NOR-gate (BS = beam splitter, PBS = polarizing beam splitter, POL = polarizer).

B and the clock. The two fibers are 75 and 350 m long, have a polarization dispersion of about 80 ps/km, and when carefully handled exhibit a polarization–extinction ratio better than 14:1. The control pulse output and the clock are directed to a correlator to measure the time shifts.

The correlation of the clock with the NOR gate output is illustrated in Figure 26. The dotted box corresponds to the clock window, and we see that C arrives within this window when no signal is present. When $A = 1$ or $B = 1$, C shifts between 2 and 3 ps out of the clock window; the shift from A is larger since C can time shift in both fibers. When $A = B = 1$, C shifts by about 4ps. The additional noise and broadening for $A = B = 1$ occurs because A and B, which are parallel polarized, interfere at the polarizing beam splitter; this interference can be avoided by using two separate beam splitters to remove A and introduce B. In this example the fan-out, or gain (= control out / signal in), is 6 and the signal energies are 5.8 pJ each. The control pulse energy in the first fiber is 54 pJ and is reduced to 35 pJ in the second fiber because of coupling losses.

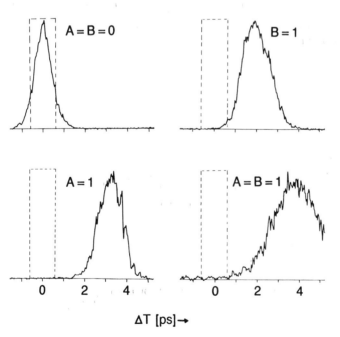

Figure 26 Cross-correlation of clock with soliton-dragging NOR gate output. The signal energy is 5.8pJ and the gain (=control out / signal in) is 6.

ULTRAFAST ALL-OPTICAL SWITCHING DEVICES

1. All-Optical Multivibrator or Ring Oscillator

To prove the cascadability and fan-out of the logic gate, we connected the NOR gate as an inverter and fed the output back to the input ($A = O$, $B =$ previous output from gate). We placed a 50:50 beam splitter at the output and sent half of the output through an adjustable, free-space, delay line to the B input. The correlator was set to the center of the clock time window. As Figure 27 shows, with the feedback blocked the output is a string of 1's. When the feedback is added, the output becomes an alternating train of 1's and 0's whose period is twice the fiber latency (1.75 μs). Although the demonstration of a submegahertz oscillator may not seem terribly impressive, this is an absolute test of both the fan-out and cascadability. After all, what could be a more stringent test of the cascadability than to force a gate to drive itself. Furthermore, in our experiment the device must have at least a fan-out of 4 since we use a 50:50 beam splitter at the output of the gate and there is an additional 3-dB loss when coupling back into the device because of our bulk optics. Thus, we demonstrate an all-optical multivibrator or ring oscillator using only a logic gate and no amplifiers.

2. Time-Domain Chirp Switch Architecture

A novel time-domain chirp switch (TDCS) architecture [24] is a generalization of fiber soliton-dragging logic gates, in which digital logic is based on time-shift keying. By using solitons, with their particle-like behavior, we can separate the frequency change due to cross-phase modulation from the phase and temporal change required for switching. Understanding the TDCS architecture permits us to (a) optimize individually the frequency change and the phase or temporal change; (b) apply our knowledge to

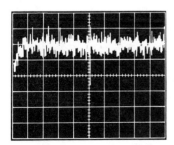
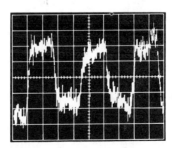

Figure 27 Output from an all-optical multivibrator, which is simply a NOR gate configured as an inverter whose output is fed back to the input. Feedback blocked on the left and feedback added on the right. The horizontal time scale is 1 μs/div.

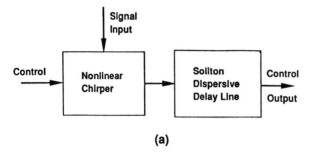

(a)

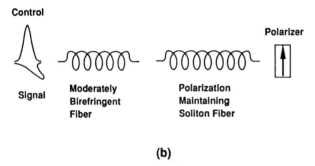

(b)

Figure 28 (a) General architecture for an all-optical time-domain chirp switch (TDCS). The signal creates a time varying index change in the nonlinear chirper, and the soliton-dispersive delay line translates the frequency shift into a time shift. (b) Soliton-dragging logic element is an all-fiber example of a TDCS.

materials other than fibers such as semiconductors and organics; and (c) lower the switching energy to levels approaching a picojoule.

As shown in Figure 28a, the TDCS consists of a nonlinear chirper followed by a dispersive delay line and has two orthogonally polarized inputs (signal and control pulses). In the absence of a signal pulse, the control pulse propagates through both sections and arrives at the output within the clock window. For a cascadable switch, the self-induced chirps on the control in both sections must balance, and the output pulse must resemble the input. Adding the signal pulse creates a time varying index change that chirps the control pulse and shifts its center frequency [22,23]. As the chirped control pulse propagates through the soliton-dispersive delay line, the frequency shift is translated into a time change.

The particle nature of solitons is crucial for separating the TDCS in two sections and relaxing the requirements on the nonlinear material. If the

ULTRAFAST ALL-OPTICAL SWITCHING DEVICES

second section were just a linear dispersive delay line (e.g., a pair of prisms or gratings), then changing a "1" to a "0" would require a frequency shift on the order of the entire spectral width. It can be shown that this is the same as requiring the interaction to result in a π phase shift within the nonlinear chirper [24]. Since a fundamental soliton acts as a particle, even a slight shift in the center frequency can cause the complete soliton to shift in time, which means that much less than a π phase shift results after the interaction between the two pulses. As in most switching configurations, we still need to obtain at least a π phase shift to change state, but the phase shift is accumulated as the frequency-shifted pulse propagates in the dispersive delay line. In other words, although all the interaction occurs in the first section, the phase shift required for switching gathers in the second section.

A simple spatial analogy that illustrates the TDCS function is illustrated in Figure 29. Each particle by itself travels along the dotted trajectories. When both particles are incident, then soliton dragging attracts the two particles and changes their angle of propagation. The change in angle occurs within the "nonlinear chirper," and just after the interaction there is insufficient shift in either pulse for switching. As the particles propagate through the dispersive delay line, however, the change in angle magnifies into a larger separation from the dotted trajectories. Therefore, the dispersive delay line acts as a "lever arm" to translate a small angle change (frequency shift in the SDLG) into a large separation from the original trajectory (time shift in the SDLG). The longer the lever arm, the smaller

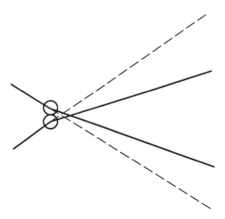

Figure 29 Spatial analogy of the time-domain chirp switch function. The dotted lines are the trajectories for either pulse alone, and the solid lines are the trajectories with both pulses present.

the change in angle through the interaction needs to be. This is why the TDCS architecture can result in low switching energies at the expense of increased latency or length of the device.

B. Soliton-Trapping Logic Gates with Output Energy Contrast

Within an ultrafast switching system we can use time-shift keying as the basis of logic. At the boundary of the system or the interface with electronics, however, we want to convert the time-shift-keyed signals to amplitude-shift keying. Soliton-trapping logic gates can perform the time-to-amplitude conversion at switching energies of ~ 42 pJ and speeds comparable with the SDLGs. The physical mechanism for soliton trapping is cross-phase modulation between orthogonally polarized pulses just as in SDLGs; but soliton-trapping logic gates rely on frequency discrimination rather than temporal shifts. Soliton trapping is a special case of soliton dragging in which the two orthogonally polarized pulses are of comparable amplitudes and the fiber has a particular range of birefringences.

Linear birefringence in optical fibers leads to pulse walk-off and broadening through polarization dispersion. Menyuk [25,26], however, showed numerically that orthogonally polarized solitons can trap one another through cross-phase modulation; thus, intensity-dependent effects can compensate for linear birefringence. For equal intensities along the two fiber principal axes, the two solitons shift their frequency in opposite directions, so that through group-velocity dispersion the soliton along the fast axis slows down, and the soliton along the slow axis speeds up. For a given birefringence and fiber length, a minimum intensity is required for the trapping.

To analyze the soliton-trapping mechanisms, we solved numerically the coupled nonlinear Schrödinger equation. In Figure 30 we plot the intensity and spectrum obtained for parameters relevant to the experiments. With a single pulse launched along the slow axis of the fiber the spectrum is symmetric, but the pulse moves with respect to $t = 0$ (Fig. 30a) because of birefringence. In Figure 30b the input is split equally between the two axes, and the two orthogonally polarized pulses trap near $t = 0$. The overlapping parts of the pulses (overlay of solid and dotted intensities in Fig. 30b) narrow because of the increased intensity-dependent phase modulation, similar to soliton compression in fibers. The individual pulses are also asymmetric, with their wings extending farthest in the direction that polarization dispersion pulls the pulse. The spectra for the two pulses shift equally and in opposite directions so that the frequency difference compensates through group-velocity dispersion for the polarization dispersion. The

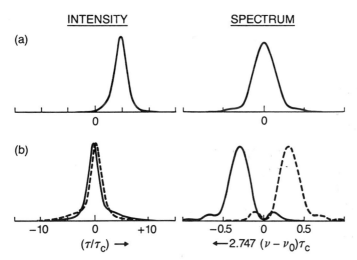

Figure 30 Intensity and spectrum obtained by numerically solving the coupled nonlinear Schrödinger equation for a normalized birefringence of $\delta = 0.517$, $L = 5.8$ soliton periods, and normalized amplitudes of (a) $A_x = 1.24/\sqrt{2}$, $A_y = 0$, or (b) $A_x = A_y = 1.24\sqrt{2}$ and $A_T = 1.24$. Standard soliton normalizations are used, as described in [2]. The solid curve corresponds to the pulse polarized along the slow axis, and the dotted curve to the pulse polarized along the fast axis. The normalizing time is $\tau_c = \tau_p/1.76$, where τ_p is the full width at half-maximum pulse width at the input.

pedestal or satellite bumps on the spectra correspond to the non-soliton part, which is stripped off through dispersion.

1. Spectral Confirmation of Soliton Trapping

In the experimental apparatus of Figure 31, $\tau \sim 300$-fs Gaussian pulses at $\lambda \sim 1.685$ μm are obtained from a passively mode-locked NaCl color-center laser. A variable attenuator is used to adjust the input power, and an isolator prevents feedback into the laser. The two polarizing beam splitters generate and recombine the two orthogonally polarized signals "A" and "B," whose amplitudes are equalized using a half-wave plate; the two beams are separated, and so we can easily block and unblock each arm. Another half-wave plate is used before the fiber to align the input polarization along the desired fiber axes. The 20 m length of fiber used has a polarization dispersion of 80 ps/km ($\Delta n \sim 2.4 \times 10^{-5}$), a zero-dispersion wavelength of 1.51 μm, and a dispersion slope of 0.05 ps/km-nm². By carefully winding the fiber on a 30-cm drum to minimize mode coupling effects, we obtained polarization extinction ratios of $\sim 20:1$.

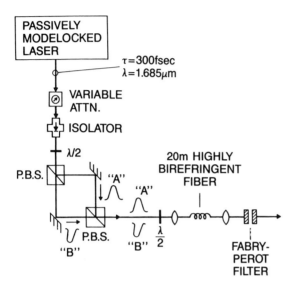

Figure 31 Experimental apparatus for testing the soliton-trapping logic gates. "A" and "B" are the orthogonally polarized inputs. (P.B.S = polarizing beam splitter, $\lambda/2$ = half-wave plate).

Figure 32 illustrates the spectral confirmation of soliton trapping. When a pulse of ~ 42 pJ energy propagates along a principal axis of the fiber, we obtain the output spectrum in Figure 32a. The bifurcated spectrum of Figure 32b results with two identical input pulses of ~ 42 pJ along each axis. For this same case, if we align a polarizer at the fiber output along the slow axis, we find the spectrum corresponding to a single trapped pulse (Fig. 32c). Comparison between Figures 32a and 32c shows that the solitons trap each other by shifting in wavelength and asymmetrizing. The spectrum in Figure 32c resembles closely the predicted spectrum from Figure 30. For the fiber dispersion of $D \sim 8.75$ ps/nm-km and the maximum separation of the spectral peaks in Figure 32b of ~ 1 THz, we find $D \times \Delta\lambda \sim 83$ ps/km, which is close to the polarization dispersion of 80 ps/km measured for this fiber.

2. Logic Gates with Frequency Filter

By adding a frequency filter at the output of the fiber, the soliton-trapping device becomes a logic gate [27]. We now add an adjustable Fabry–Perot as the frequency filter at the fiber output with sufficient free spectral range so another order is not included within the pulse bandwidth. To obtain a high contrast ratio in an inverter or exclusive-OR gate, a Fabry–Perot with 85% reflecting mirros (maximum finesse of 20) was used with the central bandpass frequency adjusted to the original center frequency of the pulses.

ULTRAFAST ALL-OPTICAL SWITCHING DEVICES 197

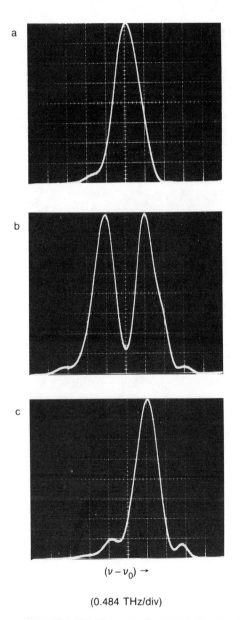

$(\nu - \nu_0) \rightarrow$

(0.484 THz/div)

Figure 32 Spectral confirmation of soliton trapping in a fiber with a polarization dispersion of 80 ps/km. (a) Pulse of ~ 42 pJ along a principal axis; (b) pulse of ~ 84 pJ at $\theta = 45°$, corresponding to two equal amplitude pulses along both axes; and (c) same as (b), with a polarizer at the fiber output aligned with the slow axis.

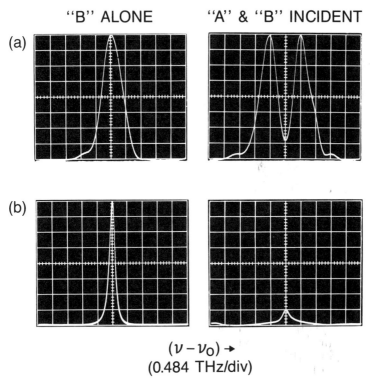

Figure 33 (a) Spectra direct from the fiber output. (b) Spectra after a Fabry–Perot bandpass filter with 85% reflecting mirrors. This corresponds to an exclusive-OR gate with an $\sim 8:1$ contrast ratio.

Figure 33b shows an exclusive-OR gate with $\sim 8:1$ contrast ratio. We obtain a large output with either "A" or "B" alone and a factor of 8 reduction in spectral intensity when both "A" and "B" are present. By placing a polarizer at the fiber output along the principal axis corresponding to "B," the exclusive-OR converts to an inverter: if "B" is left on, then we obtain NOT "A." Although the input and output center frequencies of the device coincide, the output pulses broaden to approximately the inverse of the frequency bandpass $\tau_{out} \sim 1/\Delta\nu \sim 4$ ps. It is difficult to restore these output pulses to their original width and shape.

The inverter can become a cascadable gate if we widen the filter bandpass so that the output pulses can propagate as solitons in a fiber. With 70% reflecting mirrors in the Fabry–Perot filter, a bandpass of ~ 0.58 THz was obtained, and the central bandpass was again set at the original pulse-

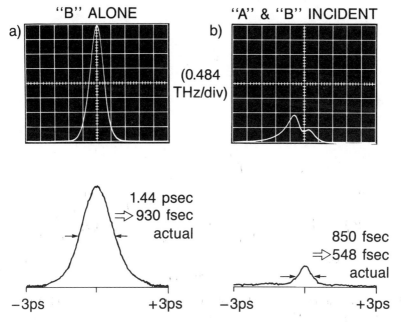

Figure 34 Output spectra and autocorrelations from an inverter (NOT "A") with 70% reflecting mirrors in the Fabry–Perot filter, and a polarizer along the "B" axis. (a) "A" blocked, and (b) "A" unblocked. Both spectra and both autocorrelations are plotted on the same scale.

center frequency. The lower finesse of the filter reduces the contrast ratio of the logic gate. Figure 34 shows the spectra and autocorrelations of the output from an inverter (we use a polarizer at the fiber output along "B" polarization, and treat "A" as the signal). The contrast ratio is $\sim 5:1$ in the autocorrelation, $\sim 4:1$ in the peak spectral intensity, and $\sim 3.2:1$ in the net energy output. With no "A" signal, the 620-fs pulses out of the fiber broaden to 930 fs after passage through the filter. When both "A" and "B" are incident on the fiber, the output has a lower intensity, and the trapped solitons are narrower in the center. Although autocorrelations do not reveal absolute timing of the pulses, the "B" alone signal comes out at a different time from the trapped solitons.

By placing a bandpass filter at the original pulse-center frequency, the soliton-trapping gates show inversion and exclusive-OR functionalities. If the filter pass band is centered ~ 0.5–1 THz away from the original pulse frequency, however, then we have an AND gate: i.e., the frequency shifts out only when both pulses are present [28]. Figure 35 shows the spectra

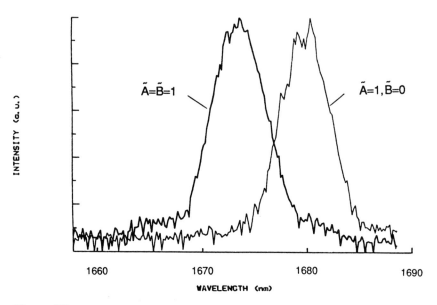

Figure 35 Digitized spectra at the fiber output after a polarizer for a single and a pair of pulses propagating in the fiber.

directly at the output of the fiber along the slow axis for a single pulse (A alone) and for a pair of temporally coincident pulses along each axis (A and B incident). Each of the pulses is ~ 42 pJ in energy, which is about three-quarters of the $N = 1$ fundamental soliton energy. This curve is similar to Figure 32 except that now a polarizer is placed at the fiber output to select only the slow axis signal, and the data is sampled and digitized for analysis. We determine the optimum center wavelength for the pass band of the frequency filter by dividing the signal for A and B by the signal for A alone. Figure 36 represents the expected contrast ratio as a function of wavelength, indicating that the contrast should exceed 20:1 for a center wavelength of the frequency filter of ~ 1672 nm. The contrast ratio between 1660 and 1670 nm appears noisy because of division by the small value of the A and B signal.

The soliton-trapping AND gate (STAG) is implemented with 85% reflecting mirrors (maximum finesse of 20) in the Fabry–Perot filter, and Figure 37 shows the spectra at the output of the STAG. The maximum contrast is ~ 22:1 at 1672 nm, which agrees with predictions, and the energy output for the on state is ~ 9.9 pJ. The polarizer introduces a 3-dB loss, which should not be fundamental, and the transmission through the Fabry–Perot adds another 3 dB of loss. Since the frequency filter with a

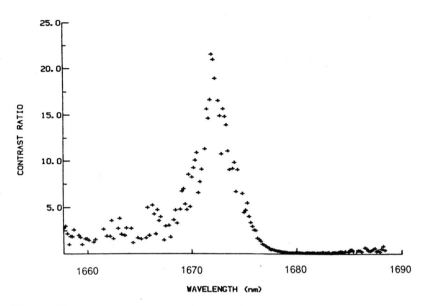

Figure 36 Expected constrast ratio for the soliton-trapping AND gate obtained by dividing the two curves (A AND B over A alone) in Fig. 35.

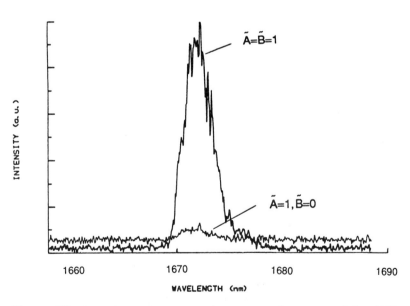

Figure 37 Experimentally measured spectra at the output of the AND gate, which consists of the fiber followed by a polarizer and Fabry–Perot frequency filter. The peak contrast ratio is 22:1, which agrees well with the values in Fig. 36. Note that we displace the origin of the two spectra simply for clarity.

pass band narrower than the pulse spectral width attenuates and broadens the pulse, this device is not cascadable because both the center wavelength and the pulse shape at the output are different from the input.

C. Cascade of Soliton-Dragging and -Trapping Gates

We cascade an SDLG with a STAG and measure an energy contrast of ~ 12:1 at the output. Therefore, we use the cascadability and fan-out of SDLGs for operations that require several levels of logic and use the energy contrast from STAGs as the interface to electronics. We also find that the experimental results are in good agreement with numerical simulations of the nonlinear Schrödinger equation.

Figure 38 shows the experimental configuration for testing the cascaded gate operation, and the inset provides a schematic of the logic circuit. A passively mode-locked color-center laser ($\tau \sim 335$ fs and $\lambda \sim 1.695$ μm) provides separate control, signal, and clock beams. The timing for the various arms is adjusted so that the control and signal are temporally coincident at the input to the SDLG, and the control and clock overlap at the input to the STAG in the absence of the signal. The SDLG consists of a 350 m

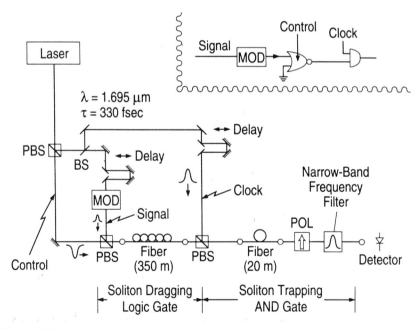

Figure 38 Experimental apparatus for testing the cascaded soliton-dragging and -trapping gates (MOD = modulator, POL = polarizer, BS = beam splitter, PBS = polarizing beam splitter). The insert shows a schematic of the logic circuit.

length of birefringent fiber surrounded by two polarizing beam splitters (PBS). The orthogonally polarized control and signal are combined at the input PBS and coupled into the fast and slow axes of the SDLG, respectively. At the output a PBS removes the signal beam and combines the orthogonally polarized clock input with the control, and both are then coupled into the STAG fiber. Because of Raman effects or soliton self-frequency shift (SSFS), the control pulse is shifted to longer wavelengths after traversing the SDLG fiber. Since anomalous group-velocity dispersion causes longer wavelengths to travel slower, we align the control along the fast axis of the STAG so that the SSFS partially cancels the birefringence and increases the interaction length. The STAG is implemented in a 20 m length of moderately birefringent fiber followed by a polarizer and a narrowband frequency filter (center bandpass wavelength of 1.688 μm, $\Delta\lambda_{FWHM} = 0.05$ Å). Soliton trapping causes the clock pulse that is along the slow axis to speed up by shifting to shorter wavelengths. By placing a polarizer along the clock axis and monitoring the up-shifted frequency, we avoid errors from SSFS or cross-Raman effects. With careful handling to minimize mode coupling effects, polarization extinction ratios of 15:1 and 70:1 were obtained for the SDLG and STAG fibers, respectively.

Figure 39a shows the cross-correlation of the control and a clock pulse at the output of the SDLG gate, and we observe that adding a 10-pJ signal pulse delays the control arrival time by 2.9 ps ($\sim 5.5\tau$). By selecting a clock window centered around the nondelayed control pulse, we obtain a NOR gate that functions as a inverter with the signal as the single data input. Observe that the 49-pJ control pulse traveling along a principal axes of the fiber broadens by 60% (from 330 to 530 fs) because the power is below the N = 1 soliton power (the soliton energy is ~ 73 pJ). In addition, the spectrum of the control pulse shifts by ~ 1 nm ($\sim 1/6$ of the FWHM spectral width), to longer wavelengths because of SSFS. Because of coupling losses between the SDLG and STAG, the control energy is lowered by 50% to 25 pJ in the STAG fiber. In Figure 39b we show the spectrum of the clock pulse after the polarizer with and without the control pulse present. The clock pulse with energy, $E_{CL} = 44$ pJ, which propagates along the slow axis of the second fiber, speeds up with the addition of the 25-pJ control pulse by shifting ~ 1 nm to shorter wavelengths. By placing a narrowband frequency filter at the fiber output and adjusting the center bandpass wavelength to the tail of the spectrum for clock alone, we obtain an AND gate that has a high output only when both inputs are temporally coincident. The ratio of the output spectrum (ratio of solid to dotted curves in Figure 39b) in Figure 39c shows an expected contrast of 13:1 at 1.688 μm.

Figure 40 illustrates the performance of the cascaded gates by displaying the detector output with the signal modulated at a 50% duty cycle. The

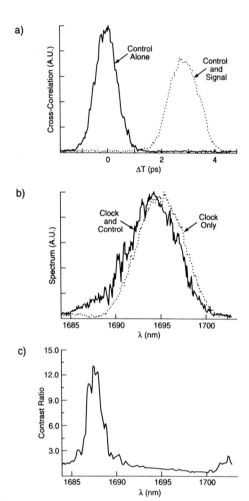

Figure 39 Cross-correlation of clock and control pulses at the output of the soliton-dragging logic gate with and without the signal pulse ($E_{CONTROL}$ = 49 pJ, E_{CLOCK} = 44 pJ, E_{SIGNAL} = 10 pJ. (b) Spectrum at the output of the soliton-trapping AND gate along the clock axis with and without the control present ($E_{CONTROL}$ = 25 pJ, E_{CLOCK} = 44 pJ in the second fiber). (c) Ratio of the two spectra in (b) that shows a peak contrast of ~ 13:1.

ULTRAFAST ALL-OPTICAL SWITCHING DEVICES

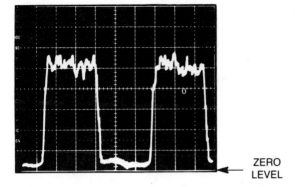

Figure 40 Output from the cascade of soliton-dragging and -trapping gates after a frequency filter centered at 1688nm. The signal is modulated at a 50% duty cycle, and the contrast is ∼ 12:1 (zero at bottom of scale).

zero level is at the bottom of the scale, and so the contrast ratio is ∼ 12:1 after the cascade of the two gates. With the signal off, the control and clock temporally coincide and trap one another, shifting the control spectrum toward shorter wavelengths. As a result, energy is transmitted through the frequency filter and detected as a "1" state. When the signal is on, the control pulse is delayed, thereby preventing soliton trapping between the clock and control. As a result, the clock pulse spectrum and amplitude remain unchanged, corresponding to a "0" state. The excess noise in the "1" state arises because of timing jitter between the two gates when both inputs are present. Timing jitter is a consequence of ±5% amplitude jitter on the laser, which was measured experimentally. This amplitude jitter leads to timing jitter because of SSFS in the SDLG fiber, and the control pulse arrival time at the STAG fiber input varies by almost ±1.5 pulse widths.

Although the previous section showed that the contrast ratio from a STAG alone can be as high as 22:1, the result is degraded because of timing jitter and coupling losses between the two gates. To understand the optimal performance of the gate pair, we numerically studied the cascaded gates using a split-step Fourier transform technique. In the SDLG, the nonlinear Schrödinger equation (NLSE) was solved for a single polarization input (control only). After proper renormalization and correction for coupling losses, the output from the SDLG was used as the control input to the STAG fiber, along with a separate clock input. The control input to the SDLG and the clock input to the STAG are assumed to be Gaussian, in accordance with earlier observations of laser-generated pulse shapes [12].

SSFS effects were included in both fibers by incorporating the term $t_d \, \delta|u|^2/\delta_t$ in the NLSE, where u is the pulse amplitude, t is the local time relative to the pulse position, and t_d is the Raman coefficient, equal to 6 fs/τ. To account for the effect of $\pm 5\%$ amplitude jitter on the cascaded performance, we averaged three cases corresponding to input intensities of $0.95|A|^2$, $1.00|A|^2$, and $1.05|A|^2$, where $|A|^2$ is the measured intensity at the input to the SDLG. Figure 41(a) shows the calculated spectral output of the STAG gate with and without the control pulse present, and the expected contrast ratio is shown in Fig. 41(b). In the numerical simulations we observe a spectral shift of 0.8 nm as a result of trapping and measure a peak contrast ratio of $\sim$ 15:1 at 1688 nm, which is in good agreement with the experimental results. To determine the optimum STAG fiber length we also include in Fig. 41(c) numerical results of the cascaded operation contrast ratio obtained at the STAG gate output as a function of STAG gate length. The simulations show that a maximum contrast ratio is obtained for a length of 20–30 m, and, thereafter, the contrast ratio decreases with increased fiber length. This decrease results from less wavelength shift from soliton-trapping and pulse shaping effects at longer lengths, which increase energy in the tails of the spectrum and reduce the obtained contrast.

IV. POTENTIAL APPLICATIONS AND TECHNOLOGICAL CHALLENGES

This chapter has reviewed some all-optical routing devices and logic gates that may be used for ultrafast serial processing. The discussion was limited to devices implemented in fibers, which generally have a long latency and must be used in pipelined, feed-forward applications. Many of these devices are at an early stage of research, and so the concepts they illustrate may be more important than the implementation details. Nonlinear guided-wave switches are based on the Kerr effect, and, if these switches are operated in the anomalous group-velocity-dispersion regime, then soliton pulses can also be supported (solitons are preferred for picosecond and femtosecond pulses). Nonlinear directional couplers may be used as a saturable absorber to remove low-level signals, and routing devices such as Kerr modulators and nonlinear optical loop mirrors may be used as demultiplexers. The digital soliton logic gates can be used in high-speed telecommunications networks, optical computing, and computer interconnection networks.

Despite the infancy of the all-optical devices, it is important to identify and direct efforts toward potential application arenas. The use of all-optical gates can enhance the system when it is the bandwidth of the switch that

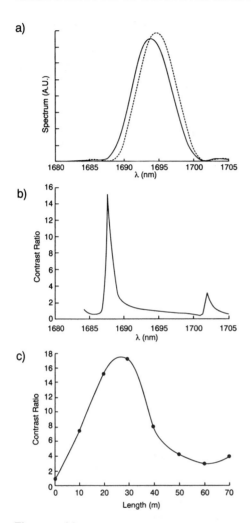

Figure 41 Calculated performance of the cascaded soliton-dragging and -trapping gates: (a) spectral output of the soliton-trapping gate (dashed: clock only; solid: clock and control); (b) contrast ratio for AND gate as a function of wavelength obtained by dividing the two curves in (a); (c) maximum contrast ratio as a function of length of the soliton-trapping fiber. In the experiments a 20-m fiber is used.

limits the performance of the sytem. Computer networks may benefit from high-speed gates for clock synchronization and communication between memory and processors, workstation and processors, or multiple-processor machines. For example, there is a growing trend toward a distributed computing environment in which every user has a workstation at their location and shares resources such as memory, math processors, mail, etc. Such an environment is not at this time limited by computing or central processing unit power. Instead, the bottleneck arises from communicating between different sectors of the network. For the distributed computing paradigm to work, there must be a high-speed, efficient, interconnecting network so every user appears to have a large virtual machine. The questions that need to be studied include the following: (a) Where are the key bottlenecks? (b) What are the constraints on the devices to be used? (c) How can ultrafast, long-latency gates help to solve some of these difficulties?

Potential applications for ultrafast gates in telecommunications networks include high-performance front and back ends of long-haul telecommunications fibers (e.g., multiplexers and demultiplexers); time slot interchangers; header reading in self-routing packet switching; and nodes in local area networks. These are all examples of time-domain switching functions where the system bottleneck is set by the bandwidth of a few critical components. Many of these applications are treated in detail in [2]. For these applications, serial processing is most natural because the system directly connects to the high-bandwidth fiber system, which is inherently serial. Furthermore, optical processing avoids the overhead and bottlenecks associated with optical-to-electronic and electronic-to-optical conversion.

Electronic logic and switching systems may be expected some day to operate at maximum speeds of $\sim$ 20–50 GHz [29]. Yet, fiber transmission systems already exceed these speeds and are limited only by the electronic demultiplexing speed at the receiver end. Considerable effort has been directed toward implementing all-optical demultiplexers based on the Kerr effect [30], four-wave mixing [31] and nonlinear optical loop mirror devices [32,33]. Current experiments in demultiplexers are targeted toward several key objectives:

1. Operating beyond 2.5 Gb/s, which is the speed for which next-generation long-haul systems are being designed, and even beyond 50 Gb/s, which should be beyond electronic speeds
2. Lowering the switching energy approaching 1 pJ/bit, since a terabit rate of data at a picojoule per bit will require lasers with a watt of average power
3. Achieving arbitrary demultiplexing of bit streams, which would warrant the use of bit-rate devices

ULTRAFAST ALL-OPTICAL SWITCHING DEVICES

Time slot interchangers (TSI) are one type of time-domain switch that are used in large telecommunications switching systems. By rearranging the temporal positions of data, a TSI reduces the size of the space-domain routing switches that are used to shuffle the data. In addition, as we use time multiplexing to achieve parallelism in time, we must also allow for communication between data streams. Communication between time-multiplexed data streams can be accomplished by use of a TSI. Just as a NOR gate provided Boolean completeness, a TSI allows for connectivity (in time) completeness. The simplest TSI problem is that two data units, A and B, are traveling down a fiber in series (not necessarily in adjacent time slots). Depending on the value of a control signal, the ordering of A and B will either remain intact (bar state) or they will interchange time slots (cross state). Two or three soliton-dragging logic gates can be used to implement an ultrafast, two-slot TSI [34,35].

Packets in a self-routing network consist of a header (address) and payload (data). Every information exchange point (node) in the network must read the header of each packet to determine its destination. By using ultrafast gates to select and decode the header, the throughput of the network could be enhanced. In lieu of ultrafast switches, researchers have proposed sending the header in a different color, at a different bit rate, over a separate network, etc. With the bit-rate switches, however, we can transmit the header in the same format as the data: This leads to the most flexible system, since sections of the data contain future routing information [2,36,37].

Local and metropolitan area networks are the next expected bottlenecks in high-speed telecommunications. As the bit rate in long-haul fiber systems increases, the bit-rate bottleneck shifts to the systems that interface to the long-distance data highways. Again, the ultrafast gates can enhance the capabilities of these networks by processing the packet header at the bit rate. For example, we have designed a 100-Gb/s soliton ring network [36,37] that may span several tens or hundreds of kilometers. The peak data rate is beyond the limits that electronic networks might be expected to achieve; e.g., this design is 1000 times faster than current high-speed FDDI (fiber-distributed data interface) ring networks. Since picosecond or femtosecond pulses will be used in this time-division-multiplexed system, soliton pulses must be used to avoid the deleterious effects from group-velocity dispersion and self-phase modulation. The soliton ring network uses digitally encoded, self-routing packet switching and is a "light-pipe" system where the data remains in optical format throughout, converting to electronics only at the host and destination nodes. This design is a pipelined, feed-forward architecture that exploits the bandwidth while downplaying the latency of the ultrafast logic gates. Also, a ring network is one potential application of all-optical gates that requires only a handful of gates per node.

The last five years have seen significant progress on ultrafast, all-optical devices, but there are several technological areas that require major breakthroughs before the field can thrive. The crucial device issue remains lowering the switching energy so that each device requires less than a picojoule net energy per bit. Novel nonlinear materials must also be studied to make compact devices with reduced latency and increased thermal stability. Perhaps the main missing component for all-optical systems is a compact laser source with average powers approaching a watt that can act as the power supply. Also absent is an all-optical random access memory, and solutions to this problem have not even been proposed. The need for random access memory in a pipelined machine may be debatable, however, since the latency in accessing the memory could be detrimental. Although commonly overlooked as an "engineering detail," accurate time-synchronization circuits are needed for bit periods approaching a picosecond. Finally, in a broader context, architectures must be crafted that use the switch bandwidth to enhance the capabilities of the system, thereby enabling this ultrafast technology to make an impact.

REFERENCES

1. N. J. Doran and D. Wood, *Opt. Lett.*, *13*:56 (1988).
2. M. N. Islam, *Ultrafast Fiber Switching Devices and Systems*, Cambridge University Press, Cambridge, 1992.
3. T. Morioka, M. Saruwatari, and A. Takeda, *Electron. Lett. 23*:453 (1987); T. Morioka and M. Saruwatari, *J. Sel. Areas Comm.*, *6*:1186 (1988).
4. S. M. Jensen, *IEEE J. Quantum Electron.*, *QE-18*:1580 (1982).
5. S. R. Friberg, A. M. Weiner, Y. Silberberg, G. Sfez, and P. W. Smith, *Opt. Lett.*, *13*:904 (1988).
6. S. Trillo, S. Wabnitz, R. H. Stolen, G. Assanto, C. T. Seaton, and G. I. Stegeman, *Appl. Phys. Lett.*, *49*:1224 (1986).
7. S. Trillo, S. Wabnitz, N. Finlayson, W. C. Banyai, C. T. Seaton, and G. I. Stegeman, *Appl. Phys. Lett.*, *53*:837 (1988).
8. A. M. Weiner, Y. Silberberg, H. Fouckhardt, D. E. Laird, M. A. Saifi, M. J. Andrejco, and P. W. Smith, *IEEE J. Quantum Electron.*, 25:2648 (1989).
9. S. Trillo, S. Wabnitz, E. M. Wright, and G. I. Stegeman, *Opt. Lett.*, *13*:672 (1988).
10. K. J. Blow, N. J. Doran, and B. K. Nayar, *Opt. Lett.*, *14*:754 (1989).
11. M. N. Islam, E. R. Sunderman, R. H. Stolen, W. Pleibel, and J. R. Simpson, *Opt. Lett.*, *14*:811 (1989).
12. M. N. Islam, E. R. Sunderman, C. E. Soccolich, I. Bar-Joseph, N. Sauer, T. Y. Chang, and B. I. Miller, *IEEE J. Quantum Electron.*, *25*:2454 (1989).

13. J. P. Gordon, *Opt. Lett.*, *11*:662 (1986).
14. F. M. Mitschke and L. F. Mollenauer, *Opt. Lett.*, *11*:659 (1986).
15. H. A. Haus and M. N. Islam, *IEEE J. Quantum Electron.*, *QE-21*:1172 (1985).
16. A. Lattes, H. A. Haus, F. J. Leonberger, and E. P. Ippen, *IEEE J. Quantum Electron.*, *QE-19*:1718 (1983).
17. K. J. Blow, N. J. Doran, B. K. Nayar, and B. P. Nelson, *Opt. Lett.*, *15*:248 (1990).
18. J. D. Moores, K. Bergman, H. A. Haus, and E. P. Ippen, *Opt. Lett.*, *16*:138 (1991).
19. H. Avramopoulos, P. M. W. French, M. C. Gabriel, and N. A. Whitaker, *IEEE Photonics Technol. Lett.*, *3*:235 (1991).
20. G. I. Stegeman and R. H. Stolen, *J. Opt. Soc. Am. B*, *6*:652 (1989).
21. M. N. Islam, C. E. Soccolich, and D. A. B. Miller, *Opt. Lett.*, *15*:909 (1990); M. N. Islam, *Opt. Lett.*, *15*:417 (1990).
22. M. N. Islam, L. F. Mollenauer, R. H. Stolen, J. R. Simpson, and H. T. Shang, *Opt. Lett.*, *12*:625 (1987).
23. M. N. Islam, C. R. Menyuk, C.-J. Chen, and C. E. Soccolich, *Opt. Lett.*, *16*:214 (1991).
24. M. N. Islam, C.-J. Chen, and C. E. Soccolich, *Opt. Lett.*, *16*:593 (1991).
25. C. R. Menyuk, *Opt. Lett.*, *12*:614 (1987).
26. C. R. Menyuk, *J. Opt. Soc. Am. B.*, *5*:392 (1988).
27. M. N. Islam, *Opt. Lett.*, *14*:1257 (1989).
28. M. W. Chbat, B. J. Hong, M. N. Islam, C. E. Soccolich, P. R. Prucnal, and K. R. German, Ultrafast soliton trapping AND-gate, *IEEE J. Lightwave Technol.*, *10*:2011 (1992).
29. P. W. Smith, *Bell System Techn. J.*, *61*:1975 (1982); R. W. Keyes, *The Physics of VLSI Systems*, Addison-Wesley, New York, 1987.
30. T. Morioka, H. Takara, and M. Saruwatari, Ultrafast, dual-path optical Kerr demultiplexer utilizing a polarization rotating mirror, *Technical Digest on Nonlinear Guided-Wave Phenomena*, Optical Society of America, Washington, D.C., 1991, vol. 15, pp. 374–377.
31. P. A. Andrekson, N. A. Olsson, J. R. Simpson, T. Tanbun-Ek, R. A. Logan, and M. Haner, *Electron. Lett.*, *27*:922 (1991).
32. B. P. Nelson, K. J. Blow, P. D. Constantine, N. J. Doran, J. K. Lucek, I. W. Marshall, and K. Smith, All-optical gigabit switching in a nonlinear loop mirror using semiconductor lasers, *Technical Digest on Nonlinear Guided-Wave Phenomena*, Optical Society of America, Washington, D.C., 1991, vol. 15, pp. 342–344.
33. N. A. Whitaker, H. Avramopoulos, P. M. W. French, M. C. Gabriel, R. E. LaMarche, D. J. DiGiovanni, and H. M. Presby, *Opt. Lett.*, *16*:1838 (1991).
34. C. E. Soccolich, M. N. Islam, J. R. Sauer, and M. Salerno, GEO-modules and all-optical time slot interchangers, OSA Proceedings on Photonic Switching (H. S. Hinton and J. W. Goodman, eds.), Optical Society of America, Washington, D.C., 1991, vol. 8. pp. 105–108.

35. M. N. Islam and J. R. Sauer, *IEEE J. Quantum Electron.*, 27:843 (1991).
36. J. R. Sauer, M. N. Islam, and S. P. Dijaili, A soliton ring network, (to be published in *IEEE J. Lightwave Technol*).
37. M. N. Islam, C. E. Soccolich, S.-T. Ho, R. E. Slusher, and J. R. Sauer, Ultrafast all-optical fiber soliton logic gates, OSA Proceedings on Photonic Switching (H. S. Hinton and J. W. Goodman, eds.), Optical Society of America, Washington, D.C., 1991, vol. 8, pp. 98–104.

5

Digital Switching Systems Based on the SEED Technology and Free-Space Optical Interconnects

H. Scott Hinton

McGill University, Montreal, Quebec, Canada

Thomas J. Cloonan, Frederick B. McCormick, Jr., Anthony L. Lentine, Rick L. Morrison, R. A. Nordin, and Gaylord W. Richards

AT&T Bell Laboratories, Naperville, Illinois

I. INTRODUCTION

Free-space digital optics is a new emerging hardware platform designed to build digital systems supporting digital signals as opposed to analog systems supporting digital signals. These digital systems are composed of optically interconnected discrete logic gates, smart pixels, and/or electronic multichip modules (MCM). This connection-intensive hardware can then be used to match the massive interconnection requirements of switching fabrics and other high-performance digital systems. This chapter will begin with a discussion of the benefits of free-space digital optics. This will be followed by an overview of free-space digital optical hardware, including reviews of the self-electrooptic effect device (SEED) technology, spot array generation, optical interconnects, and optical hardware modules. Finally, there will be a discussion of free-space interconnection networks with an emphasis on extended generalized shuffle networks.

II. BENEFITS OF FREE-SPACE DIGITAL OPTICS

Several advantages of photonics technology can be used to extend the performance and capability of the existing electronics technology. These advantages include (a) lower communication energy, (b) lower on-chip power dissipation per pin-out, (c) lower skew, (d) large interconnection density medium, and (e) new connection-intensive architectures. Each of these strengths will be discussed in more detail below.

A. Lower Communication Energy

One of the attributes of optics that has not yet been exploited is the lower energy requirement for communicating logical signals from one integrated circuit (IC), multichip module (MCM), or printed circuit board (PCB) to another [1–3]. The signal energy required to move an electrical signal from one point to another depends on whether the signal propagates on a lossy resistive lumped element line or a lossless terminated transmission line. For the case of a properly terminated transmission line the signal energy per bit E_s is given by

$$E_S \text{ (properly terminated)} > \frac{\tau V^2}{R}$$

where V is the logic-level voltage, R is the characteristic transmission line impedance, and τ is the pulse width of a bit of information. As an example, a 1-V, 1-ns pulse on a 100-Ω transmission line requires at least 10 pJ of energy per bit.

When the electrical line length is less than the phase velocity times the signal rise time ($L \leq v_p \tau_r$), the electrical interconnect can be treated as an unterminated lumped element line instead of a properly terminated transmission line. In this case, the amount of energy stored in the charging capacitor is $C_{tot} V^2 / 2$, while the same amount of energy is dissipated in the lumped inductance; thus,

$$E_S \text{(lumped element line)} > C_{tot} V^2$$

where the total capacitance C_{tot}, is given by $C_{tot} = Lc_L + C_p + C_g$ where c_L is the capacitance per unit length of the interconnecting line, C_p is the pad capacitance, and C_g is the gate capacitance. As an example, a 1-ns signal traveling on a 1-mm unterminated lumped element line requires 140 fJ of energy (see Fig. 1).

For the optical case, the minimum energy required is a function of the quantum efficiency of the photodetector β, the wavelength of the light λ, the voltage V required to be developed across a resistor R in series with a photodetector, the capacitance of the diode C_D, and the efficiency of the interconnect η_i. Assuming that $RC_D > \tau$, then

$$E_S \text{ (optical)} > \frac{hcC_D V}{\eta_i \beta \lambda e}$$

SEED TECHNOLOGY

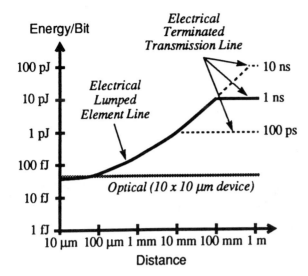

Figure 1 Minimum required communication energy as a function of distance for different bit durations.

where $h = 6.626 \times 10^{-34}$ J-s, $c = 2.998 \times 10^8$ m/s, and $e = 1.602 \times 10^{-19}$ C. When $\lambda = 850$ nm, $\beta = 1$, $\eta_i = 1.0$, and $C_D = 11.5$fF (115 aF/μm^2), there is a minimum optical energy per bit of approximately 58 fJ.

Optical interconnection provides an energy advantage, because it sends signals as bunches of photons. Hence, there are no lines to charge. This feature of optics has been called *quantum impedance conversion*. One consequence of this effect is that, beyond a certain "break-even" distance, optics requires less energy than electrical connections. The practical break-even length depends on how good the optical interconnect technology is.

To take full advantage of quantum impedance conversion requires that we can make small optoelectronic devices that are efficient at low power levels. The photodetectors have to be small (e.g., 10×10 μm), and they must be integrated right beside the electronic circuits. With small integrated modulators and detectors, the break-even length could be a few hundred microns.

B. Lower On-Chip Power Dissipation Per Pin-Out

Although new electronic technologies are increasing the number of pin-outs per chip, one fundamental limiting factor will be the power dissipated on the chip [4]. The by-product of lower communication energy is that there will be less power dissipated on-chip per pin-out. Here optical inter-

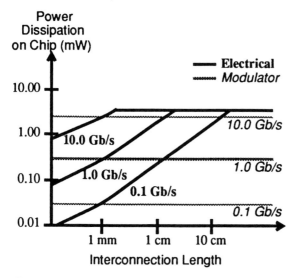

Figure 2 A comparison between electrical and optical interconnection. (See Table 2.) [4]

connects may have a distinct advantage over their electrical counterparts. Figure 2 shows the results of this comparison.

The parameters used to create this figure are shown in Table 1. This figure illustrates the on-chip power dissipation, required from the source, versus the interconnection length, with the bit rate, and the source type: electrical (e.g., C4 or flip-chip bonding), laser (e.g., surface emitting laser), or modulator (e.g., symmetric self-electrooptic effect device) as parameters.

For the electrical case, the flat part of the curves are associated with electrical lines that are long enough to be treated as properly terminated transmission lines. In this case the dissipated power is given by

$$P_D \text{ (properly terminated)} > \frac{(1/\eta_e - 1) V^2}{R}$$

For short electrical line lengths the electrical interconnect can be treated as an unterminated lumped element line instead of a properly terminated transmission line; thus,

$$P_D \text{ (lumped element)} > \frac{(1/\eta_e - 1) C_{tot} V^2}{\tau_r}$$

as shown by the curved lines in Figure 2. The capacitance C_{tot} is the sum of the transmission line capacitance, the pad capacitance, and the gate

SEED TECHNOLOGY

Table 1 Parameters for Figure 2

Parameter	Electrical	Modulator
Received detector voltage	1 V	1 V
Electrical driver efficiency (η_e)	0.75	
Transmission line capacitance (c_L)	1 pF/cm	
Input gate capacitance (C_g)	20 fF	20 fF
Input pad capacitance (C_p)	20 fF	
Photodetector capacitance (C_d)		5 fF
Photodetector responsivity ($\mathcal{R}$)		0.8 A/W
Interconnect efficiency (η_i)		0.8
Modulator efficiency (η_m)		0.5
Load resistance capacitance (C_L)	5 fF	5 fF

Source: [4].

capacitance. As an example, a 1-Gb/s signal (τ_r = 200 ps) traveling on a 1-mm unterminated lumped element line dissipates approximately 0.2 mW. On the other hand the power dissipated on-chip when driving a 1-mm properly terminated transmission line is 3.3 mW, assuming a 100-Ω line.

The detector model used by both the modulator portions of the curve is a resistor in series with a photodetector. The voltage across the resistor is the input to a gate. The resistor is chosen to guarantee that the rise time is equal to the fall time of an incident pulse; thus,

$$R = \frac{V}{C_{tot}\, dv/dt} \cong \frac{\tau_r}{C_{tot}}$$

The optical power required by the photodetector for a 1-V signal is

$$P_{opt} > \frac{C_{tot}\, dv/dt}{\mathcal{R}} \cong \frac{VC_{tot}}{\mathcal{R}\tau_r}$$

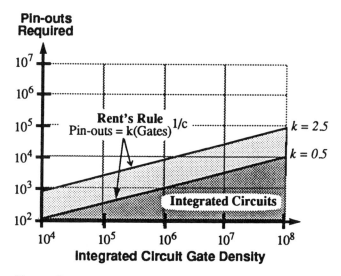

Figure 3 System and device pin-out requirements and minimum required system ICs as a function of the integrated circuit gate density (assuming $c = 1.79$).

Finally, for the case of the modulator, the absorbed power is given by

$$P_D(\text{modulator}) > \frac{(1/\eta_m - 1)P_{opt}}{\eta_i}$$

$$> \frac{(1/\eta_m - 1)VC_{tot}}{\mathcal{R}\tau_r\eta_i}$$

This implies that the dissipated power at 1 Gb/s ($\tau_r = 200$ ps) for a modulator is 0.2 mW.

With increasing interconnection length, the electrical and modulator interconnection schemes cross over near 1 mm, at a 1 Gb/s bit rate. For interconnection lengths greater than 1 mm, optics (modulators and low-loss interconnection) are more efficient. For lengths less than 1 mm, electrical techniques are more efficient. Increasing the length even further, the electrical-based interconnection schemes cross over near 1 cm. Therefore, one may conclude that optics can reduce the total dissipated power of a digital system when interconnecting both chip-to-chip and substrate-to-substrate. Hence, in systems where high density I/Os are needed with a high percentage of the I/O required to be active at one time, optics can provide a performance advantage, with respect to power dissipation, beginning at the chip-to-chip interconnection level.

The need for a large number of *connections* both at the device and system level in future high-performance digital systems is another driving

SEED TECHNOLOGY

force behind the interest in the photonic technology. As the integrated circuit gate density has increased, so has the need for pin-outs or connections, as illustrated in Figure 3. In this figure the left axis projects the pin-outs required by future ICs, assuming the empirical Rent's rule will continue to be valid [5,6]. Rent's rule is given by

Pin-outs = $k(\text{Gates})^{1/c}$

where Gates is the number of digital gates, k is a constant that depends on the ability to share signal lines (for high performance applications, $k = 2.5$), and c is another constant within the range of 1.5–3.0 (1.79 appears to be the best match for high-performance packages with 10 to 100,000 digital gates). Note that for ICs with 1-M digital gates, the package will require between 1000 and 10,000 pin-outs. This suggests the future need to thermally manage 10–100 W of on-chip heat generated for electrical chip-to-chip communication. Reducing this an order of magnitude through optical communication could be a significant improvement for future high-performance systems.

C. Lower Skew

When building large systems that require any form of synchronization, a limiting system parameter is skew, δ_s, which is the maximum difference in propagation delay between parallel channels [4]. This skew can be caused by either dielectric or dimensional variations in the system interconnects (see Fig. 4).

The skew due to dielectric variations is given by

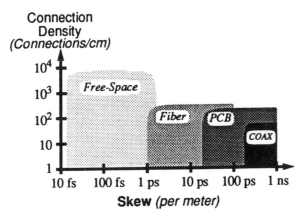

Figure 4 A comparison between electrical and optical skew and connection density.

$$\delta_s \text{ (dielectric)} = \frac{\sqrt{\epsilon_r}\Delta}{c} = \frac{n\Delta}{c}$$

where ϵ_r is the relative permittivity, n is the index of refraction of the dielectric, Δ is the manufacturing tolerance of the ϵ_r or n of the dielectric, and c is the speed of light. Examples of the skew for printed circuit boards (PCB) include the dielectric FR4, given by δ_s (FR4) = 70–140 ps with ϵ_r = 4.6 and Δ = 0.01–0.02 and the dielectric teflon, given by δ_s(teflon) = 55–110 ps when ϵ_r = 2.8 and Δ = 0.01–0.02. The result of tigher manufacturing tolerances for silica fiber yields a lower skew of δ_s(fiber) = 4.8–9.6 ps when n = 1.45 and Δ = 0.001–0.002. The skew for free-space interconnection can be controlled within several wavelengths for aberration-controlled telecentric imaging systems. Interconnections based on holographic interconnects and/or nontelecentric systems result in skews that increase with the field size of the image.

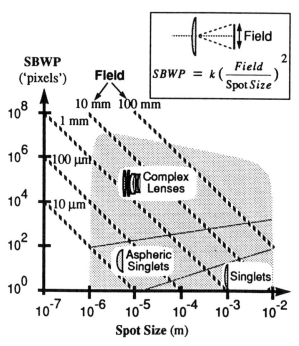

Figure 5 Space–bandwidth product versus spot size (assuming < 1 in. diameter lenses, λ = 850 nm).

D. Large Interconnection Density Medium

Another strength of the optical domain is the spatial bandwidth available through either classical bulk optics, conventional holography, or computer generated holography. The maximum number of elements or pixels that can be supported by an optical system is referred to as its space–bandwidth product (SBWP) or the degrees of freedom of the system [7]. The SBWP for several types of lens systems are shown in Figure 5. If each pixel can be equated to a pin-out, then device pin-outs greater than 10^4 can be achieved.

E. New Connection-Intensive Architectures

Using this enhanced connection and communications capability of optics, new *connection-intensive* digital processing systems can be designed that maximize connections between the processing nodes (smart pixels), ICs, and MCMs rather than minimizing them as in current *connection-constrained* electronic systems. This provides the opportunity of avoiding communication bottlenecks associated with connection-constrained architectures (e.g., buses), thereby leading to new high-performance terabit aggregate capacity digital systems (see Fig. 6).

These new connection-intensive systems are normally composed of concatenated two-dimensional optoelectronic integrated circuits (2-D OEICs) interconnected with either bulk optics or holograms [8]. These 2-D OEICs

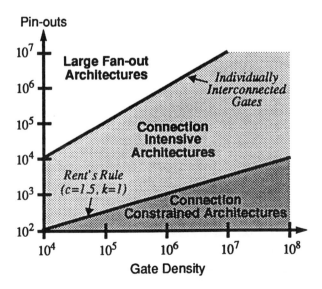

Figure 6 System architectural design space.

are a mixture of electronic and optical devices and can be thought of as electronic chips with photonic I/O. When a 2-D OEIC is composed of a two-dimensional matrix of functional circuits (smart pixels) [8], each requiring photonic I/O, it can be referred to as a smart pixel array. This unlikely mixture of electronic and optical devices is designed to take advantage of the strengths in both the electrical and photonic domain. The photonic devices include detectors to convert the signals from the previous 2-D OEIC to electronic form and either modulators, surface emitting lasers, or LEDs to optically transfer the results of the electronically processed information to the next 2-D OEIC. The electronics does the intelligent processing on the data, and the photonic devices provide the connectivity. Finally, high fan-out architectures could even further utilize the potential connectivity offered by the optical domain.

As the bit rates and clock rates of digital systems continue to increase, it forces PCBs, MCMs, and even ICs to be capable of supporting aggregate capacities in excess of 1 Tb/s. This is especially true in connection-intensive systems such as switching fabrics. Figure 7 illustrates the aggregate capacity

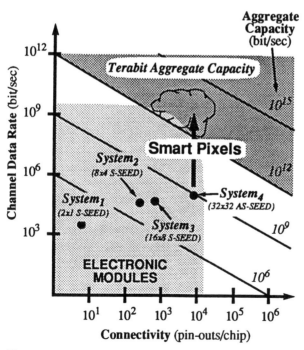

Figure 7 Device aggregate capacity.

as a function of connectivity (pin-outs per chip) and per channel data rate (bits per second). The capabilities of electronic modules (PCBs, MCMs, and ICs) are located in the lower left corner of the figure. The upper right is the high-performance region supporting greater than a terabit aggregate capacity that can be accessed though smart pixels and free-space optical interconnects. Also included in this design space is the performance of the first four system demonstrations by AT&T.

III. FREE-SPACE DIGITAL OPTICAL HARDWARE

The hardware required to implement a digital switching fabric based on free-space digital optics is composed of four basic components [9]. The first is the active device such as a symmetric SEED (S-SEED) array or an array of smart pixels. For modulator-based devices, an optical clock or power source is required in the form of a spot array generator. The optical interconnect hardware functionally interconnects the active devices. Finally, a beam combination unit is required to combine the signal inputs from the previous devices and the optical clock, and then to direct the output from the device to the next stage of the network. An example of each of these components is described in more detail below.

A. SEED Technology

The self-electrooptic effect device (SEED) technology is based on multiple quantum well (MQW) modulators [10]. Multiple quantum wells consist of thin, alternating layers of narrow and wide bandgap materials such as GaAs and AlGaAs. Because of confinement of carriers in the quantum wells, the absorption spectrum shows distinct peaks, which are termed exciton peaks. When an electric field is applied perpendicular to the plane of the quantum wells, the positions of the peaks shift as illustrated in Fig. 8. This electroabsorption mechanism is called the quantum confined Stark effect (QCSE) [11,12], and it is strong enough that a 1 μm thick multiple quantum well stack can have changes in absorption coefficient of a factor of two or so for a 5-V change across the stack. By placing the multiple quantum well material in the intrinsic region of a reverse-biased p-i-n diode, the resulting device can modulate light in response to a change in voltage. The same device can also detect light.

The evolution of the SEED technology includes the Resistor SEED [13–15], the symmetric SEED (S-SEED) [16], the Logic SEED (L-SEED) [17], and currently the integration of field effect transistors (FETs) with MQW modulators (FET-SEED) [18]. This section will give a brief description of the basic operation of some of these devices.

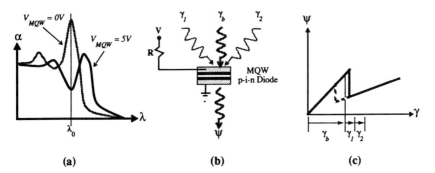

Figure 8 (a) Absorption spectra of MQW material for both 0 and 5 V; (b) schematic of MQW p-i-n diode; (c) input/output characteristics of MQW p-i-n diode.

1. The Resistor SEED

The characteristic curve shown in Figure 8c results when this MQW material is placed in the intrinsic region of a p-i-n diode and electrically connected to a resistor as shown in Figure 8b. When the incident intensity γ_i is low, there is no current flowing through the p-i-n diode or resistor, and thus the majority of the voltage is across the p-i-n diode. If the device is operating at the wavelength λ_0, the device will be a low absorptive state. As the incident intensity increases, so does the current flowing in the p-i-n diode; this in turn reduces the voltage across the diode, which increases the absorptioon and current flow. This state of increasing absorption creates the nonlinearity in the output signal, ψ, shown in part (c) of the figure. Optical logic gates can be formed by biasing the R-SEED close to the nonlinearity γ_b and then applying lower-level data signals γ_1 and γ_2 to the device.

2. Symmetric SEED

The S-SEED, which behaves like an optical inverting S–R latch, is composed of two electrically connected multiple quantum well (MQW) pin diodes as illustrated in part (a) of Figure 9. In this figure, the device inputs include the entering signals, γ_i (set) and $\bar{\gamma}_i$ (reset), and the clock signals. To operate the S-SEED, the γ_i and $\bar{\gamma}_i$ inputs are also separated in time from the clock inputs as shown in part (c) and (d). The γ_i and $\bar{\gamma}_i$ inputs, which represent the incoming data and its complement, are used to set the state of the device. When $\gamma_i > \bar{\gamma}_i$, the S-SEED will enter a state where the lower MQW p-i-n diode will be transmissive, forcing the upper diode to be absorptive. When $\gamma_i < \bar{\gamma}_i$, the opposite condition will occur. Low switching intensities are able to change the device's state when the clock signals are

SEED TECHNOLOGY

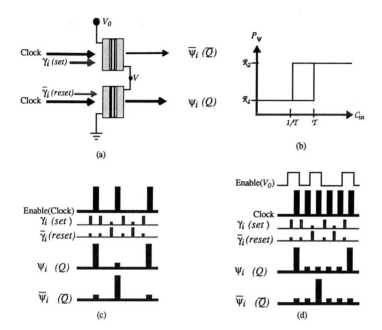

Figure 9 Symmetric self-electrooptic effect device (S-SEED): (a) S-SEED with inputs and outputs; (b) power transfer characteristics; (c) optically enabled S-SEED; and (d) electrically enabled S-SEED.

not present. After the device has been put into its proper state, the clock beams are applied to both inputs. The ratio of the power between the two clock beams should be approximately one, which will prevent the device from changing states. These higher energy clock pulses will transmit the state of the device to the next stage of the system. Since the γ_i and $\bar{\gamma}_i$ inputs are low-intensity pulses and the clock signals are high-intensity pulses, a large differential gain may be achieved. This type of gain is referred to as time-sequential gain.

The operation of an S-SEED is determined by the power transfer characteristic shown in Figure 9b [19]. The optical reflected power (Ψ_i), when the clock signal is applied, is plotted against the ratio of the total optical signal power impinging on the set and reset windows (when the clock signal is not applied). Assuming the clock power is incident on both signal windows, the output power is proportional to the reflectivity, $\mathcal{R}_i$. The ratio of the input signal powers is defined as the input contrast ratio $C_{in} = P_\gamma/P_{\bar{\gamma}}$. As C_{in} is increased from zero, the reflectivity of the bottom diode switches from a low value, $\Re_1$, to a high value, $\Re_2$, at a C_{in} value approximately equal to the ratio of the absorbances of the two optical windows: $\mathcal{T} = (1 - \Re_1)/(1 - \Re_2)$. Simultaneously, the reflectivity of the top diode switches from $\Re_2$ to $\Re_1$. The return transition point (ideally) occurs when $C_{in} = (1 - \Re_2)/(1 - \Re_1) = 1/\mathcal{T}$. The ratio of the two reflectivities, $\Re_2/\Re_1$, is the output contrast C_{out}. Typical measured values of the preceding parameters include $C_{out} = 3.2$, $\mathcal{T} = 1.4$, $\Re_2 = 50\%$ and $\Re_1 = 15\%$ [19].

The operation of an S-SEED as a 2-module (switching node) can be accomplished by either optically or electrically enabling the individuals S-SEEDs. To optically enable an S-SEED array, a spatial light modulator can be used to select which S-SEEDs receive clock pulses. If an S-SEED receives a clock pulse, the information previously latched into the device will be transferred to the next stage of the network. If no clock pulse is received, the information cannot be transferred. This is illustrated in Figure 9c. On the other hand, the S-SEEDs can also be electrically enabled by controlling the voltage applied to the devices as shown in part (d). If the appropriate voltage is present, the S-SEEDs behave as previously described and the information will be transferred. If no voltage is present, both MQW p-i-n diodes will become absorptive, preventing the stored information from transferring to the next stage.

3. FET-SEEDs

To take further advantage of the spatial bandwidth available in the optical domain, integrated electronic circuits need to be integrated with optical detectors (inputs) and modulators or microlasers (outputs). This mixture of the processing capabilities of electronics and the communications capabilities of optics will allow the implementation of connection-intensive architectures with more complex nodes than simple switches. In addition, the gain provided by the electronic devices should allow higher speed operation of the nodes.

Integrated quantum well modulators with GaAs field effect transistors (FETs) have been demonstrated [20]. The GaAs FETs were fabricated on top of a p-i-n diode structure grown p-side down. A circuit diagram and layer structure of a demonstrator are shown in Figure 10. Optical switching energies of less than 100 fJ have been observed [21].

Over the past five years there has been considerable progress at AT&T in developing the SEED technology. It began in 1987 with the first 2 × 1 array of S-SEEDs and has evolved to the point where 32K arrays of S-SEEDs have been fabricated. Also the switching energies have decreased from 100s of pJ with the first S-SEEDs to less than 100 fJ in the most recent FET-SEEDs in the same period of time. This progress is illustrated in Fig. 11 [22].

SEED TECHNOLOGY

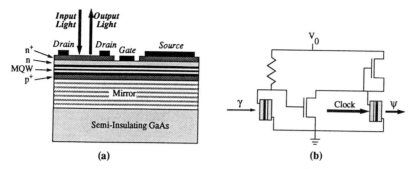

Figure 10 F-SEEDs: (a) layer structure and (b) simple demonstrator circuit.

B. Spot Array Generation

The two-dimensional arrays of SEED technology based switching nodes that have been previously discussed require an optical power supply to clock the devices [9]. The generation of 2-D arrays of uniform intensity spots requires two basic components. The first is a high-power, single-frequency, diffraction-limited laser that can provide the appropriate power per pixel required to meet the system speed requirements. The second component in a spot array generator requires some mechanism to equally and uniformly divide the power from the laser and distribute it among the smart pixels. The amount of optical power required by the laser source, P_{tot}, is given by

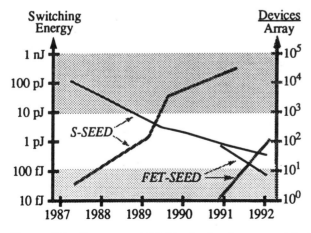

Figure 11 Progress at AT&T in the development of the SEED technology.

$$P_{tot} = E_{sw} \frac{1}{t_{sw}} \frac{1}{T_{sys}} N$$

where E_{sw} is the device switching energy, t_{sw} is the device switching time, T_{sys} is the system optical transmission, and N is the number of smart pixels. As an example, to power 1000 smart pixels with switching energies of 1 pJ/smart pixel at a bit rate of 100 MHz, assuming a 10-dB loss in the optical interconnect, will require $P_{tot} = 1$ W.

There have been several different approaches to the distribution of optical power between the smart pixels. They are listed in Table 2 [23]. The approach that has received the most attention is Fourier-plane spot array generation using phase gratings to uniformly distribute the optical power to the smart pixels. These phase gratings are made by etching glass with a repetitive multilevel pattern. For the case of a binary phase grating (BPG), there would be two thicknesses of glass. This grating is illuminated by a plane wave from a laser source as illustrated in Figure 12. The light transmitted through the grating is transformed at the back focal plane of a lens, which is the output plane of the spot array generator and the location of a smart pixel array.

Table 2 Methods of Spot Array Generation

Category	Type	Size	Efficiency	Uniformity	References
Fourier-plane	Dammann [24]	$\leq 1 \times 9$	50–70%		[25]
		$\leq 1 \times 21$	60–70%		[26]
		64×64	64%	±10%	[27,28]
		201×201		±7%	[29]
	Multilevel Dammann	5×5	63%	±5%	[30]
		4×4	70.6%	±1%	[31]
	Kinoform	$\leq 1 \times 101$	80–90%		[32]
		1×9	92%	±7%	[33]
	Microlens array	4800 beams	Lenslet dependent	Poor at boundaries	[34]
Fresnel-plane	Talbot	100×100	≤100%	Similar to input beam	[35-37]
Image-plane	Phase contrast imaging	32×32 number of cells	≤100%	Similar to input beam	[38]
	Microlens array	Lenslet number	≤100%	Similar to input beam	[39-41]
Cascade-type	Dammann multiple imaging	81×81	25%	±29%	[42]

SEED TECHNOLOGY

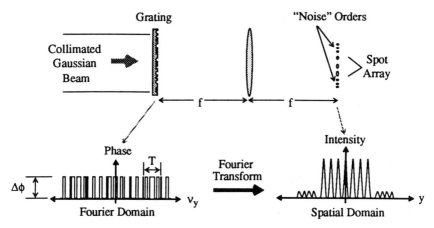

Figure 12 Fourier-plane spot array generation.

An example of the spot array generation hardware used by AT&T's $System_4$ is shown in Figure 13 [27,45]. In this subsystem, the laser drive electronics and the thermoelectric cooler are used to control the hand-picked 60-mW, semiconductor 850-nm laser. The output light is initially collimated and then circularized by the Brewster telescope. The analyzer and 1/4-wave plate are used as an isolator to reduced the destabilizing effects of back reflections of the laser. The collimated light then passes through the Risley prisms, which are used to register the spot arrays on the SEED device photosensitive windows. Finally, the light passes through a 64

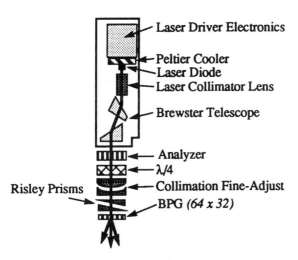

Figure 13 $System_4$ spot array generation.

× 32 Dammann grating, which in conjunction with the objective lens of system (see Fig. 16) redistributes the input optical power into 2048 uniform spots of light.

C. Optical Interconnects

When an optical signal leaves a smart pixel it must be directed to the smart pixels in the next stage of the system. The mechanism used to create this path for information from one stage to the next provides the role of electrical wire used in electronic systems. These optical interconnects can be provided through bulk optics or holography.

The interconnect used for $System_4$ is used to create a 3-D network composed of a 2-D $N \times M$ network replicated X times to create X parallel $N \times M$ networks [27,45]. This is equivalent to an $N \times M$ network that is X bits deep. This type of network can be implemented with a simple 2-D interconnect. An example of the interconnect is shown in Figure 14, where the output of each node is directed to the pupil-plane, where a binary phase grating splits the signal into three equal parts. These copies of the original signal are then directed to the inputs of the next stage of the network. Part (b) of the figure illustrates how a Banyan network (thick lines) can be created using these interconnects. The thick lines represent active connections between nodes in $stage_i$ and nodes in $stage_{i+1}$; the thin lines represent connections between stages in the network that are blocked by placing metal masks in front of the optical windows of specific nodes in $stage_{i+1}$. If all three signals created by the interconnect are used, instead of just two

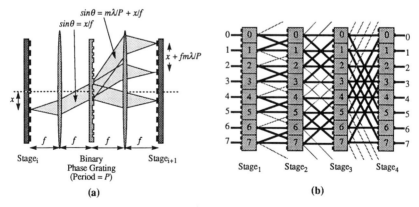

Figure 14 Pupil-plane interconnection: (a) 1 × 3 BPG interconnect; (b) Banyan network using 1 × 3 interconnects.

SEED TECHNOLOGY

as described above, a TIADM network can be implemented. Note that the splitting angle of the BPG is different for each stage in the network. The Banyan network [47] shown in Figure 14 is an example of a 16 × 32 network in which each channel is 32 bits deep.

D. Beam Combination

Free-space photonic switching systems based on SEED technology arrays require that each device must be able to receive two input signals plus a clock [48,49]. In addition, the reflected output signal must be directed from the device to the next stage of the switching fabric. The major constraints of this problem are that the spots must be small ($< 5 \mu$m), requiring a high-resolution, low-aberration objective lens, and that the entering signals must not interfere at the device's optical window. A simple but lossy method of combining two inputs signals, the clock, and the output signals is through the use of a polarization beam splitter (PBS) and a partially reflecting mirror, M (50:50), as shown in Figure 15 [43–45]. The input image enters the beam combination unit s-polarized; thus, it is reflected up through a $\lambda/4$ plate to the partially reflecting mirror M. Half of the power is reflected, passing through the $\lambda/4$ plate again, which will rotate the image to p-polarized light. This reflected image will then pass through the polarization beamsplitter, PBS, through another $\lambda/4$ plate, the objective lens, L, and onto the array of switching nodes. For the case of S-SEED nodes, the information present on the input image is latched into the S-SEEDs. The circularly polarized clock can then pass through all the elements, be modulated by the information stored in the switching node array, and then be

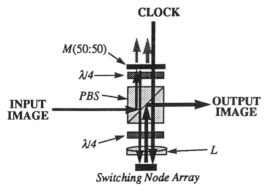

Figure 15 Lossy beam combination.

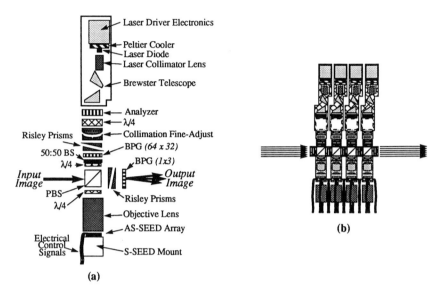

Figure 16 (a) Optical hardware module; and (b) four concatenated OHMs.

reflected out of the beam combination unit to provide the output image to the next part of the system.

E. Optical Hardware Module

By combining all of the piece parts that have been discussed, an optical hardware module (OHM) can be created to provide the optical hardware necessary for each stage of a multistage network as shown in Figure 16a [45]. These OHMs can then be concatenated together to create multistage interconnection networks for switching applications or pipeline structures for a computing environment as illustrated in Figure 16b.

IV. FREE-SPACE INTERCONNECTION NETWORK

The main objective of the free-space technology is to exploit the spatial bandwidth (pin-outs or connections) available in the optical domain [8]. This has allowed designers to look for multistage interconnection networks (MINs) that are connection intensive as opposed to maximizing the available bandwidth in a limited number of connections (temporal bandwidth). The devices used in free-space system proposals have generally been digital

SEED TECHNOLOGY

and not analog, which implies that the design issues for large systems will be significantly different. As an example, digital systems regenerate the signal at each gate, and so system signal-to-noise ratio and loss are not important system or network design issues.

The performance of a switching fabric is related to its connection capability, that is, the number of available paths or connections from any input channel to any output channel. Figure 17 shows a line of connection capability that extends from partial connectivity to redundant strictly nonblocking connectivity and some of the known fabric topologies [50]. The line can represent point-to-point networks or broadcasting networks. At the extreme left of the line is no network at all: the case of isolated users with no communication. As we move to the right, the networks allow partial connections between users, but any two users chosen at random have a probability less than one of being connected. At full connectivity, any one input may connect to any one output, but some connections are not possible if other connections already exist in the network. If the network is rearrangeably nonblocking, any point may be connected to any other point provided we are allowed to reroute some existing connections. All permutations of inputs to outputs can be established by a rearrangeably nonblocking network. Wide-sense nonblocking networks guarantee that a connection between an idle input and an idle output will never be blocked provided that the order in which the connections are established follows the right algorithm. Strictly nonblocking net-

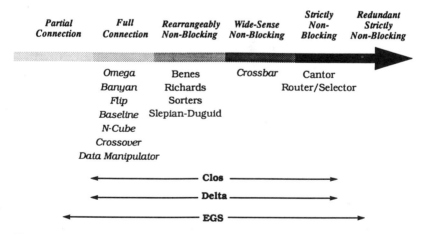

Figure 17 Connection capability of several known interconnection network topologies.

works also guarantee that a connection between an idle input and an idle output will never be blocked, but any available path can be used for any connection. On the far right of the connectivity continuum are networks that have more than enough elements to be strictly nonblocking; these networks are designed to maintain an acceptable level of connectivity even when elements within the network fail. A more detailed discussion of the fabric topology of the extended generalized shuffle network (EGS) is given below.

A multistage interconnection network (MIN) is an interconnection of *stages* of switching nodes, where a stage is a set of *identical* switching nodes as shown in Figure 18 [51]. Let S_i, $i = 1, \ldots, s$, denote the *i*th stage of an *s*-stage MIN, where S_i contains r_i nodes, each having n_i inputs and m_i outputs. The $N = r_1 n_1$ inlets of the switching nodes of S_1 are the N inlets of the MIN and the $M = r_s m_s$ outlets of the switching nodes of S_s are the M outlets of the MIN. For $i = 2, \ldots, s$, the inlets of the switching nodes of S_i are connected by links *only* to outlets of the switching nodes of S_{i-1}, and the outlets of the switching nodes of S_i are connected by links *only* to inlets of the switching nodes of S_{i+1}. So that all stage i outlets can be connected on a one-to-one basis with all stage $i + 1$ inlets, it is required that $r_i m_i = r_{i+1} n_{i+1}$, for $1 \leq i \leq s - 1$. Thus, N-input, M-output MINs contain multiple stages of nodes (node-stages), and each consecutive pair of node-stages is connected by the links within a link-stage.

Nodes actively route data, while links passively transport data from one node-stage to the next. Several different types of digital switching nodes are illustrated in Figure 19. The triplet notation shown in this figure represents the following: (number of inputs [n_i], number of outputs [m_i], capacity of the node [c_i]) [52]. This third parameter indicates the number of channels that can be actively passed through the node at a given time. The (2, 2, 2) node has two inputs and two outputs with the capability of having both inputs and outputs simultaneously active at any time, $c_i = 2$. This node

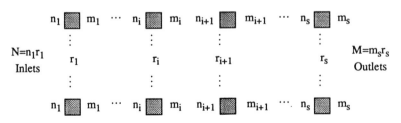

Figure 18 Multistage interconnection network.

SEED TECHNOLOGY

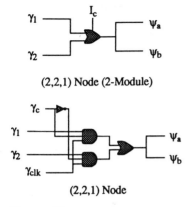

Figure 19 Switching nodes.

is topologically equivalent to a directional coupler, but since it is composed of digital gates it does not have the optical transparency of its analog counterpart. The (2, 2, 1) node in the center of the figure has two inputs and two outputs, although both outputs contain the same information. Hence, the node has $c_i = 1$. In this node the input AND gates select which input channel can pass its contents to the outputs. These nodes work well in networks that have been designed to guarantee that only one input to a given node can be active at any time. Such networks include dilated Benes [53], Ofman [54], and EGS networks [55]. Finally, there is the 2-module. This node also requires that the network guarantee that only one node input is active at any time but is more restrictive than the previously described node, $c_i = 0$. Since this node cannot block signals entering on the inputs, it requires that no signal be present on the unused input line.

An EGS network [51,55,56] is an MIN with a particular specifed sequential interconnection pattern. A MIN G is defined to be an EGS network if either of the following two conditions holds.

Condition 1: For every S_i or G there exists a one-to-one mapping ϕ_i from a node in stage $S_i \in$ the integer set $\{0, 1, \ldots, r_i-1\}$ onto the integer set $\{0, 1, \ldots, r_i-1\}$ such that, for $i = 1, 2, \ldots, s - 1$, switching node $\alpha \in S_i$ connected to switching node $\beta \in S_{i+1}$ if and only if

$$\phi_{i+1}(\beta) \in \{[\phi_i(\alpha)m_i + o_i] \bmod r_{i+1} : o_i \in \{0, 1, \ldots, m_i - 1\}\}.$$

Condition 2: For every S_i of G there exists a one-to-one mapping ψ_i from a node in stage $S_i \in$ the integer set $\{0, 1, \ldots, r_i - 1\}$ onto the integer set

$\{0, 1, \ldots, r_i - 1\}$ such that, for $i = 2, 3, \ldots, s$, switching node $\alpha \in S_i$ is connected to switching node $\gamma \in S_{i-1}$ if and only if

$$\psi_{i-1}(\gamma) \in \{[\psi_i(\alpha)n_i + \iota_i] \mod r_{i-1} : \iota_i \in \{0, 1, \ldots, n_i - 1\}\}.$$

These two conditions may be equivalent in some cases, but in general they are not. The two conditions assert interconnection properties for G from two different perspectives. For two stages i and $i + 1$, Condition 1 defines an interconnection pattern from the perspective of the switching nodes in stage i, whereas Condition 2 does the same from the perspective of stage $i + 1$.

Condition 1 essentially says that a MIN G is an EGS network if one can label the switching nodes in each stage i from 0 through $r_i - 1$ so that switching node 0 in stage i is connected to switching nodes 0 through $m_i - 1$ in stage $i + 1$, switching node 1 in stage i is connected to switching nodes m_i through $2m_i - 1$ in stage $i + 1$, and so on, where each succeeding switch node in stage i is connected to the next m_i switching nodes in stage $i + 1$ in a cyclic fashion; i.e., when the last switching node $(r_{i+1} - 1)$ in stage $i + 1$ is reached, the next switching node to be connected starts over again at 0.

Similarly, Condition 2 essentially says that a MIN G is an EGS network if one can label the switching nodes in each stage i from 0 through $r_i - 1$ so that switching node 0 in stage i is connected to switching nodes 0 through $n_i - 1$ in stage $i - 1$, switching node 1 in stage i is connected to switching nodes n_i through $2n_i - 1$ in stage $i - 1$, and so on, where each succeeding switch node in stage i is connected to the next n_i switching nodes in stage $i - 1$ in a cyclic fashion; i.e., when the last switching node $(r_{i-1} - 1)$ in stage $i - 1$ is reached, the next switching node to be connected starts over again at 0.

EGS networks, then, are a broad class of MINs that do not place any restrictions on the number of nodes within the node-stages or on the number of node-stages within the MIN. In addition, EGS networks do not require the nodes in the network to have any particular functionality, and so acceptable EGS networks can be designed using nodes with n inputs and m outputs and type α, where n and m can be any positive integers and $\alpha = 2$ for full capacity nodes ($c = n$), $\alpha = 1$ for capacity 1 nodes ($c = 1$), $\alpha = 0$ for n-module nodes ($c = 1$). By definition, EGS networks have link-stage connections that are sequentially interconnected, or they are isomorphic to such networks.

The EGS networks described in this paper will be limited to a small subset of the general EGS class of networks. For positive integers n, k, F, and s this subset will have $N = n^k$ inputs, a fan-out section composed of $1 \times F$ nodes, a switching section composed of s stages of (n, n, α) nodes, a fan-

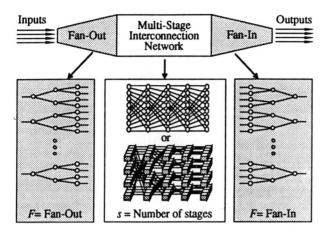

Figure 20 Components of extended generalized shuffle network.

in section composed of $F \times 1$ nodes, and $N = n^k$ outputs as shown in Figure 20. As a result, every node-stage in the switching section contains NF/n (n, n, α) nodes.

An important parameter required in the design of EGS networks is the probability of blocking of the network, $P(B)$. For EGS networks, the variables N, F, s, n, and α can be used to approximate $P(B)$ and then represent and EGS with the sextuplet $[N, F, s, P(B), n, \alpha]$. This approximation is given by

$$P(B) = \left[1 - \left(1 - \frac{n-1}{Fn^{\alpha-1}} \right)^{s-\alpha+1} \right]^{Fn^s/N}$$

This equation can be manipulated to give

$$\kappa = \ln\left(-Fn^s \ln\left[1 - \left(1 - \frac{n-1}{Fn^{\alpha-1}} \right)^{s-\alpha+1} \right] \right)$$

where κ is equal to

$$\kappa = \ln(N) + \ln\left[\ln\left(\frac{1}{P(B)}\right) \right]$$

From this equation we can plot the relationship between s, F, and κ. This is illustrated in Figure 21. From this figure, it can be observed that a network with a given N, $P(B)$, n, and α can have a range of shapes varying from short and wide (s = small, F = large) to long and thin (s = large, F = small).

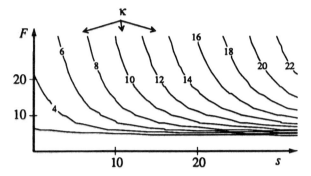

Figure 21 Relationship between EGS variables, $[N, F, s, P(B), 2, 0]$.

An EGS network using (n, n, α) nodes is strictly nonblocking if

$$\frac{Fn^s}{N} + (n-1)^2 \left\lfloor \frac{n^{s-2(\alpha-1)}}{FN} \right\rfloor \geq$$
$$\left\lceil n^{(s-\alpha)/2} \left(2n^{s/2} + (n-1)(1-s_2) \right) \right\rceil +$$
$$2 \left\lfloor [(n-1)(s-k+1-\alpha) - 1]\frac{n^{s+1-\alpha}}{N} \right\rfloor + 1$$

where $0 \leq s \leq 2k - 2$, s_2 denotes the remainder when dividing $s + \alpha$ by 2, and $N = n^k$.

If the fan-out/fan-in sections are implemented using smart pixels acting as binary splitters/combiners, respectively, the total number of node-stages (T) required in an EGS network is $T = s + 2 \log 2^F$. The number of nodes per stage is NF/n, and thus the size of the smart pixel node arrays is $\sqrt{NF} \times (\sqrt{NF})/n$ [57,58].

The strictly nonblocking (SNB) regions of an EGS network as a function of F and s are illustrated in Figure 22 [58]. In this figure the corresponding node array size is shown below each F value. This F versus s plot indicates that for SNB operation larger values of F will typically require smaller values of s, and smaller values of F will typically require larger values of s. The optimum operating point using current devices (which tend to have smaller sizes) is indicated by the black dot in Figure 22, and it illustrates that a SNB EGS network with $N = 256$ inputs will require 19 device arrays of size 128×64. The optimum operating points have been calculated for SNB EGS networks of various sizes (N) and are shown in Table 3 [58].

Not all switching applications require SNB operation. Many applications can tolerate small amounts of blocking within the network. The hardware costs of EGS networks can be decreased if the application will allow non-

SEED TECHNOLOGY

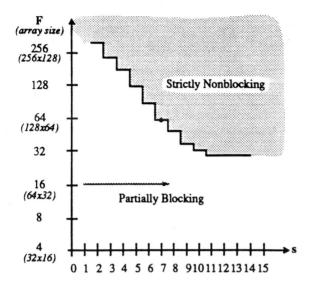

Figure 22 Required F versus s for [256, F, s, $P(B)$, 2, 0] SNB EGS networks.

Table 3 Hardware requirements for [N, F, s, $P(B)$, 2, 0] EGS networks

Network Inputs (N)	Array Size	Required Arrays	
		SNB	$P(B) < 10^{-8}$
16	8 × 16	10	10
32	16 × 32	13	12
	8 × 16		16
64	32 × 64	15	14
	16 × 32		14
128	32 × 64	18	15
256	64 × 128	19	17
	32 × 64		17
512	64 × 128	23	18
1024	128 × 256	24	20
	64 × 128		19
2048	256 × 512	26	21
	128 × 256		21
4096	256 × 512	28	22

zero blocking probabilities, because F and s can be reduced from the SNB values shown in Table 3. Simulations were used to determine the blocking probability as a function of s with fixed N and F. The results for a network with $N = 256$ and $F = 16$ (shown by the gray line in the blocking region of Fig. 22) are plotted as the dark line in Figure 23. In this figure the blocking probability drops rapidly as s is increased, as was evident in the 3-D view of Figure 21. If a $P(B)$ of 10^{-8} is acceptable, then an [256, 16, 9, 10^{-8}, 2, 0] EGS network can be used. This corresponds to a network with $T = 17$ node-stages of size 64×32, thus providing a savings in hardware cost over the SNB network shown in Table 3.

The $P(B)$ simulations were also used to study the effects of faulty nodes on the operation of an EGS network. The gray line plots in Figure 23 show that the blocking probability will increase as faulty nodes are added to the MIN section of the EGS network, but the increases are relatively small. As a result, a few extra stages can be added to the network during the design phase to permit the network to tolerate these faults and still maintain acceptable blocking probabilities. Catastrophic faults can still occur within the network, and the only solution to this problem is to add redundancy. For a photonic free-space system, as will be described later, the amount of

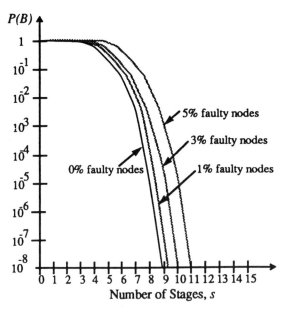

Figure 23 Blocking probability $P(B)$ versus s for [256, 16, s, $P(B)$, 2, 0] EGS network.

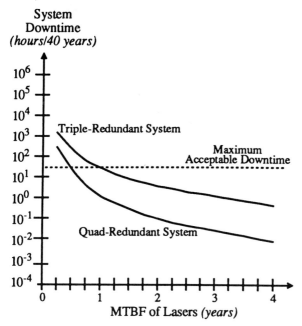

Figure 24 System down time versus MTBF of lasers in [256, 16, 9, 10^{-8}, 2, 0] EGS network.

required redundancy is related to the mean time between failures (MTBF) of the lasers. For an [256, 16, 9, 10^{-8}, 2, 0] network, the system down time was calculated as a function of the MTBF of the lasers. Figure 24 shows that triple- or quad-redundancy may be required to yield acceptable system down times (< 2 h in 40 years) for both computing and telecommunications applications [57,58].

To control the network, there needs to be a method of choosing a path through the network from any inlet **x** to any outlet **y**, where $\mathbf{x} = \{i_k \ldots x_2 x_1\}$ and $\mathbf{y} = \{y_k \ldots y_2 y_1\}$ represent the input and output addresses of the network, respectively [55]. A path **p** of the Fn^s/N possible paths through the network will pass through $node_i$ of the ith stage, where

$$node_i(\mathbf{x}, \mathbf{p}, \mathbf{y}) = \left\lfloor \frac{Fn^s\mathbf{x} + N\mathbf{p} + \mathbf{y}}{n^{s+1-i}} \right\rfloor \mod((NF)/2)$$

For the case where $n = 2$, this equation reduces to

$$node_i(\mathbf{x}, \mathbf{p}, \mathbf{y}) = \left\lfloor \frac{F2^s\mathbf{x} + N\mathbf{p} + \mathbf{y}}{2^{s+1-i}} \right\rfloor \mod((NF)/n)$$

Observe, in this special case, the numerator can be represented by

$$F2^s\mathbf{x} + N\mathbf{p} + \mathbf{y} \Rightarrow \underbrace{x_k \ldots x_2 x_1}_{\text{Inlet Address}} \underbrace{p_q \ldots p_2 p_1}_{\text{Path Number}} \underbrace{y_k \ldots y_2 y_1}_{\text{Inlet Address}}$$

where the term $F2^2$ shifts the binary location of the inlet address $\log_2 F + s$ bit positions to the left. On the other hand the multiplier N of the second term shifts the path number, $\mathbf{p}$, $\log_2 N$ positions to the left. The denominator of the above equation creates a shifting window that indentifies the traversed node, node_i, in stage i, as illustrated below:

$$\left\lfloor \frac{F2^s\mathbf{x} + N\mathbf{p} + \mathbf{y}}{2^{s+1-i}} \right\rfloor_{\mod(NF/2)} \Rightarrow x_k \ldots x_2 x_1 p_q \ldots \underbrace{p_2 p_1 y_k \ldots y_2 y_1}_{\substack{\text{window} \\ \log_2 n + \log_2 F - 1 \\ \text{node}_i}} \quad \underset{s+1-i}{\leftrightarrow}$$

The window, of width $\log_2 N + \log_2 F - 1$, identifies the location of the node in stage i. The window shifts with each stage in the network $s + 1 - i$ positions to the right. Methods have been devised that determine the dis-

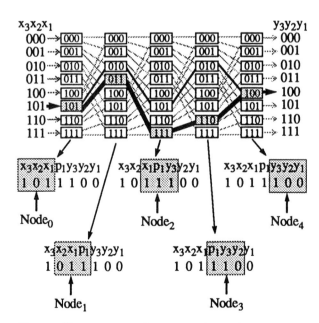

Figure 25 Routing of $[8, 2, 3, P(B), 2, \alpha]$ EGS networks.

SEED TECHNOLOGY

Figure 26 Picture of 16 × 32 (32 bit wide) free-space EGS network.

joint paths throughout the network allowing simple and fast path hunting algorithms and hardware [55]. A simple example is shown in Figure 25. In this figure there is path set up between inlet $x = 101$ and outlet $y = 100$. Note how the window shifts for each stage and how the location of the node is found in the window. Also, note how the bit to the left of the window indentifies which input will receive the signal (0 = upper input, 1 = lower input) and the bit to the right of the window tells which output will send the signal to the next stage (0 = upper output, 1 = lower ouput).

A picture showing the optical hardware that demonstrated the first six stages of a 16 × 32 EGS network is shown in Figure 26 [45].

V. CONCLUSION

This chapter has briefly reviewed the free-space digital optical technology as a new emerging hardware platform designed to build digital systems supporting digital signals. It began with a discussion of the benefits of free-space digital optics. This was followed by an overview of free-space digital optical hardware, including reviews of the SEED technology, spot array generation, optical interconnects, and optical hardware modules. Finally, there was a discussion of free-space interconnection networks with an emphasis on extended generalized shuffle networks.

REFERENCES

1. D. A. B. Miller, Optics for low-energy communication inside digital processors: Quantum detectors, sources and modulators as efficient impedance converters, *Opt. Lett.*, *14*:146 (198a).

2. M. R. Feldman, S. C. Esener, C. C. Guest, and S. H. Lee, Comparison between optical and electrical interconnects based on power and speed considerations, *App. Opt.*, 27: (1988).
3. H. H. Arsenault, T. Szoplik, and B. Macukow, *Optical Processing and Computing*, Academic Press, 1989, pp. 1–31.
4. R. A. Nordin et al., A systems perspective on digital interconnection technology," *J. Lightwave Technol.*, *10*: 811 (1992).
5. Rent never published. See B. J. Landman and R. L. Russo, pin vs. block relationships for partitions of logic graphs, *IEEE Trans. Computers*, *C20*:1469 (1971).
6. R. R. Tummala and E. J. Rymaszewki, *Microelectronics Packaging Handbook*, Van Nostrand Reinhold, New York, 1989.
7. J. W. Goodman, , *Introduction to Fourier Optics*, McGraw-Hill, New York 1968.
8. H. S. Hinton, Architectural considerations for photonic switching networks, J. Sel. *Areas Comm.*, *6*:1209 (1988).
9. H. S. Hinton, Photonic switching fabrics, *IEEE Comm. Magazine,* *28*:71 (1990).
10. H. S. Hinton and A. L. Lentine, Multiple quantum-well technology takes SEED, *Circuits and Devices,* March 1993, pp. 12–18.
11. D. A. B. Miller et al., Bandedge electro-absorption in quantum well structures: The quantum confined stark effect, *Phys. Rev. Lett.*, *53*:2173 (1984).
12. D. A. B. Miller et al., Electric field dependence of optical absorption near the bandgap of quantum well structures," *Phys. Rev. Lett.*, *B32*:1043 (1985).
13. D. A. B. Miller et al., Novel hybrid optically bistable switch: The quantum well self-electro-optic effect device, *Appl. Phys. Lett.*, 45:13 (1984).
14. D. A. B. Miller et al., The quantum well self-electro-optic effect device: Optoelectronic bistability and oscillation, and self-linearized modulation, *IEEE. J. Quantum Electron.*, *QE-21*:1462 (1985).
15. D. A. B. Miller, Quantum well self electro-optic effect devices, *Opt. Quantum Electron.*, *22*:561 (1990).
16. A. L. Lentine et al., Symmetric self-electro-optic effect device: Optical set–reset latch, differential logic gate, and differential modulator/detector, *IEEE. Quantum Electron.*, 25:1928 (1989).
17. A. L. Lentine et al., Logic self-electro-optic effect devices: Quantum-well optoelectronic multiport logic gates, multiplexors, demultiplexors, and shift registers, *IEEE J. Quantum Electron.*, *28*:1539 (1992).
18. D. A. B. Miller, M. D. Feuer, T. Y. Chang, S. C. Chunk, J. E. Henry, D. J. Burrows, and D. S. Chemla, Field-effect transistor self-electrooptic effect device: Integrated photodiode, quantum well modulator and transistor, *IEEE Photonics Technol. Lett.*, *1*:62 (1989).
19. F. B. McCormick, F. A. P. Tooley, T. J. Cloonan, J. L. Brubaker, A. L. Lentine, R. L. Morrison, S. J. Hinterlong, M. J. Herron, S. L. Walker, and J. M. Sasian, Experimental investigation of a free-space optical switching network by using symmetric self-electro-optic-effect devices, *Appl. Opt., 31*:5431 (1992).

20. T. K. Woodward, L. M. F. Chirovsky, and A. L. Lentine, Operation of integrated GaAs–AlGaAs FET-SEEDs: Modulators with gain, IEEE LEOS Annual Meeting, 1991, postdeadline paper PD10 (1991).
21. A. L. Lentine et al., 4 × 4 arrays of FET-SEED embedded control 2 × 1 switching nodes, IEEE LEOS Summer Topical Meeting on Smart Pixels, Santa Barbara, CA, 1992, postdeadline paper.
22. P. J. Anthony, Progress in the SEED Technology, OSA Annual Meeting, Albuquerque, NM, 1992.
23. N. Streibl, Beam shaping with optical array generators, *J. Mod. Opt.*, *36*:1559 (1989).
24. H. Dammann and K. Gortler, High-efficiency in-line multiple imaging by means of multiple phase holograms, *Opt. Comm.*, *3*:312 (1971).
25. H. Dammann and E. Klotz, Coherent optical generation and inspection of two-dimensional periodic structures, *Optica Acta*, *24*:505 (1977).
26. U. Killat, G. Rabe, and W. Rave, Binary phase gratings for star couplers with high splitting ratio, *Fiber Integrat. Opt.*, *4*:159 (1982).
27. R. L. Morrison, S. L. Walker, T. J. Cloonan, F. A. P. Tooley, F. B. McCormick, J. M. Sasian, A. L. Lentine, J. L. Brubaker, R. J. Crisci, S. J. Hinterlong, and H. S. Hinton, Diffractive optics in a free-space digital optical system, Diffractive Optics: Design, Fabrication, and Applications, New Orleans, LA, 1992, pp. 28–30.
28. R. L. Morrison, Symmetries that simplify the design of spot array phase gratings, *J. Opt. Soc. Am. A* (1992).
29. M. R. Taghizadeh, J. Turunen, B. Robertson, A. Vasara, and J. Westerholm, Passive optical array generators, 1991 Topical Meeting on Optical Computing, Salt Lake City, UT, 1991, pp. 148–151.
30. S. J. Walker and J. Jahns, Arrays generation with multilevel phase gratings, *JOSA*, *7*:1509 (1990).
31. R. L. Morrison and S. L. Walker, Progress in diffractive phase gratings used for spot array generation, Proceedings of the Topical Meeting on Optical Computing, 1991, pp. 144–147.
32. J. Turunen, A. Vasara, and J. Westerholm, Kinoform phase relief synthesis: A new method, *Opt. Eng.* (1989).
33. D. Prongue, H. P. Herzig, R. Dandliker, and M. T. Gale, Optimized kinoform structures for highly efficient fan-out elements, *Appl. Opt.*, *31*:5706 (1992).
34. N. Streibl, U. Nolscher, J. Jahns, and S. J. Walker, Array generation with lenslet arrays, *Appl. Opt.*, *30*:2739 (1991).
35. J. A. Thomas, *Binary Phase Elements in Photoresist,* Masters Thesis, University of Erlangen, 1989.
36. A. W. Lohmann and J. A. Thomas, Realization of an array illuminator based on the Talbot effect, *Appl. Opt.*, *29*:4337 (1990).
37. A. Kolodziejczik, *Opt. Comm.*, *59*:97 (1986).
38. A. W. Lohmann, J. Schwider, N. Streibl, and J. Thomas, Array illuminator based on phase contrast, *Appl. Opt.*, *27*:2915 (1988).
39. A. C. Walker, M. R. Taghizadeh, J. G. H. Mathew, I. Redmond, R. J. Camp-

bell, S. D. Smith, J. Dempsey, and G. Lebreton, Optically bistable thin film interference devices and holographic techniques for experiments in digital optics, *Opt. Eng.*, 27:38 (1988).
40. M. R. Taghizadeh, I. Redmond, A. C. Walker, and S. D. Smith, Design and construction of holographic optical elements for photonic switching applications, *Photonic Switching* (T. K. Gustafson and P. W. Smith, eds.), Springer-Verlag, New York 1988, pp 111–114.
41. A. Lohmann and F. Sauer. Holographic telescope arrays, *Appl. Opt.*, 27:3003 (1988).
42. F. B. McCormick, Generation of large spot arrays from a single laser beam via multiple imaging with binary phase gratings, *Opt. Eng.*, 28:299 (1989).
43. F. B. McCormick, F. A. P. Tooley, J. M. Sasian, J. L. Brubaker, A. L. Lentine, T. J. Cloonan, R. L. Morrison, S. L. Walker, and R. J. Crisci, Parallel interconnection of two 64 × 32 symmetric self-electro-optic effect device arrays, *Electron. Lett.*, 27:1869 (1991).
44. F. B. McCormick, F. A. P. Tooley, J. L. Brubaker, J. M. Sasian, T. J. Cloonan, A. L. Lentine, R. L. Morrison, R. J. Crisci, S. L. Walker, S. J. Hinterlong, and M. J. Herron, Design and tolerancing comparisons for S-SEED-based free-space switching fabrics, *Opt. Eng.*, 31:2697 (1992).
45. F. B. McCormick, T. J. Cloonan, F. A. P. Tooley, A. L. Lentine, J. M. Sasian, J. L. Brubaker, R. L. Morrison, S. L. Walker, R. J. Crisci, R. A. Novotny, S. J. Hinterlong, H. S. Hinton, and E. Kerbis, A six-stage digital free-space optical switching network using S-SEEDs, *Appl. Opt.*, to appear.
46. L. R. Goke and G. J. Lipovski, Banyan networks for partitioning multiprocessor systems, Proceedings of the First Annual Symposium on Computer Architecture, 1973, pp. 21–28.
47. M. E. Prise, M. M. Downs, F. B. McCormick, S. J. Walker, and N. Streibl, Design of an optical digital computer, *Optical Bistability IV* (W. Firth, N. Peyhambarian, and A. Tallet eds.), Les Editions de Physique, 1988, pp. C2-15–C2-18.
48. F. B. McCormick and M. E. Prise, Optical circuitry for free space interconnections, *Appl. Opt.*, 29:2013 (1990).
49. J. R. Erickson and H. S. Hinton, Optically transparent systems, *An Introduction to Photonic Switching Fabrics* (H. S. Hinton, ed.), Plenum Publishing, to be published in 1993.
50. H. S. Hinton, J. R. Erickson, T. J. Cloonan, and G. W. Richards, Space-division switching, *Photonics in Switching II* (J. E. Midwinter, ed.), Academic Press, to be published in 1993.
51. G. M. Masson, G. C. Gingher, and S. Nakamura, A sampler of circuit switching networks, *Computer*, June 1979, pp. 32–48.
52. Padmanabhan and A. N. Netravali, Dilated networks for photonic switching, *IEEE Trans. Comm.*, COM-35:1357 (1987).
53. C.-T. Lea, Bipartite graph design principle for photonic switching systems, *IEEE Trans. Comm.* (1990).
54. G. W. Richards, U.S. Patent Numbers 4,993,016 and 4,991,168.

55. T. J. Cloonan, G. W. Richards, A. L. Lentine, F. B. McCormick, H. S. Hinton, and S. J. Hinterlong, A complexity analysis of smart pixel switching nodes for photonic extended generalized shuffle switching networks, *J. Quantum Electron.* (1993).
56. T. J. Cloonan, G. W. Richards, F. B. McCormick, and A. L. Lentine, Extended generalized shuffle network architectures for free-space photonic switching, OSA Proceedings on Photonic Switching (H. S. Hinton and J. W. Goodman, eds.), Optical Society of America, Washington, D.C., 1991, Vol. 8, pp. 43–47.
57. T. J. Cloonan, G. W. Richards, A. L. Lentine, F. B. McCormick, and J. R. Erickson, Free-space photonic switching architectures based on extended generalized shuffle networks, *Appl. Opt.*, *31*:7471 (1992).

6
Free-Space Holographic Grating Interconnects

Abdellatif Marrakchi
Siemens Corporate Research, Inc., Princeton, New Jersey

Kasra Rastani
Bellcore, Red Bank, New Jersey

I. INTRODUCTION

As current communication and computing systems grow more complex, efficient and fast routing of information between elements such as gates, chips, boards, and networks becomes crucial. Electronic interconnects are fast approaching their limits, and many have suggested that optical techniques would be a viable alternative [1–5]. At a higher hierarchical level of interconnects (board-to-board), the limited bandwidth in the current electrical transmission lines is the primary limiting factor. At a lower level (chip-to-chip, or intrachip), other factors such as switching speed, power dissipation, packing density, and immunity to electromagnetic interference become dominant.

Most of the real estate in a chip is taken by the interconnect lines. To reduce the area that they occupy, one could scale down their size. Although there are some advantages in using small linewidths (submicron lines), one quickly realizes that there are also limits to this exercise since the speed will then be limited by the interconnect delay [1]. An alternative to scaling is to find a technology that would allow the interconnects to be brought out of the chip. This will enable the integration of more functionality within the chip itself. Optical techniques are attractive since

1. Photons do not interact, thereby eliminating the electromagnetic interference problem.
2. The three-dimensional nature of optics allows parallel access of two-dimensional structures.
3. One can foresee having programmable optical interconnects using dynamic elements, a task that is at best difficult for electronics.

Routing optical beams by diffraction from multiplexed volume holograms has been suggested for interconnecting information nodes [6–14]. In this chapter, we analyze issues relevant to the free-space interconnect system shown schematically in Figure 1. In such a configuration, optical beams in an input array are routed to different locations of an output array using diffraction from gratings recorded in a holographic medium. For clarity, only two input beams are shown in the figure, which also illustrates fan-out (beam splitting among several output directions) and fan-in (beam combining along one output direction). At the output stage, one may have detector arrays or inputs to the next stage of a cascaded system. Therefore, the development of two-dimensional arrays of devices (lasers, spatial light modulators, detectors) is essential to the realization of this architecture.

The centerpiece of the configuration shown in Figure 1 is the holographic medium in which the gratings are recorded. A detailed characterization of such media is beyond the scope of this chapter. It is important to note, however, that fixed holographic gratings have been recorded in photochemical materials such as dichromated gelatin and silver halide [15], and

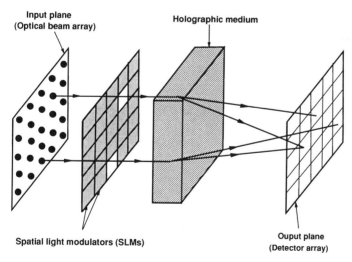

Figure 1 Schematic of a holographic grating interconnect. The spatial light modulator controls the array of input beams.

that dynamic gratings have been recorded in photorefractive materials such as lithium niobate (LiNbO$_3$), barium titanate (BaTiO$_3$), bismuth silicon oxide (Bi$_{12}$SiO$_{20}$), gallium arsenide (GaAs), and many more [16].

Following a background on grating diffraction in Section II, and on the photorefractive effect in Section III, we concentrate on the peripheral components of a holographic interconnect system. These devices are also critical in other free-space interconnect architectures. In Section IV, fixed beam array generators are considered, whereas programmable devices are analyzed in Section V. A detailed characterization of a reconfigurable holographic grating interconnect system is offered in Section VI. Here, emphasis is put on interconnects with analog weights for neural networks, although the same architecture can be used for other applications. In Section VII, we attempt to shed some light on issues such as grating erasure, beam fan-in, and grating degeneracies that limit the performance of a holographic interconnect system. Finally, some concluding remarks are given in Section VIII.

II. BACKGROUND ON GRATING DIFFRACTION

A. Thin Versus Thick Diffraction

Interaction of optical waves with gratings has been treated at different levels of complexity [17,18]. Our objective in providing a background on diffraction from gratings in the following sections, both in the Raman–Nath and Bragg regimes, is to facilitate their assessment as they apply to optical interconnects. Hence, we will limit the discussion to reviewing basic principles.

A diffraction grating element has either surface relief rulings or a modulation of the bulk refractive index. Considering a lossless grating with spatial period Λ_G (wavenumber $K_G = 2\pi/\Lambda_G$), its dielectric constant (ϵ) is real, periodic, and of the form

$$\epsilon(\mathbf{r}) = \epsilon_2 + \Delta\epsilon \cos(\mathbf{K}_G \cdot \mathbf{r}) \tag{1}$$

in which ϵ_2 is the average dielectric constant of the grating and $\Delta\epsilon$ is its modulation. The parameter $\mathbf{r}$ is a vector defining the spatial coordinate. A phase grating can also be characterized by a modulation of the refractive index n of the material (which is related to ϵ by $\epsilon = n^2$):

$$n = n_2 + \Delta n \cos(\mathbf{K}_G \cdot \mathbf{r}) \tag{2}$$

The diffraction characteristics depend on whether the grating is "thin" or "thick." It is widely recognized that the Raman–Nath (thin diffraction regime occurs when the relevant parameters satisfy the inequality [19]

$$\frac{\Delta n}{n_2} \leq \left(\frac{\Lambda_G \cos \theta_i}{\pi d}\right)^2 \tag{3}$$

in which Δn is the refractive index modulation, n_2 is the background refractive index, θ_i is the angle of incidence, and d is the grating thickness. From this relation other equivalent criteria can be deduced [20,21]. It is generally accepted that for $d/\Lambda_G \ll 10$, diffraction follows the Raman–Nath (or thin grating) regime of approximation, with many diffracted orders. This is briefly considered in the next section. When $d/\Lambda_G \gg 10$, the diffraction is known to be in the Bragg regime, with only one diffracted beam. This is presented in Section II. C using the coupled-mode theory for thick gratings.

B. Raman–Nath Diffraction

In this section, we briefly analyze diffraction from a thin grating under the Raman–Nath regime of approximation. The multiple transmitted and reflected diffracted beams are shown in Figure 2. The appropriate refractive indices are shown in this figure for the case $n_2 > n_3 > n_1$. The incident beam with wavevector $\mathbf{k}_1$ makes an angle θ_i with the grating direction. This beam is refracted (following Snell's law) into the grating region. The radii of the left and right semicircles are equal to the wavenumbers (magnitude of wavevectors) of the reflected and transmitted waves; i.e., $k_1 = 2\pi n_1/\lambda_0$ and $k_3 = 2\pi n_3/\lambda_0$, respectively, in which λ_0 is the free-space wavelength. The spacing between the horizontal dashed lines is equal to the magnitude of $\mathbf{K}_G$.

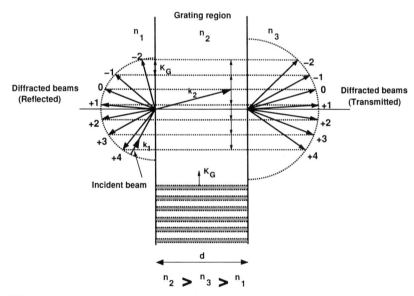

Figure 2 Phase matching diagram for grating diffraction in the Raman–Nath regime. The grating thickness is d.

Following the wavevector matching condition (e.g., $\mathbf{k}_2 + \mathbf{K}_G = \mathbf{k}_{+1}$), the angle θ_m of the mth transmission diffracted order is given by

$$\Lambda_G(n_1 \sin \theta_i - n_3 \sin \theta_m) = m\lambda_0 \tag{4}$$

The angle that the mth order makes upon reflection can be determined from a relationship similar to Eq. (4) by replacing n_3 by n_1.

In the Raman–Nath approximation, the dephasing between the spatial harmonics is neglected; hence, an infinite number of diffracted waves are allowed to propagate. Under this assumption, the diffraction efficiency (η_m) of the mth order is given by an integer-order ordinary Bessel function of the first kind:

$$\eta_m = J_m^2(2\gamma) \tag{5}$$

in which the argument is $\gamma = \pi \Delta n\, d / \lambda_0 \cos \theta_i$ for an incident beam with TE polarized light (electric field perpendicular to the plane of incidence), and $\gamma = \pi \Delta n d \cos 2\theta_i / \lambda_0 \cos \theta_i$ for TM polarized light. Assuming an infinite number of diffracted orders, conservation of energy is satisfied from $\sum_{n=-\infty}^{\infty} J_m^2(2\gamma) = 1$. It should be pointed out that observing large numbers of diffracted orders does not necessarily indicate Raman–Nath diffraction satisfying Eq. (5), but rather it could indicate an intermediate regime between Raman–Nath and Bragg.

C. Bragg Diffraction

In this section, we present the results of the coupled-wave theory of diffraction for volume phase gratings in the Bragg regime. A more in-depth analysis of this regime is appropriate here, since thick gratings lend themselves better to optical interconnect applications owing to their narrow angular and spectral selectivities. Besides the coupled-wave theory presented here, other methods such as the modal analysis [22,23] and the beam propagation method [24] have been reported. For a rigorous coupled-wave analysis of uniaxial anisotropic diffraction gratings, the reader is referred to [25], of which an extension to biaxial anisotropic gratings is presented in [26].

1. Isotropic Diffraction

A thick grating can be visualized as a lamination of thin ones. In this case, diffraction at each layer undergoes repeated scattering by other layers, leading to interference. This interference is generally destructive unless the angle of incidence satisfies the Bragg condition:

$$2\Lambda_G \sin \theta_B = \frac{\lambda_0}{n_2} \tag{6}$$

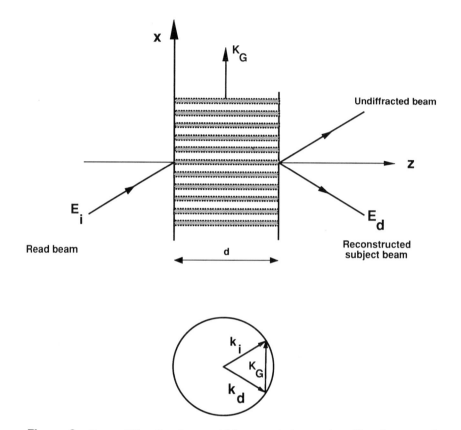

Figure 3 Bragg diffraction from a thick transmission grating. The phase matching diagram is depicted at the bottom of the figure.

in which the angle of incidence θ_B is measured within the grating region. Figure 3 shows a transmission grating and the associated incident, undiffracted, and diffracted beams. The phase matching diagram is shown in the bottom of the figure. The grating wavevector is parallel to the X axis (representing the medium's boundary). The angle between each beam and the Z axis is θ_B, since we have assumed equal refractive indices for both the incident and the transmission regions. The expression of Eq. (6) can be derived from Eq. (4) by assuming only one diffracted beam ($m = 1$) and equal angles of incidence and diffraction.

Kogelnik derived a detailed theoretical model of diffraction for volume gratings [27]. The most relevant result to our discussion is the diffraction efficiency of a phase grating, given by

$$\eta = \frac{\sin^2(\xi^2 + v^2)^{\frac{1}{2}}}{1 + \xi^2/v^2} \tag{7}$$

In this expression, ξ and v are given by

$$\xi = \Delta\theta k_2 d \sin\theta_B \tag{8}$$

$$v = \frac{\kappa d}{\cos\theta_B} \tag{9}$$

in which $\Delta\theta$ is the misalignment of the incident beam from the Bragg angle θ_B, $k_2 = 2\pi n_2/\lambda_0$, and κ is the coupling constant; $\kappa = \pi \Delta n/\lambda_0$ for a TE wave, and $\kappa = -\pi \Delta n \cos 2\theta_B/\lambda_0$ for a TM wave.

The preceding results reveal that the diffraction efficiency in a lossless thick grating is cyclic as a function of $\Delta n\, d$ and theoretically can reach unity. In reality, however, a combination of loss through the medium and small achievable Δn limit the efficiency to lower values. In addition, this efficiency is quite sensitive to misalignment from the Bragg incidence angle and to differences between recording and readout wavelengths. These characteristics make diffraction from volumetric gratings attractive for optical interconnects because they allow control over how much and where light is diffracted.

2. Anisotropic Diffraction

If a holographic medium has linear and/or circular birefringence, the polarization of the diffracted wave may be different from that of the incident wave. This is known as anisotropic diffraction. To illustrate how anisotropy can change the characteristics of diffraction, such as location of diffracted peaks and angular selectivity, consider the case depicted in Fig. 4, which represents the cross section of the wave surface of a material with circular and linear birefringence [28]. Reading out the grating with a beam having wavevector $\mathbf{k}_{R1}$ or $\mathbf{k}_{R4}$ would diffract the orthogonal polarization (which is phase matched to the grating wavevector $\mathbf{K}_G$), whereas $\mathbf{k}_{R2}$ or $\mathbf{k}_{R3}$ would diffract the same polarization. Hence, by changing the angle of incidence two peaks could be diffracted at an angle that is different from the Bragg incidence given in the isotropic case. This is clearly illustrated in Figure 5, showing the diffraction efficiency as a function of Bragg detuning.

The limitation on the number of interconnects that can be realized will ultimately be determined by many factors that include the diffraction regime (thin or thick), the diffraction mode (isotropic or anisotropic), in addition to variables such as optical wavelength, crystal thickness, and refractive index modulation depth, among others. Some of these apsects will be further developed in Section VII.

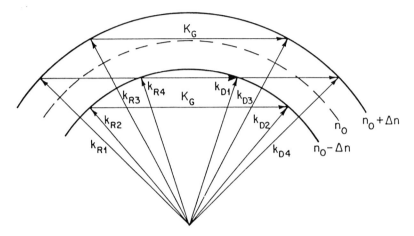

Figure 4 Cross section of the wave surface of a photorefractive bismuth silicon oxide (BSO) crystal showing the four eigenmodes of diffraction. The double shell is a consequence of optical activity.

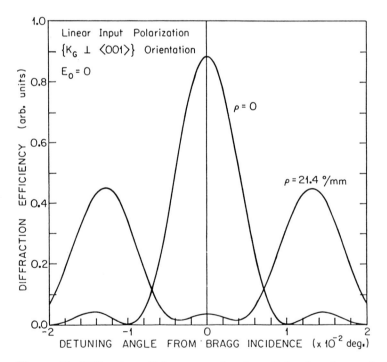

Figure 5 Diffraction efficiency as a function of the detuning angle (diffusion limit). The parameter ρ indicates whether the crystal is optically active.

III. BACKGROUND ON THE PHOTOREFRACTIVE EFFECT

A. Basic Principles

The purpose of this section is to give enough background to the reader in order to understand the mechanisms that give rise to the photorefractive effect. There are basically two materials characteristics that are necessary for this effect to take place: (a) the material has to be photoconductive (with donor and trap centers); and (b) it should exhibit an electrooptic effect. Other physical properties might also be present in some materials, such as the optical activity found in crystals of the sillenite family, or the beam fanning exhibited by some ferroelectric materials. These have found application in signal-to-noise enhancement of holographically recorded images [29], or in self-pumped phase conjugation [30], but they are not essential to the photorefractive effect itself.

Diffraction due to the photorefractive effect cannot be explained with the traditional picture of the optical nonlinear wave mixing in media with $\chi^{(3)}$ susceptibility tensors, despite some analogies in their theoretical formulation [31,32]. The nonlinearities of the refractive index are so small that generally they can be detected only when high-power laser sources are utilized (typically in the MW cm^{-2} range, compared with mW cm^{-2} for the photorefractive effect). In addition, the space-charge density buildup in photorefractive media is an integrative process [33], whereas stimulated and parametric interactions occur essentially instantaneously and are primarily limited by the response time of the atomic or molecular arrangement of the medium. Consequently, the photorefractive effect is bound to involve distinct physical mechanisms during its formation.

When two mutually coherent beams are allowed to interfere within the volume of photoconductive and electrooptic materials that have donor and trap sites, free carriers are nonuniformly generated by absorption in the bright areas, and they are redistributed by diffusion and/or drift under the influence of an externally applied electric field until trapped at an acceptor level in darker areas. This process creates a charge density (or dipole distribution) that replicates the incident intensity pattern, either in phase when the diffusion process dominates, or phase shifted by an angle Ψ when an external field is applied to the crystal (see Fig. 6). A space-charge field E_{sc} builds up in the bulk of the material, which is $(\pi/2 + \Psi)$ phase shifted with respect to the intensity pattern, because of the divergence relationship between the charge density and the field. The presence of this electric field translates into a refractive index modulation through the linear electrooptic (Pockels) effect, and thus records a volume phase hologram. A typical value for the refractive index modulation is of the order of 10^{-5}.

Charge transport under nonuniform illumination in photorefractive materials has been recently described utilizing two separate models. The hop-

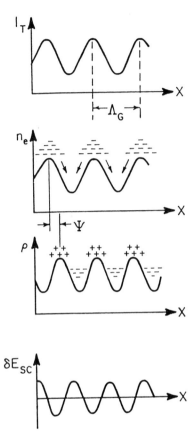

Figure 6 Transformation process of a sinusoidal intensity distribution into a refractive index change.

ping model [34] assumes that carrier transport occurs by hopping from a donor center to a neighboring empty trap. The band transport model [35] assumes that charges are excited in the bright regions and migrate in the conduction band by diffusion and/or drift, before being permanently trapped in the dark regions. In both descriptions, a space-charge density results in an electric field E_{sc}, which in turn modulates the refractive index via the Pockels effect. In general, the space-charge field is out of phase with the incident illumination when the crystal has a nonlocal or noninstantaneous response [36].

An elementary phase grating is recorded in a photorefractive material when two beams with vertical polarizations and respective intensities I_1 and

I_2 are allowed to interfere within its bulk. The incident intensity distribution is then given by

$$I_T(x) = I_G (1 + m_G \cos K_G x) \tag{10}$$

in which $I_G = I_1 + I_2$ is the total incident intensity, $m_G = 2(I_1 I_2)^{\frac{1}{2}}/(I_1 + I_2)$ is the fringe modulation ratio, and $K_G = 2\pi/\Lambda_G$ is the grating wavevector amplitude. The steady state space-charge field E_{sc} has a component in phase with the illumination [35],

$$A = E_d \frac{(E_0/E_d)}{(1 + E_d/E_q)^2 + (E_0/E_q)^2} \tag{11}$$

and a $\pi/2$ phase shifted component,

$$B = E_d \frac{1 + (E_d/E_q) + (E_0^2/E_d E_q)}{(1 + E_d/E_q)^2 + (E_0/E_q)^2} \tag{12}$$

in which E_0 is the externally applied electric field, E_d is the diffusion field, and E_q is the maximum space-charge field that can be induced in the bulk of the material (limited by the trap density). This maximum corresponds to a complete separation of positive and negative charges by one spatial period Λ_G. The total space-charge field amplitude, for a negligible dark conductivity and weak beam coupling (i.e., m_G is not affected by the writing process), is given by $E_{sc} = m_G(A^2 + B^2)^{\frac{1}{2}}$, or

$$E_{sc} = m_G E_d \frac{[1 + (E_0/E_d)^2]^{\frac{1}{2}}}{[(1 + E_d/E_q)^2 + (E_0/E_q)^2]^{\frac{1}{2}}} \tag{13}$$

and the phase shift ψ between the incident intensity distribution and the holographic grating by $\tan \psi = B/A$, or

$$\tan \psi = \frac{1 + E_d/E_q + E_0^2/E_d E_q}{E_0/E_d} \tag{14}$$

Hence, the total (first harmonic) space-charge field with its spatial dependence can be written as

$$E_{sc}(x) = E_{sc} \cos \left(\frac{2\pi x}{\Lambda_G} + \psi \right) \tag{15}$$

In the diffusion regime, when no external electric field is applied to the crystal ($E_0 = 0$), the phase shift ψ is equal to $\pi/2$, and the refractive index is modulated only through the diffusion field.

Figure 7 shows E_{sc}/E_d as a function of the applied field normalized to the maximum space-charge field E_0/E_q, for three typical values of the fringe

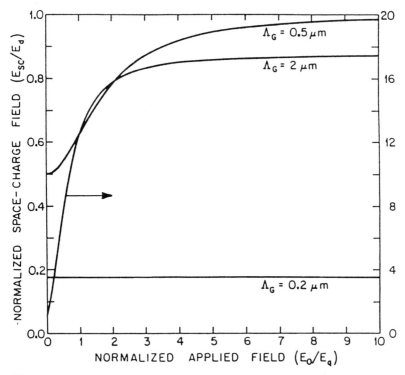

Figure 7 Normalized space-charge field (E_{sc}/E_d) as a function of the normalized applied field (E_0/E_q). The curves are plotted for three typical values of the grating fringe spacing.

spacing. The total space-charge field is maximized for large fringe spacings and applied fields. The numerical values used in conjunction with these plots are $E_d = 0.16\,\Lambda_G^{-1}$ V cm^{-1} and $E_q = 7 \times 10^7\,(\Lambda_G)$ V cm^{-1}.

Figure 8 shows the phase shift ψ between the holographic grating (i.e., the sinusoidal distribution of the refractive index) and the incident illumination pattern, as a function of the applied field normalized to the maximum field E_q. When the diffusion process dominates, ψ is equal to $\pi/2$ independent of the relative magnitudes of the different electric fields involved. A slight variation from this condition with an externally applied field, however, could have large implications on the phase shift, depending on the ratio E_d/E_q. The $\pi/2$ phase shift represents the optimum condition for energy transfer between the two writing beams (as will be explained in the section on self-diffraction), and gives efficient beam coupling through the component B of the space-charge field [34–38]. It is interesting to note that

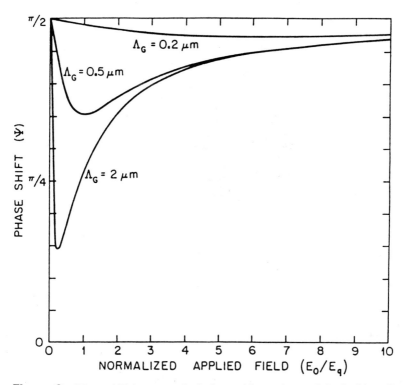

Figure 8 Phase shift between the holographic grating and the incident illumination pattern as a function of the normalized applied field. Curves for three typical values of the fringe spacing are shown.

for a large E_0 (compared with E_q), the phase shift tends toward a $\pi/2$ value, which means that energy transfer should be optimized with respect to the phase shift and also with respect to the magnitude of the shifted component of the field.

B. Photorefractive Grating Diffraction

A sinusoidal intensity distribution incident on a photorefractive crystal, such as bismuth silicon oxide (or BSO) for example, generates the space-charge field given by Eq. (15), which in turn induces a modulation of the refractive index along a principal axis, via the linear electrooptic (Pockels) effect, such that

$$\Delta n(x) = \tfrac{1}{2}n_2^3 r_{41} E_{sc} \cos(K_G x + \psi) \qquad (16)$$

in which r_{41} is the electrooptic coefficient. This modulation represents a volume phase grating in the bulk of the crystal, which will diffract a probe beam incident at an optimum angle and polarized to "sense" the refractive index spatial variation. In the case of photoconductive materials, the readout process is destructive due to charge relaxation, unless the two writing beams are kept incident on the medium, or unless the diffraction process is so efficient that the transmitted and diffracted beams write in turn a second grating from which self-diffraction will occur. This section assumes that the probe beam only reads out the holographic grating and does not itself affect the refractive index, because of its weakly absorbed wavelength or its small intensity compared with the writing beams. The diffraction efficiency η is defined as the ratio of the diffracted intensity to the transmitted intensity without the grating.

The mathematical derivation of the diffraction efficiency involves solving the wave equation. In this process, the dielectric tensor is transformed into its diagonal form in the polarization eigenmode frame, and only the transmitted and one diffracted beam are allowed to propagate in the material under the Bragg condition. In this case, the general expression for the diffraction efficiency η as derived by Kukhtarev et al. [35] is then

$$\eta = \frac{2\beta}{1+\beta} \frac{\exp(\Gamma d/2)[\cosh(\Gamma d/2) - \cos(\delta A d)]}{1 + \beta \exp(\Gamma d)} \quad (17)$$

in which β is the ratio of the writing beam intensities ($\beta \leq 1$), $\Gamma = -2\delta B$ is the energy transfer gain in two-wave mixing, $\delta = \pi n_2^3 r_{41}/\lambda_0 \cos\theta_i$ is the coupling constant, and A and B are the space-charge field amplitudes given by Eqs. (11) and (12). In these expressions, d is the grating thickness, λ_0 is the readout wavelength, and θ_i is the incidence angle. For a weakly absorbed readout beam, with corresponding intensity absorption coefficient α at the wavelength λ_0, Eq. (17) is simply multiplied by the term $\exp(-\alpha d/\cos\theta_i)$.

Since the diffraction efficiency in BSO crystals is only of the order of a few percent (under normal experimental conditions), η can be written as

$$\eta = \frac{\beta}{(1+\beta)^2} (\delta E_d d)^2 \frac{(1 + E_0/E_d)^2}{(1 + E_d/E_q)^2 + (E_0/E_q)^2} \quad (18)$$

in which the contributions of the different fields are emphasized. Measurement of the diffraction efficiency as a function of the fringe spacing yields valuable information about parameters that control the space-charge field amplitude. One such example is the trap density N_A that can be deduced from the saturation diffraction efficiency in the drift limit, which is reached when the applied field is equal to the maximum space-charge field at a

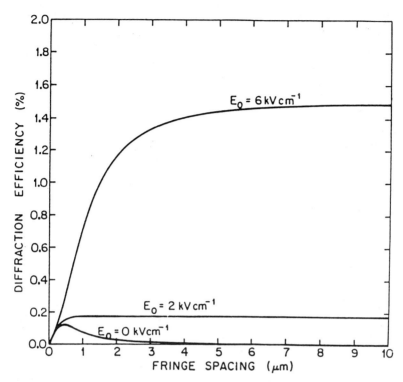

Figure 9 Diffraction efficiency of a photorefractive phase grating as a function of fringe spacing, for three values of the applied field E_0.

particular fringe spacing. To minimize the error in the determination of the trap density, caused by the uncertainty on the effective field inside the material, it would be preferable to fit the calculated and measured diffraction efficiencies in the diffusion limit. With this technique, Huignard et al. [39] measured $N_A = 10^{15}$ cm^{-3}. Figure 9 is a plot of the diffraction efficiency η given by relation (18), as a function of the fringe spacing Λ_G, for three different values of the externally applied electric field ($E_0 = 0, 2$, and 6 kV cm^{-1}).

Assuming equal writing beam intensities ($\beta = 1$) and small coupling ($\Gamma d \ll 1$), the general expression of the diffraction efficiency reduces to the well-known relation

$$\eta = \sin^2 \frac{\pi d \, \Delta n}{\lambda_0 \cos\theta_B} \tag{19}$$

derived by Kogelnik in the coupled-wave theory [27], for a readout beam incident at the Bragg angle θ_B. A deviation from this angle decreases the diffracted intensity. The full width at half-maximum (FWHM) of the diffraction efficiency curve as a function of the angle of incidence of the readout beam defines the angular selectivity $\Delta\theta_B = \Lambda_G n_2/d$, which is about 4 mrad for $\Lambda_G = 3$ μm and $d = 2$ mm.

In the four-wave mixing configuration (simultaneous readout with a counterpropagating beam at the same wavelength as the writing beams), a phase-conjugate beam is diffracted in real-time, with a reflectivity proportional to η for small beam coupling. A detailed analysis of both configurations can be found in [16].

C. Photorefractive Grating Self-Diffraction

Self-diffraction refers to the process whereby the two writing laser beams, which interfere to form the photorefractive grating, diffract from the forming grating, thereby modifying the interference fringe profile deeper within the crystal [35,37,38]. This effect modifies both the modulation depth of the grating and the phase of the fringe system. The result is a temporal and spatial evolution of the grating strength and phase, which eventually stabilizes in the steady state limit to a spatially varying grating strength and phase.

One manifestation of self-diffraction effects is energy coupling between the two recording laser beams, in which the intensity of one of the beams can be amplified at the expense of the intensity in the second beam. This phenomenon has been extensively investigated in the literature [16]. Maximum energy coupling occurs when a 90° phase shift exists between the optical interference pattern and the resulting space-charge field. This optimum phase shift occurs naturally when recording stationary interference patterns in the diffusion regime, but typical space-charge fields are low in this regime. The photosensitivity can be significantly enhanced by applying a bias electric field (as shown in Fig. 7), but the phase relationship between a stationary light interference pattern and the resulting space-charge field is generally not optimal for energy coupling.

At least two techniques have been demonstrated for enhancing the self-diffraction process for energy coupling applications. One technique is to Doppler shift one of the recording laser beams, for example, by reflecting this beam off a piezoelectrically driven constant velocity mirror, as shown in Figure 10 [40–42]. A Doppler shift causes the light interference pattern to translate with speed u, which modifies the phase between the optical interference and space-charge field profiles. One particular grating speed exists that asserts the 90° phase shift needed for maximum energy coupling.

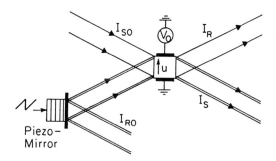

Figure 10 Optical arrangement for Doppler-enhanced two-beam coupling. A moving grating is generated by shifting the piezoelectrically driven mirror.

An additional benefit of the Doppler technique is a significant increase in the magnitude of the space-charge field. A second technique for enhancing energy coupling is to record a stationary interference pattern with a rapidly reversing applied bias field [43]. It is important to note that self-diffraction effects can also manifest themselves as a change in the polarization states of the two writing beams. This is particularly noticeable in optically active photorefractive materials of the sillenite family.

It was mentioned previously that the exponential energy transfer gain as defined by Kukhtarev et al. [35] is given by $\Gamma = -2\delta B$, in which B is the $\pi/2$-shifted component of the space-charge field. It is, however, more informative to determine the effective gain, which is the ratio of the signal beam with the optical pump present to the signal in the absence of the pump [37]. This gain (γ) is derived in a straightforward manner from the coupled wave equations mentioned previously, by calculating the intensity of the different beams at the exit of the crystal and taking the relevant ratio:

$$\gamma = \frac{(1 + \beta) \exp(\Gamma d)}{1 + \beta \exp(\Gamma d)} \tag{20}$$

in which β is the ratio of the writing intensities and d the interaction thickness. In the case of enhanced beam coupling, analytic expressions for the field B, assuming a single mobile charge species and a single trap site, have been derived in [40] and refined by Valley [41] and Refregier et al. [42].

D. Self-Pumped Phase Conjugation

As is suggested by its name, self-pumped phase conjugation is the process by which the phase conjugate of a beam incident upon a photorefractive

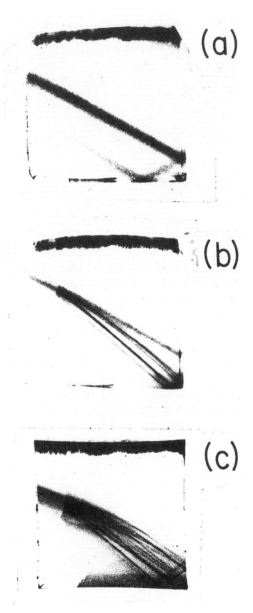

Figure 11 Photomicrographs of a barium titanate crystal acting as a self-pumped phase conjugator. The pumping waves are self-generated from the single incident beam, which enters near the top of the crystal's left face. The dark horizontal line across the top of the photographs is light scattered from a crystal face that was

crystal is self-generated. It is generally observed in ferroelectric materials such as $BaTiO_3$. This phenomenon was explained by Feinberg as illustrated in the photographs of Figure 11 [30]. All the necessary beams are induced inside the crystal. The incident beam impinges on the material in such a way as to strike a corner of the crystal. After a few moments, beam fanning starts to develop, which leads to total internal reflections along opposite directions of the corner. This generates counterpropagating pump waves as required in a typical phase conjugator. These multiple waves then interfere to create the appropriate gratings that diffract back the incident beam. High reflectivities close to 70% have been achieved.

IV. FIXED BEAM ARRAY GENERATORS

A. Dammann Gratings

The Dammann grating is a two-dimensional structure that diffracts a collimated beam into many angularly multiplexed beams [45,46]. In a manner shown in Figure 12, an array of focused beams is generated from the diffracted orders at the back focal plane of a Fourier transform lens. In most applications, it is desirable to generate an array of beams with equal intensities. Much of the recent research has been in designing such arrays.

The transmittance function $T(x, y)$ of a two-dimensional Dammann grating is a separable function in x and y, i.e., $T(x, y) = t(x)g(y)$. In addition, the desired array has usually a fourfold rotation symmetry about the z axis (direction of light propagation), which translates into $t(x) = g(y)$. Therefore, we will only consider the $t(x)$ function, which is periodic, symmetric, and binary [47]. In a basic period, there are $2N$ transitions for an array size of $2N + 1$. Assuming a period of 1 ($x_N = 0.5$), the transmittance function $t(x)$ over a half-period is of the form

$$t(x) = \sum_{s=0}^{N} (-1)^s \text{rect}\left(\frac{x-(x_{s+1}+x_s)/2}{x_{s+1}-x_s}\right), \quad 0 \le x \le 0.5 \quad (21)$$

---

damaged during crystal poling. The c-axis is directed from top to bottom. (a) The incident beam is an ordinary ray, and the stimulated gain is below threshold. No pumping waves are formed and no phase-conjugate beam is generated. (b) The incident beam is an extraordinary ray, and the stimulated gain is above threshold. Loops of light containing the pumping waves are seen between the incident beam and the lower right-hand corner of the crystal. (c) The incident beam contains a complex image and many loops form inside the crystal.

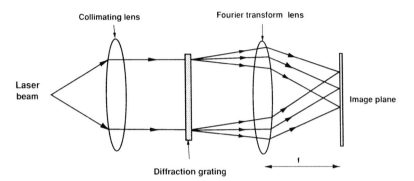

Figure 12 Spot array generation by diffraction from a Dammann grating and Fourier transformation by a lens.

A single period of a Dammann grating with $N = 4$ is shown in Figure 13. The transmittance function is shown in Figure 13a, and the corresponding $T(x, y)$ is shown in Figure 13b. For a binary phase grating, the phase difference between $t(x) = 1$ and $t(x) = -1$ is π.

With the appropriate assumptions (Fraunhoffer diffraction and infinitely long grating), one can calculate the amplitudes of the diffracted orders and find that they are given by

$$a_0 = 4 \sum_{s=1}^{N} (-1)^{s+1} x_s + (-1)^{N+1} \tag{22}$$

$$a_m = \frac{2}{m\pi} \sum_{s=1}^{N} (-1)^{s+1} \sin(2\pi m x_s), \quad m \neq 0 \tag{23}$$

In most cases of interest, the objective is to determine the variables x_1, $x_2, \ldots, x_N$ such that the intensities of all the diffracted orders are equal. Different algorithms have been reported for determining these variables [46,47]. Convergence of these algorithms and the exponential growth of their complexity with the array size, however, are major drawbacks of Dammann gratings.

Once the design of a Dammann grating is completed, its fabrication is relatively straightforward. The objective is to pattern steps (or grooves) in a substrate that induce a π phase shift on an incident wave with wavelength λ_0. This requires step heights that are equal to $\lambda_0/2 \, \Delta n$, where Δn, is the refractive index change across the step. Dammann gratings can be fabricated with techniques such as photolithographic patterning of a photosensitive film, thin-film deposition, or selective etching or milling of a substrate. As an illustrative example, use of Dammann gratings fabricated in silicon

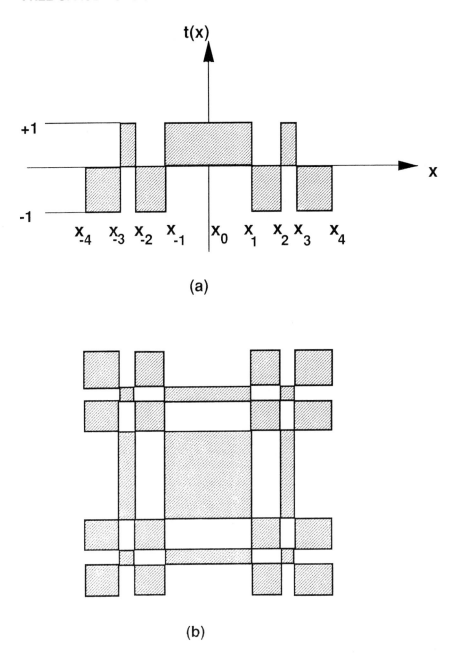

Figure 13 (a) Transmittance function of a Dammann grating $t(x)$. (b) Corresponding two-dimensional Dammann gratings. Only one period is shown in both drawings.

nitride that generate a two-dimensional array of 201 × 201 beams has been reported in [48].

B. Fresnel Microlens Arrays

Diffractive binary, multiphase, and blazed Fresnel microlenses have been reported as means of generating arrays of coherent optical beams [49–53]. Fabrication is straightforward for binary elements, but the difficulty increases rapidly for multiphase and blazed devices. The geometry of a binary amplitude Fresnel microlens (BA-FML) is shown in Figure 14a. This lens consists of alternating opaque and transparent zones. In the case of a binary phase Fresnel microlens (BP-FML) all the zones are transparent with a π phase shift between adjacent ones. The radius of the mth zone R_m is related to that of the first zone by $R_m^2 = mR_1^2$.

The transmittance function of a BP-FML is

$$t(R) = \sum_{n=-\infty}^{\infty} a_n \exp\left(i\pi n \frac{R^2}{R_1^2} \right) \tag{24}$$

where a_n, the amplitude of the nth Fourier component, is

$$a_n = \frac{1}{2R_1^2} \int_0^{2R_1^2} t(R) \exp\left(-i\pi n \frac{R^2}{R_1^2} \right) d(R^2) \tag{25}$$

Calculating the diffraction characteristics of such a transmittance reveals that the lens acts as both a converging and diverging element with foci located at

$$f_n = -\frac{R_1^2}{n\lambda_0} \quad \text{with } n = \pm 1, \pm 3, \pm 5, \ldots \tag{26}$$

in which λ_0 is the optical wavelength. The intensity at these foci, for a binary phase lens, is given by

$$I_n = \text{sinc}^2 \frac{n}{2} = \left(\frac{\sin(n\pi/2)}{n\pi/2} \right)^2 \quad \text{with } n = \pm, \pm 3, \pm, \ldots \tag{27}$$

Diffraction into the primary focus ($n = -1$) is therefore 41%, and for the secondary foci the efficiencies reduce as $1/n^2$. In the context of holographic interconnects, multiple-order focusing of these elements can have deleterious effects of erasing primary gratings and recording unwanted secondary ones. To alleviate these problems, multiphase Fresnel microlenses with deemphasized higher-order foci can be used at the cost of increased fabrication difficulty.

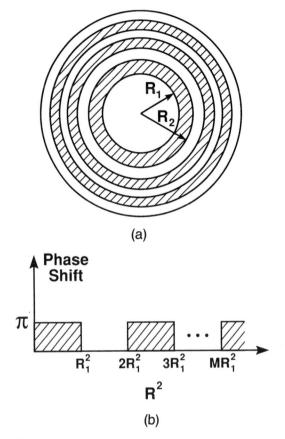

Figure 14 (a) Schematic diagram of a binary amplitude/phase Fresnel lens. In the amplitude lens every other zone is opaque, while in the phase lens they are all transparent with an induced π phase shift between the neighboring zones. (b) The phase shift induced by zones of a binary phase Fresnel lens as a function of R^2. The period of this function is $2R_1^2$. The amplitude of the transmittance function is unity.

1. Design and Fabrication Issues

Three important parameters govern the design of an FML array (FMLA): the operating wavelength, the desired focal length of each element, and the aperture size. As an illustrative example, consider a focal length of 20 mm for $\lambda_0 = 0.514$ μm. For such a lens, the radius of the first zone is 101 μm, and an aperture of 1.2 mm would allow for 35 zones. To design an 8 × 8 array with 100% fill factor would require a 9.6 mm aperture, which is within reasonable limits of a practical system. A 32 × 32 array, on the other hand, having a 38.4 mm square aperture would be cumbersome for some systems. Trade-offs will have to be factored in the design. Given a fixed wavelength, the lens aperture size (number of zones) and its focal length can be changed to accommodate large arrays. Reducing the number of zones, however, has the disadvantage of increasing the focal spot size. Reducing the focal length allows for larger packing density by reducing the radii of the zones, but the focal length must be large enough to allow room for the incorporation of elements such as spatial light modulators.

The fabrication steps for a BP-FMLA are similar to the ones described previously for a Dammann grating. The processing sequence for the ion-milling technique is shown in Figure 15. A pattern of Fresnel microlenses is formed in a photoresist layer coating the substrate. The sample is then ion milled to generate grooves in the exposed area. The exposure to the ion beam is controlled to achieve the desired groove depth. For a glass substrate (refractive index $n_2 = 1.5$), the depth is 0.51 μm for $\lambda_0 = 0.514$ μm. At the end of this process, the residual photoresist is removed. An optical micrograph of a section of an 8 × 8 BP-FMLA ion-milled on a glass substrate is shown is Figure 16. This array has the characteristics described above (20 mm focal length, 1.2 mm center-to-center spacing). A scanning electron micrograph of one element is shown in Figure 17, revealing uniform surfaces and sidewalls, which are necessary to minimize unwanted scattering.

An array of 8 × 8 focused spots generated by diffracting a collimated argon ion laser beam from the above microlenses is shown in Figure 18. A scan across four of the focused spots (Fig. 19) shows uniformly diffracted intensities. A magnified view of a single focused spot is shown in Figure 20. The diffraction pattern follows a $sinc^2$ distribution caused by diffraction from the square aperture of the lens. The beam spot size (measured to $1/e^2$ point) is about 12 μm, close to the theoretical 9 μm. The measured diffraction efficiency is about 30%, slightly less than the predicted value of 41%. Unwanted scattering due to sidewall nonuniformities and statistical variations in step heights is a major cause of this reduction [54].

While a different array size would require a whole new computation of the Dammann pattern, in the case of Fresnel zone plates changing the size of the structure could simply be done by adding or removing a row and/or column of lenses.

C. Holographic Optical Elements

Diffractive elements that are holographically recorded in a photosensitive medium are known as holographic optical elements (HOEs). A lens, for example, can be recorded as shown in Figure 21. In this configuration, the

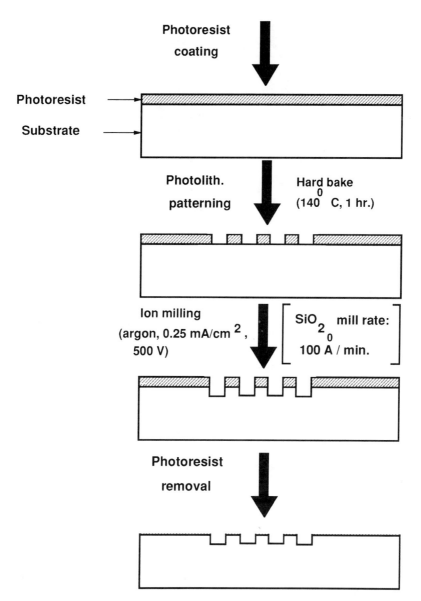

Figure 15 Sequence for patterning and fabrication of BP-FMLAs using selective ion beam milling of a substrate such as SiO_2.

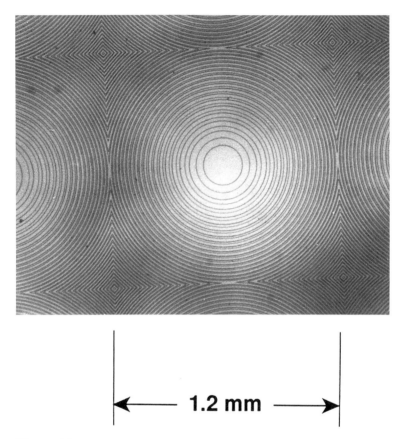

Figure 16 An optical micrograph of a BP-FMLA fabricated on SiO_2 through ion beam milling. The array consists of 8 × 8 elements, each with a 1.2-mm square aperture.

recorded pattern is the interference between a spherical object beam from a lens and a collimated reference beam. A conjugate of the spherical beam is reconstructed when the hologram is exposed to the conjugate of the reference beam. An array of such holographic lenses can be fabricated by multiple exposure and translation of the medium (see Fig. 21). Hence, when such a holographic element is exposed to a broad readout beam (conjugate of the reference beam) an array of focused spots can be generated. HOEs recorded in dichromated gelatin have been used to generate large arrays of uniform focused beams with efficiency as high as 97% [55].

FREE-SPACE HOLOGRAPHIC GRATING INTERCONNECTS

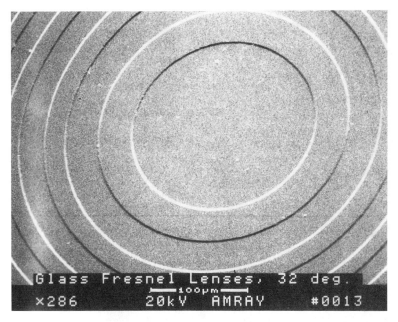

Figure 17 A scanning electron micrograph (SEM) of the BP-FML with higher magnification reveals surface uniformity.

In Figure 12, we showed the use of Dammann gratings with a Fourier lens to generate an array of spots from a single beam. Such an array can also be recorded holographically by interference with a reference beam. In the reconstruction process, the hologram can be used without the Fourier lens [56], simplifying the device structure. Similarly, the Fresnel zone plates and other diffracting structures could be holographically recorded.

V. PROGRAMMABLE BEAM ARRAY GENERATORS

A. Spatial Light Modulators

In the previous section, we presented ways of generating fixed arrays of optical beams from a single source for interconnect applications. Programmable arrays require means of affecting the characteristics of the individual elements. Spatial light modulators (SLMs) are devices suitable for this purpose. Both electrically and optically addressed modulators are now commercially available. Optical interconnects would typically require a densely packed array with more than 100×100 pixels in a 1-cm^2 device. Therefore, the addressing scheme is an important issue. For example, in a

1.2mm

Figure 18 An 8 × 8 array of focused beams ($\lambda_0 = 0.514$ μm) generated by a glass BP-FMLA.

100 × 100 array, 10^4 lead wires are needed for electrical direct addressing of the individual elements. This could prove difficult with current technologies. An alternative approach is matrix addressing, which would require only 200 wires for the preceding example, at the expense of speed. In addition to speed, sensitivity, switching energy, contrast ratio for binary modulators, and number of grey levels for analog ones, are also important considerations.

Electrooptic modulation using single crystals has been studied extensively [57,58]. These crystals behave as programmable wave plates. Thin films of lead lanthanum zirconate titanate (PLZT) have also been used with silicon (Si) driving transistors [59,60]. Both electrically addressed and opti-

FREE-SPACE HOLOGRAPHIC GRATING INTERCONNECTS 277

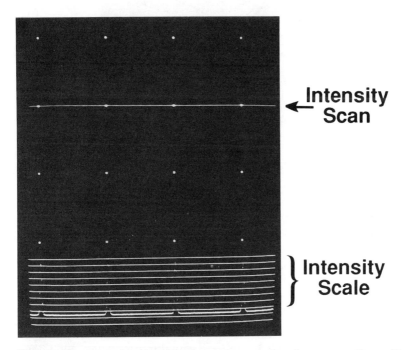

Figure 19 Scan of the intensity across four generated spots revealing uniformity.

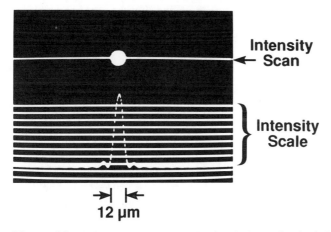

Figure 20 Diffraction pattern at the focal plane of a single binary phase Fresnel microlens (BP-FML). The pattern is characteristic of the diffraction from a square aperture.

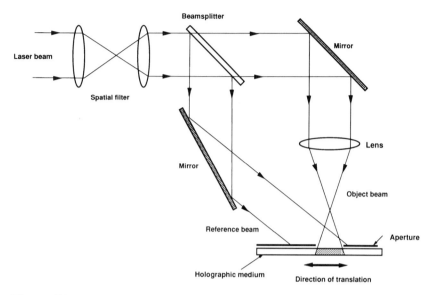

Figure 21 Recording of a holographic optical element (HOE). An object beam consisting of a diverging spherical wave, and a collimated reference beam are used. Sequential recordings and translations are necessary to make array of lenses.

cally addressed SLMs have been fabricated with this technology [59]. Other modulator types include the deformable mirrors [61] and multiple quantum well (MQW) structures [62,63]. This list, however, is by no means exhaustive. In the following, we will concentrate on the liquid crystal technology because of its mature state [64–67].

The geometry of an SLM using twisted nematic liquid crystals is shown in Figure 22. A thin layer (5–10 μm) of liquid crystal is sandwiched between two glass plates. The device has an input polarizer, and an orthogonal analyzer at the output. The orientation of the molecules, represented by the direction of the long axis (also called the director), is such that these directors are orthogonal at the two glass plate interfaces. This forces the molecules to form a spiral in the bulk of the crystal (hence the name twisted nematic). An incident light beam polarized along the director at the entrance of the device locks onto the molecules such that the emerging light is orthogonally polarized (see Fig. 22) and hence passes through the second polarizer. Thin films of indium tin oxide (ITO) coatings on the liquid crystal side of each of the glass plates are used to apply a voltage to the cell (these electrodes are not shown in Fig. 22). The applied electric field forces the molecules to more or less line up with it, destroying the twisted format. The polarization of the incident beam is therefore unaltered in this case,

FREE-SPACE HOLOGRAPHIC GRATING INTERCONNECTS

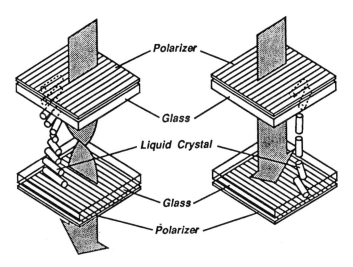

Figure 22 Geometry and operation of a twisted nematic liquid crystal SLM.

and the analyzer blocks the beam. Analog intensity modulation can be achieved with varying field amplitudes. This structure switches in the millisecond range and has extinction ratios that are better than 200:1.

One SLM implementation that uses optical addressing is schematically shown in Figure 23. Elements of this device are also sandwiched between two glass plates with transparent electrodes. A hydrogenated amorphous silicon (Si:H) film (3 μm thick) forms a PIN photodetector on one glass plate. A highly reflective layer separates the Si:H photodetector and a 1 μm thick ferroelectric liquid crystal (FLC) layer. A voltage is applied across the transparent electrodes. This voltage drops across the FLC in that region of Si:H that is illuminated with an address beam. In this region, the FLC rotates the polarization of the incident read beam, and the output polarizer therefore extinguishes the read beam. In the region of the Si:H that is not illuminated with the addressing beam, no field is transferred to the FLC. Here, the read beam is unaltered upon reflection and hence not extinguished by the polarizer. An example of the above implementation achieving eight grey levels and fast response times of the order 1–100 μs has been demonstrated [67].

B. Fresnel Microlens Arrays

Programmable beam arrays have been generated using addressable liquid crystal spatial light modulators and two-dimensional arrays of Fresnel microlenses [68]. Four specific configurations with various addressing tech-

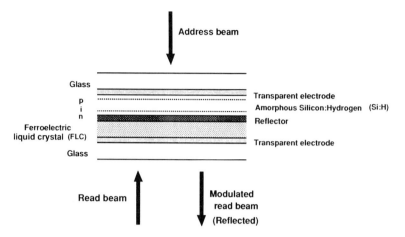

Figure 23 Schematic of an optically addressed SLM based on hydrogenated silicon photodetectors (Si:H) and ferroelectric liquid crystals (FLC). The optical address beam is incident from the top of the structure. The read beam, incident from the bottom, is modulated upon retroreflection.

niques were investigated: an optically addressed light modulator, a matrix-addressed pixellated (television) display with thin-film transistors, and two modulators that are electrically addressed through specially designed electrode patterns. Implementations with binary amplitude and binary phase Fresnel microlenses have been discussed, identifying programmable features and resolution limits of each type of modulator. In the following, we describe one of the alternative designs developed in [68] as an illustrative example.

A custom liquid crystal device with its electrodes patterned as Fresnel zone plates would enable the generation of beam arrays. Such a design allows for individual addressing of the elements in the array. The cell consists of a thin layer of nematic liquid crystal confined between two parallel glass plate surfaces, separated by 10 μm. In this geometry, the device acts as an electrically controlled phase plate. The optical path length is changed by reorienting the molecules with an applied electric field, thus causing the effective index to change since the two indices parallel and perpendicular to the long axis of the molecules are different. This device could be fabricated to optimize either phase (cell with liquid crystal molecules parallel to each other) or amplitude (twisted cell) modulation.

A Fresnel phase microlens is created by inducing a π phase difference between two adjacent zones in a parallel cell. The zones are directly etched on the transparent electrodes, which allows for an electrical control of the

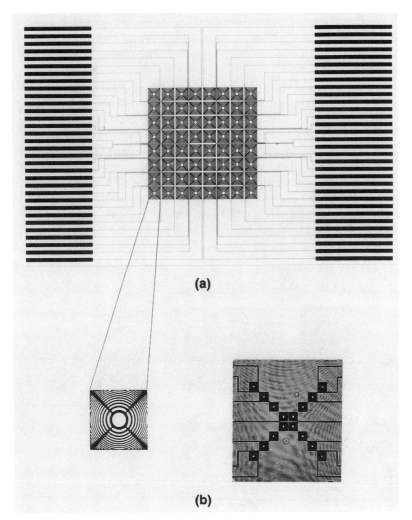

Figure 24 (a) Electrode mask used to pattern the ITO electrode. (b) Illustration of the programmability afforded with a phase modulator.

focusing capability of the cell. Figure 24a shows the electrode mask used to pattern the indium tin oxide (ITO) electrode. The applied voltage reaches all the dark zones in each microlens, because they are interconnected by lines extending diagonally from the central zone to the edge of the lens pattern. The microlenses are placed 1.2 mm center-to-center and have 15 bright zones. Figure 24b illustrates the capability of programming this device. In this figure, a voltage is applied to the pixels along the diagonals of an 8 × 8 array while the others are grounded. In this experiment a phase device was used, which explains the bright background. An average efficiency of 26% was reached with the 514-nm line of an argon laser.

The device described above is polarization sensitive, which could be a drawback for some applications. Polarization independence can be achieved by forcing the directors of the molecules in the adjacent zones to be perpendicular to each other [69]. At the other interface the molecules are homeotropic (i.e., perpendicular to the surface). During the fabrication process, the orientations of molecules in neighboring zones are made orthogonal by controlling the alignment layer. This was done by first rubbing the entire plate in one direction, and then photolithographically defining the zones of the Fresnel microlens (which covered every other zone with photoresist). The substrate was subsequently rubbed in the perpendicular

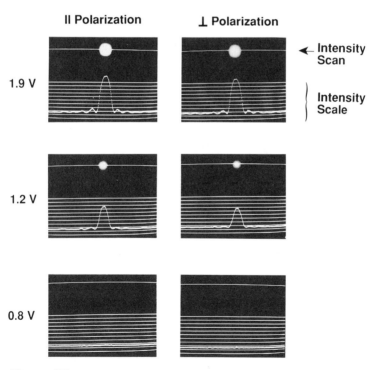

Figure 25 Intensity scan at the focal plane of a single element for orthogonal polarizations.

FREE-SPACE HOLOGRAPHIC GRATING INTERCONNECTS

direction in the exposed zones. After this step, the photoresist was removed and the substrate was cleaned. A 5-μm cell was then filled with the liquid crystal and tested. An array of focused beams was generated from an argon ion laser beam by these microlenses. By varying the applied voltage, one changes the focused beam intensity, as shown for two orthogonal polarizations in Figure 25. The diffraction efficiency into the focus as a function of the applied voltage is shown in Figure 26, confirming polarization independence.

C. Semiconductor Laser Arrays

Recent advances in two-dimensional vertical cavity surface emitting laser diode arrays (VC-SELDAs) provide exciting new possibilities in the field of optical interconnects [70–74]. A schematic representation of such a device is shown in Figure 27. Each resonator consists of an active region sandwiched between two mirror stacks [72]. The active region is made of InGaAs/GaAs quantum wells and the mirrors are composed of alternate AlAs/GaAs $\lambda/4$ layers. The bottom mirror is Si n-doped and the top mirror is Be p-doped to facilitate electrical contacts. A protective buffer layer of GaAs separates the bottom mirror stack from the substrate. To apply the drive current, gold contacts on the top mirrors are used. Above threshold, the laser beam is partially transmitted through the bottom mirror. Use of

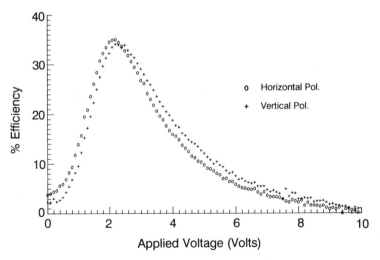

Figure 26 Plot of the conversion efficiency as a function of the voltage applied to the liquid crystal cell (horizontal polarization correponds to circles, and vertical to crosses).

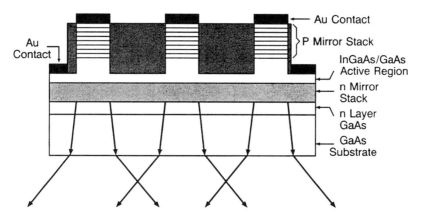

Figure 27 Schematic of an implementation of vertical cavity surface emitting laser arrays. The active medium consists of InGaAs/GaAs multiple quantum wells. Top and bottom mirrors are p-type and n-type AlAs/GaAs λ/4 layers. The laser beam is emitted through the substrate.

the InGaAs active layer ensures that the lasing wavelength (980 nm) is sufficiently away from the absorption band edge of the GaAs substrate.

Laser cavities with dimensions of 10×10 up to 100×100 μm^2 have been fabricated [72]. The threshold current for a 20×20 μm^2, for example, is about 4.2 mA, and a maximum output power of 11 mW has been achieved at a quantum efficiency of 20%. Better designs and fabrication techniques are expected to reduce the threshold currents and increase the quantum efficiencies. Individual laser response times are about 10 ns, providing significant modulation speeds. Progress has also been reported in both matrix addressing [73] and individual addressing [74] of surface emitting laser diode arrays.

In the geometry of VC-SELDAs discussed above, the beam exiting each resonator diffracts from the aperture. This diffraction can lead to crosstalk between neighboring elements in a densely packed array. Corrective measures can be made owing to the unique structures of these devices. For example, one can integrate microlenses to collimate, bend, or focus the individual beams as illustrated in Figure 28. The integration of binary phase Fresnel microlenses with VC-SELDAs has been reported [75]. These binary phase Fresnel lenses have been fabricated with the selective ion beam milling on the VC-SELDA substrate. A scanning electron micrograph of a section of the lens array (32×32) is shown in Figure 29a. Figure 29b is a higher-magnification view of one element, revealing surface uniformity. The fabricated zone step height in GaAs (refractive index $n_2 =$

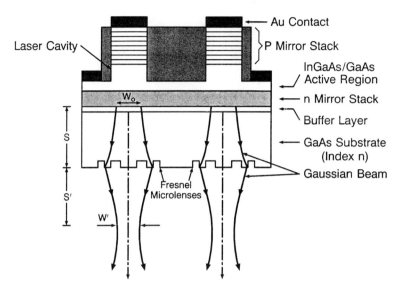

Figure 28 Schematic for the integration of microlenses with surface-emitting laser arrays.

3.5) is 0.19 μm, which yields a π phase shift at the laser wavelength of λ_0 = 980 nm. Each lens has a focal length of 108 μm and aperture of 80 μm. The center-to-center spacing is 100 μm. A 2 × 2 array of focused spots generated by the same number of laser/microlens elements is shown in Figure 30.

VI. RECONFIGURABLE HOLOGRAPHIC GRATING INTERCONNECTS

A. A General Photorefractive Interconnect System

Photorefractive materials can be used to store holographic gratings that implement interconnection links between individual processing elements of two distinct planes. The combined parallelism of optics and storage capacity of these recording media allow for a potentially large number of independent interconnections to be set up by multiplexing multiple gratings [76–78]. There are two general classes of holographic grating interconnects (or HGIs) [79]: the allocated volume (AV) HGIs, in which a distinct spatial region is allocated to the interconnections associated with each input beam, and the shared volume (SV) HGIs, in which all interconnections share the entire volume of the holographic medium. This is schematically shown in

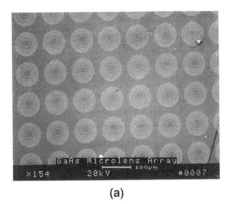

(a)

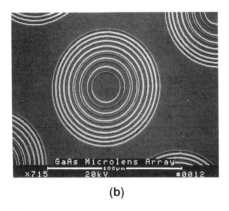

(b)

Figure 29 (a) A scanning electron micrograph of a section of the binary phase Fresnel lens array (32 × 32) ion milled on the exit interface of the substrate with laser array. (b) A scanning electron micrograph of one microlens with larger magnification reveals the surface uniformity.

Figure 31. In the AV mode, two subconfigurations exist. In the first (top right of Fig. 31), only one bulk single crystal is used. Since the diffraction characteristics can be modified by externally applying an electric field, however, it would not be possible to affect individual links in this case (unless a reflection-type arrangement with pixellated electrodes is used). This problem is alleviated by the second subconfiguration, which is made of a mosaic of crystals, each with its individual control voltage (top left of Fig. 31). In the SV mode, the grating efficiency cannot be controlled by an external field for each interconnect.

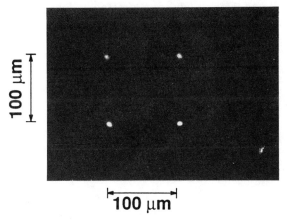

Figure 30 A 2 × 2 section of the array of beams generated by the laser arrays with integrated Fresnel microlenses.

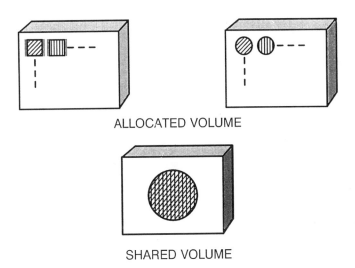

Figure 31 Multiplexing geometries for photorefractive interconnect gratings.

In the design of a programmable holographic interconnect system, there are many fundamental issues, of which a few are addressed in detail in this chapter. In a telecommunications environment, the optical interconnect is either "on" or "off"; in a neural network, it can have any analog value. Hence, the choice of a recording technique that allows variable interconnection strengths (or weights) would give more flexibility to system designers. An additional requirement is the capability of continuously modifying these weights with minimal crosstalk. It is also desirable in some telecommunication systems to have point-to-point and broadcast capabilities, but in addition to these requirements fan-in is necessary in neural network applications. Some other issues related to the implementation of holographic interconnects are currently under investigation, as evidenced by the numerous recent publications [80–89].

The proposed scheme for a reconfigurable holographic grating interconnect system constructed around a photorefractive crystal is shown in Figure 32. The purpose is to connect a matrix of sources in plane P_{IN} with a matrix of detectors in plane P_{OUT}, with fan-in and fan-out capability. In order to write the respective gratings that will form the optical links, matrices of mutually coherent control sources are needed in the planes P_{C1} and P_{C2}. All the beams from P_{C1} represent the object wave in the traditional sense of holography, and each beam from P_{C2} represents a reference wave. Each interconnection set in this coherent system, defined by a specific configuration of P_{C1} and one reference beam from P_{C2}, has to be recorded separately

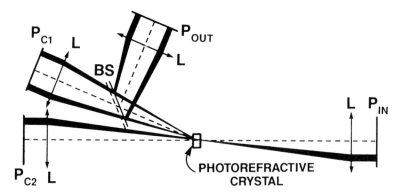

Figure 32 Schematic of a holographic interconnect system. The photorefractive crystal is placed in the Fourier plane of lenses L. Beams from P_{C1} and P_{C2} write the interconnection gratings that link sources from P_{IN} to detectors in P_{OUT}. The beam splitter (BS) separates recording from diffracted beams.

FREE-SPACE HOLOGRAPHIC GRATING INTERCONNECTS

from all the others in order to generate multiple optical links without appreciable crosstalk. As an illustrative example, the dark areas in Figure 32 show a set of interconnections. Two gratings are written in the bulk of the photorefractive material by interfering two sources from P_{C1} with one from P_{C2}. The beam incident on the crystal from P_{IN} satisfies the Bragg condition and diffracts simultaneously from these two gratings, thereby connecting itself with two different locations in the output plane. This situation would correspond to a fan-out configuration.

In the following, we analyze grating erasure in photorefractive materials for the implementation of optical interconnects with analog weights. The effects of both intensity and phase modulation of the writing beams on the diffraction efficiency and on the response time are considered. The phase modulation analysis includes the double-exposure and the time-average techniques.

1. Intensity Erasure

When two coherent beams are allowed to interfere within the bulk of a photorefractive material, the resultant intensity distribution is mapped onto a corresponding refractive index modulation owing to charge generation, transport, and trapping. If the writing beams are plane waves and the recording is assumed to be in the linear regime, the resulting index modulation is described by Eq. (16), in which the space-charge field is a function of the incident intensities through its dependence on m_G. For a thick grating, diffraction will be at maximum intensity for a light beam that satisfies the Bragg condition. The amplitude of the diffracted optical field is a linear function of this modulation depth m_G, to the first-order approximation and for isotropic diffraction. Consequently, the strength of a holographic interconnection can be controlled by varying the intensity of the recording beams. This is indeed verified in Figure 33, which shows the normalized diffraction efficiency as a function of the square of the modulation depth. The results obtained in this figure and in the following experiments are for a bismuth silicon oxide (BSO) crystal that is $10 \times 10 \times 2$ mm^3, with the smallest dimension being the thickness. A holographic grating with a fringe spacing of 1.8 μm is recorded at saturation in the bulk of this material by interfering two plane waves originating from an argon laser operated at a wavelength of 514 nm. The diffraction efficiency is probed with a He–Ne laser beam incident at the Bragg angle. No voltage is applied to the crystal, since optimizing the diffraction efficiency is not our primary objective.

Such control over the diffraction efficiency with varying modulation depth of the incident light distribution can easily be extended to multiple gratings multiplexed within the bulk of the same material. Assuming independent gratings and neglecting cross-terms that would arise if the nonlin-

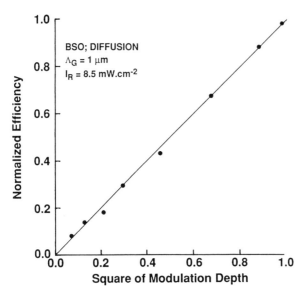

Figure 33 Normalized diffraction efficiency as a function of the square of the grating modulation depth.

earity of the recording were to be taken into accout, the modulation depth for each interconnection is given by

$$m_{Gij} = \frac{2(I_{1i}I_{2j})^{\frac{1}{2}}}{\sum_{ij}(I_{1i} + I_{2j})} \tag{28}$$

in which the subscripts i and j refer to the appropriate writing beams. It is evident from this equation that under special simplifying assumptions, multiple gratings can be individually controlled. The main drawback of such a technique, however, derives from the dependence of the photorefractive response time on the modulation depth itself. Decreasing the intensity of one of the writing beams and keeping the other one unchanged would decrease the overall intensity incident upon the crystal. Since the response time of the material is itself inversely proportional to the total incident intensity (to the first order of approximation), given a constant laser power, the time required to write a given interconnection grating depends on the value of the strength itself. This is illustrated in Figure 34, showing the response time for writing an elementary grating to saturation as a function of the modulation depth. Hence, one would need a fairly accurate tracking and control of each grating exposure, for all sets of interconnections, therefore leading to a complex and impractical system.

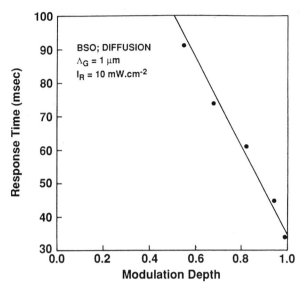

Figure 34 Response time of grating recording in BSO as a function of modulation depth.

2. Double-Exposure Erasure

Gabor et al. demonstrated that the superposition of two π phase-shifted holograms can be used to perform coherent subtraction (or addition) of two-dimensional transparencies [90]. This technique has also been successfully applied to the partial erasure of information contained in a transparency, in the presence of other holograms that are angularly multiplexed outside the Bragg selectivity bandwidth of each other [91]. Here, we extend the principle of coherent erasure by double exposure of phase-shifted holograms to the case of elementary gratings that implement real-time optical interconnections in photorefractive materials. The effect of continuously varying the phase shift on the diffraction efficiency is quantified and shown to be applicable to the implementation of interconnects with programmable variable weight [92].

The double-exposure technique is a two-stage process in which a phase shift is induced on one of the control beams between recordings. The conventional ways of inducing this phase shift are either to electrooptically phase modulate the beam, or reflect it off a piezoelectrically driven mirror [40]. In the following experiments, we use a mirror mounted on a stack of piezoelectric ceramics. In this technique, following the first recording, a second phase-shifted grating is written in the same material, at the same spatial frequency. Hence, assuming an identical modulation at saturation, the composite grating is described by

$$\Delta n(x) = \Delta n_s [\cos(K_G x) + \cos(K_G x + \Phi)] \qquad (29)$$

which can be expressed as

$$\Delta n(x) = 2\Delta n_s \cos\left(\frac{\Phi}{2}\right) \cos\left(K_G x + \frac{\Phi}{2}\right) \qquad (30)$$

in which Φ is the phase shift between the two gratings. An analysis of Eq. (30) reveals that the spatial frequency of the composite grating is not altered, that the diffracted beam will suffer a phase variation, and that most importantly, the overall modulation depth of the grating is multiplied by $\cos(\Phi/2)$. Consequently, since the diffraction efficiency is proportional to the square of this modulation depth (to the first-order approximation), the diffracted intensity will vary as $\cos^2(\Phi/2)$, thereby yielding a variable interconnection between the readout source and the detector plane. Note that when Φ is equal to π, the two gratings cancel each other as previously reported in [90,91].

In our experiment, one of the plane waves is reflected off a piezo-mirror driven by a square-wave voltage whose period is long compared with the material's response time in order to facilitate measurements. The phase difference between the two recorded gratings is dictated by the voltage-induced displacement of the piezo-mirror. The composite grating is read out in real-time with a He–Ne laser incident at the Bragg angle, although for phase matching over a wider spatial bandwidth it would be preferable to use the same wavelength as for writing. In this particular experiment, the photorefractive recording is again performed in the diffusion regime; i.e., no external field is applied to the crystal. In such a configuration, the overall diffraction efficiency is quite small (about 10^{-3}). Asymmetries in the erasure were observed when an electric field was externally supplied to the material.

Figure 35 shows oscilloscope traces of the diffracted intensity (η) and the relative phase (Φ) between the two gratings for three sets of experimental values. As expected, when the second grating is superimposed on the first one, the diffracted intensity decreases by an amount that is related to the magnitude of the phase shift between the two recordings. From top to bottom, the applied voltage to the piezo-mirror induces a shift that is less than, equal to, or larger than π, respectively. Consequently, the diffracted intensity at the minimum point decreases to a value that is equal to zero for $\Phi = \pi$ (Fig. 35b), before increasing again. Since the writing beams are continuously on, the diffraction efficiency grows back up to its saturation value after erasure.

FREE-SPACE HOLOGRAPHIC GRATING INTERCONNECTS

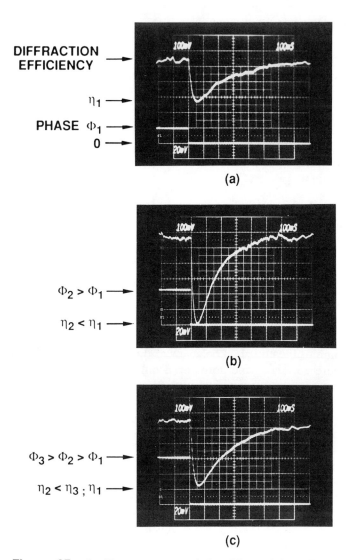

Figure 35 Oscilloscope traces of the diffracted intensity and the phase shift between the recorded gratings. The magnitude of the shift is increasing from top to bottom.

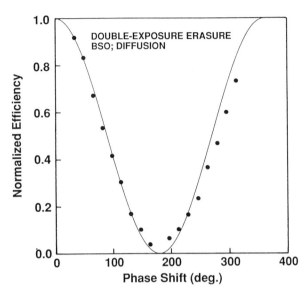

Figure 36 Normalized diffraction efficiency as a function of the phase shift in units of degrees between the two gratings recorded with a double-exposure technique.

The confirmation of the $\cos^2(\Phi/2)$ behavior of the diffracted intensity is illustrated in Figure 36. Here, the normalized diffraction efficiency after erasure, i.e., the minimum intensity in Figure 35, is plotted as a function of the phase shift between the two gratings. Theory and experiment are in very good agreement. A variable and controllable interconnection strength is thus possible with this technique. Obviously, linearity is still an issue, although this problem can be alleviated by having an adequate nonlinear relationship between applied voltage and phase shift induced by the modulator.

Even though there are two competing mechanisms, i.e., recording of the second grating and erasure of the first one, it appears that the coherent erasure process has a response time that is much faster than that of the space-charge field buildup [93]. Intuitively, this could be explained by the more effective erasure with an adequate light distribution. Under normal circumstances, grating decay is induced by uniform illumination throughout the whole crystal. For an identical total intensity, however, by choosing a distribution that perfectly matches the phase grating but is spatially shifted, the trapped charges are more effectively reexcited during erasure, due in part to a larger local optical power density. In addition, contrary to a technique that relies on varying the intensity of the writing beams, the

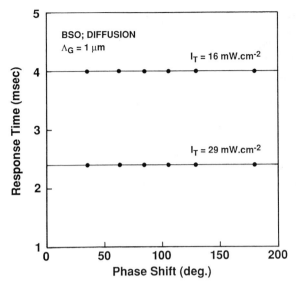

Figure 37 Response time as a function of phase shift in a double-exposure configuration, for two different values of the total incident intensity.

erasure time constant is independent of the phase shift between the two gratings (Fig. 37), leading to an identical exposure time for all recordings and hence to a simpler system. The speed improvement is clearly evident in Figure 37, which shows a response time of about 2.5 ms at a total incident intensity of 29 mW cm^{-2}, as compared with 30 ms for a total intensity of 20 mW cm^{-2} in Figure 34.

In a practical interconnection scheme, many gratings will share the same volume in the photorefractive crystal. It is thus necessary to find out whether erasing one grating will affect the others. In one experiment illustrating fan-out, two gratings with an angular separation of 4 mrad are recorded in the BSO crystal, still in the diffusion regime. Since the Bragg selectivity for the He–Ne beam is not critical with this small angular separation, two waves are diffracted. During recording, one of the writing beams is phase shifted while the efficiency is continuously monitored. The angle of incidence of the readout beam is such that equal intensity is diffracted along both directions. The result in Figure 38 shows oscilloscope traces of the diffracted intensity in each beam and the relative phase shift. In this particular experiment, erasure of one grating does not affect the other. For larger angular separations between the writing beams, we have confirmed the results reported in [91] that the different gratings do not affect each

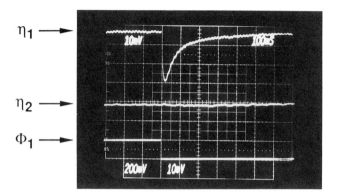

Figure 38 Oscilloscope traces of the diffracted intensity in a fan-out situation and of the phase shift of one of the writing beams.

other either. Whenever several input sources have to be directed to the same output, however, coherent diffraction might become a problem if different diffracted beams affect the overall intensity through either constructive or destructive interference [94]. If such is the case, spatially separated gratings, fast sequential reading, use of incoherent sources for readout, or phase encoding could solve this problem [95]. This issue is addressed later in this chapter.

3. Time-Average Erasure

The double-exposure technique is now extended to the case of multi-exposure or time average over the response of the material. This type of interferometric arrangement has been applied in the past to the nondestructive testing of two-dimensional vibrating structures [29,96,97]. Here, by keeping control over the time varying phase of the writing beams, we show that one can implement interconnections with variable diffraction efficiency.

Assuming a time varying phase for one of the writing beams of the form

$$\Phi(t) = \frac{4\pi a}{\lambda_0} \sin(\Omega t) \tag{31}$$

in which a is the amplitude of the modulation (or mirror displacement), λ_0 the wavelength, and Ω the temporal frequency, the intensity distribution incident upon the photorefractive crystal is given by

$$I_T(x, t) = I_1 + I_2 + 2(I_1 I_2)^{\frac{1}{2}} \cos[K_G x + \Phi(t)] \tag{32}$$

in which I_1 and I_2 are the respective intensities of the writing beams, and K_G is the grating wavevector. Owing to the finite response time of the crystal (τ), which is assumed to be much larger than the period of the phase modulation ($\tau \gg 2\pi/\Omega$), the modulation depth of the induced refractive index change becomes

$$m_G = \frac{2(I_1 I_2)^{\frac{1}{2}}}{I_1 + I_2} J_0\left(\frac{4\pi a}{\lambda_0}\right) \tag{33}$$

in which J_0 is the zeroth-order Bessel function. Consequently, since the diffraction efficiency is proportional to the square of this modulation depth (to the first-order approximation), the diffracted intensity will vary as $J_0^2(4\pi a/\lambda_0)$, thereby forming a variable interconnection between the readout source and the detector plane.

As in the double-exposure technique, the following experiments have been performed with a bismuth silicon oxide crystal. The experimental setup is identical to the one described previously, except for the piezo-mirror drive voltage, which is now a sine wave. This voltage has a much shorter period compared with the material's response time, which is about 30 ms for a total intensity of 20 mW cm^{-2} and a unity modulation ratio. The amplitude of the phase modulation is dictated by the voltage-induced displacement of the piezo-mirror. The recorded grating is read out in real-time with a He–Ne laser incident at the Bragg angle. In this particular experiment, the photorefractive recording is again performed in the diffusion regime.

The confirmation of the $J_0^2(4\pi a/\lambda_0)$ behavior of the diffracted intensity is illustrated in Figure 39. The normalized diffraction efficiency is plotted as a function of the modulation factor $(4\pi a/\lambda_0)$. Theory and experiment are in very good agreement. The slight shift between the two curves is attributed to the nonlinear response of the piezoelectric stack to relatively large voltages (500 V for a 1 μm displacement). In any case, this result demonstrates that a variable and controllable interconnection strength is possible with this technique.

Here again, the behavior in the presence of many gratings sharing the same volume in the photorefractive crystal is analyzed in an experiment similar to the one performed for the double-exposure case to illustrate fan-out. The result in Figure 40 shows oscilloscope traces of the diffracted intensity in each beam and the phase modulation of one of the writing beams. The diffraction efficiency (η_1) in the first beam is affected by the amplitude (a_1) of the phase modulation in a way that the diffracted intensity is lower than when such a modulation is absent. In this particular experiment, partial erasure of one grating does not affect the other, and hence

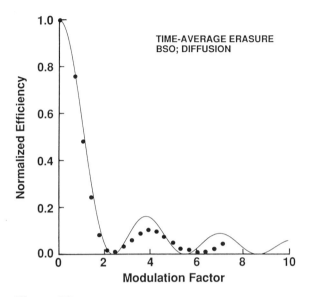

Figure 39 Normalized diffraction efficiency as a function of the phase modulation factor (described in the text).

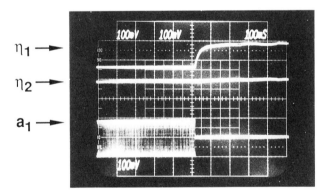

Figure 40 Oscilloscope traces of the diffracted intensity in a fan-out situation with two recorded gratings, of which one is phase modulated according to the lower trace.

control over a given interconnection can be performed without significant crosstalk. It is important to note that the speed improvement achieved with the double-exposure technique is lost, but implementation of an interconnect system with multiple gratings is made easier, since the writing beams can remain turned on during the whole recording process.

It appears then that the time-average technique is most amenable to system implementation, since it is not necessary to clock the system as is the case for intensity and double-exposure erasure. This photorefractive interconnect system with coherently erasable synapses (or PISCES) is detailed in Figure 41. The system components include arrays of sources and detectors, and spatial light modulators (both amplitude and phase), which have been described earlier in this chapter.

B. Optical Interconnects Using Correlation

Once it is realized that wave mixing in photorefractive materials is essentially a multiplicative function of two or more optical fields, designing systems that perform all sorts of parallel processing, and in particular two-dimensional interconnections, becomes an easy task. Recent examples are

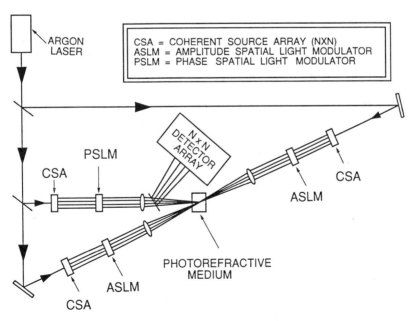

Figure 41 PISCES, or a photorefractive interconnect system with coherently erasable synapses.

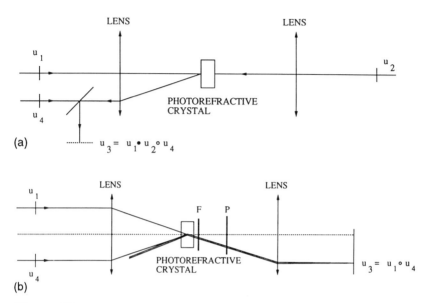

Figure 42 Two-dimensional optical correlator based on (a) four-wave mixing (after [98]), and (b) two-wave mixing (after [99]). The open circle denotes the convolution function, and the closed one denotes the product.

illustrated in Figure 42. In both systems the material records the product of the Fourier transform of the fields u_1 and u_2. In Figure 42a recording and readout of the hologram are achieved using the same wavelength (four-wave mixing geometry) [98], whereas in Figure 42b a different wavelength is used [99]. The diffracted optical field is proportional to the product of the readout wave and the convolution of the two recording beams. Consequently, these optical processors can be used to perform pattern recognition tasks. Recently however, J. Ford et al. have shown dynamic optical interconnects with similar experimental arrangements [100]. In the following we briefly summarize their results.

Figure 43 illustrates the concepts behind optical interconnects using correlation techniques. One of the input fields is an $N \times N$ array of square pixels ($\Delta \times \Delta$). This array is denoted $g(x, y)$. The other input field is the interconnection tensor $W(x, y)$, also arranged in an ($N \times N$) array of $N \times N$ pixel subarrays. Each subarray contains the interconnection information for a single output position. Figure 43 shows the case of a 3×3 interconnect system with a 1-to-1 interconnect pattern $W(x, y)$, since only one pixel is activated in any subarray. Assuming optimal experimental conditions, the output image $c(x, y)$ is defined as the cross-correlation of $g(x, y)$ with

FREE-SPACE HOLOGRAPHIC GRATING INTERCONNECTS

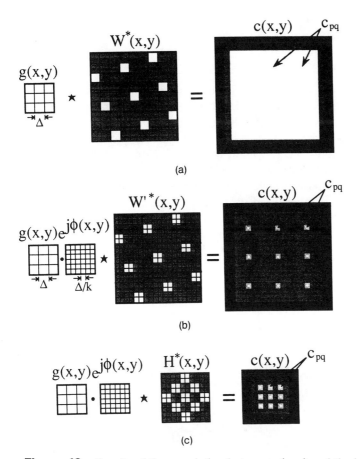

Figure 43 Results of the correlation between $g(x, y)$ and the interconnect control tensor $W(x, y)$: (a) the output is imbedded in noise; (b) noise reduction with phase coding of $g(x, y)$; (c) compression of $W(x, y)$ into $H(x, y)$.

the complex conjugate of $W(x, y)$. As depicted in Figure 43a, the correlation peaks are imbedded in noise and can barely be distinguished. Multiplying $g(x, y)$ by a phase code before the correlation, however, improves the result at the expense of an increased requirement on the space–bandwidth product of the tensor $W(x, y)$. This requirement can be eased using compression techniques. In Figure 43c the interconnection pattern $W(x, y)$ is replaced with a function $H(x, y)$ in which each of the subarrays has been shifted so that its center is separated by exactly Δ from its neighbors. The interconnect system described above has been implemented with a $BaTiO_3$ photorefractive crystal.

C. Optical Interconnects Using Two-Wave Mixing

Energy transfer in two-wave mixing experiments was described in Section III. C. Here we show how this principle can be applied to optical interconnects. Since the process is dynamic (both writing beams have to be present for the coupling to take place), it is not possible to use the shared volume configuration for interconnecting many inputs. Allocated volume configurations, however, would allow such multiple interconnects. Figure 44 shows the experimental setup for connecting one input to a two-dimensional array of outputs [101]. Maximum intensity transfer occurs when the phase shift between the recording intensity pattern and the phase grating is exactly $\pi/2$. This happens naturally when recording is done in the diffusion regime. If one of the writing beams is phase shifted, however, a moving grating is

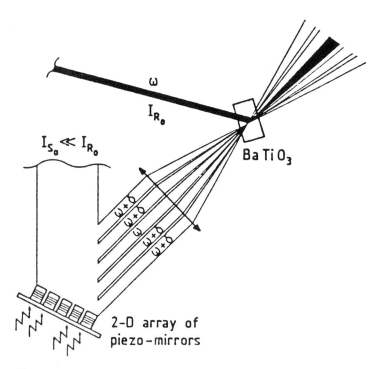

Figure 44 Optical interconnects based on two-wave mixing. Addressing is achieved with an array of piezoelectrically driven mirrors.

generated instead, and the optimum $\pi/2$ phase shift requirement is spoiled if the grating moves at the right speed. The optimum velocity is determined by the response time of the photorefractive material. The input is self-diffracted to several output locations in a controllable fashion by addressing the array of piezo-mirrors. In this setup, the incident beam is split into two mutually coherent beams. A strong pump beam is sent straight to the $BaTiO_3$ crystal, and a weaker beam is expanded to illuminate a 4×4 array of piezo-mirrors. The crystal is placed at the focal plane of the lens to allow for maximum overlap of the various beams.

The operation of this interconnect system is quite simple. If no voltage is applied to the piezoelectrically driven mirrors, beam coupling is optimized and the intensity is transferred from the pump beam to the individual pixels. Addressing is achieved by exciting the appropriate phase modulator so that beam coupling does not occur for that particular beam direction. The extension of this technique to multiple inputs using the allocated volume configuration has been demonstrated in an efficient optical matrix-vector multiplier [102].

D. Optical Interconnects Using Self-Pumped Phase Conjugation

In all the previously described systems, to each input/output pair we have associated a grating for the interconnect. When the number of interconnects grows large, severe constraints are placed on the system because of the Bragg degeneracy. The gratings have to be multiplexed very tightly, and the Bragg selectivity is not sufficient to avoid crosstalk between neighboring interconnects. One alternative approach is to angularly and spatially distribute the gratings in the volume of the photorefractive material [103].

In the self-pumped phase conjugation configuration, the incident beam fans into a broad wave that is subject to total internal reflection and generates its own phase conjugate beam. In the overlapping region, numerous gratings are formed that diffract along the same direction the incoming optical beam. If the input is an array of illuminated apertures, the phase-conjugated beam will exactly mimic this array. An experimental demonstration of this connectivity is shown in Figure 45. The top photograph shows the input array of circular apertures, and the middle photograph shows the conjugated beam (apertured by the system's optical components). When the input is shifted by half of the grid period, the diffracted array disappears, which verifies the spatial distribution of the gratings (bottom of figure). In addition, if after recording only one pixel is turned on, this pixel is globally interconnected to all other locations. To connect it only to a few outputs, these would have to be activated during recording. Partial erasure has also been demonstrated [103].

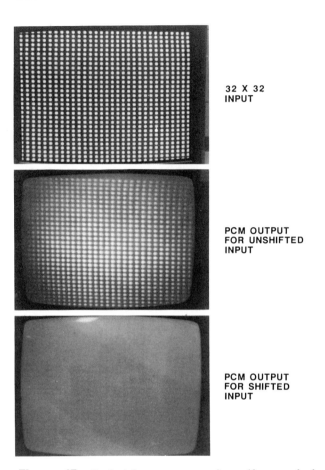

Figure 45 Optical interconnects using self-pumped phase conjugation: (top) input mask; (middle) conjugated image; (bottom) shifted input.

VII. LIMITATIONS OF PHOTOREFRACTIVE-BASED INTERCONNECTS

A. Grating Erasure

The physical mechanism of holographic recording in photorefractive materials makes it difficult to multiplex several holograms. An incident beam redistributes the charges from their traps, thereby weakening the previously recorded space-charge field and the grating strength. When multiplexing several holograms using sequential recording to avoid crosstalk, the diffraction efficiency of the latter holograms is larger than that of the first

FREE-SPACE HOLOGRAPHIC GRATING INTERCONNECTS

ones because of erasure. To achieve uniform diffraction, an exposure schedule has to be found that relies on the response time of the material used. With this technique, 500 holograms were recently recorded in $LiNbO_3$ with uniform efficiency [104]. After recording all patterns, this information has to be fixed in the material in order to avoid erasure during subsequent readout [105].

In what follows, we analyze the grating erasure problem in an architecture that is devised to minimize this erasure during both recording and readout [106]. The discussion is for a modest implementation in terms of number of interconnects involved. This architecture uses angularly multiplexed plane waves for reference beams and spatially multiplexed input beams to record routing holograms as shown in Figure 46. The read beams are conjugates of the reference beams. They reconstruct conjugates of the object beams, which are reflected by a beam splitter (BS) toward a detector array. Spatial light modulators (SLMs) control the appropriate beam

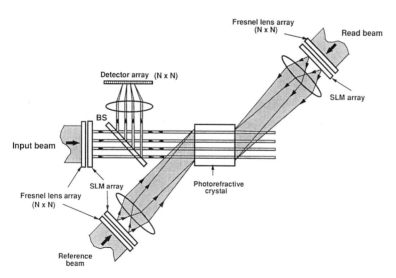

Figure 46 A schematic diagram of a dynamic interconnect system based on reconfigurable volume holograms in photorefractive crystals. Recording of gratings is done betwen spatially multiplexed pencil-like beams of an input array and angularly multiplexed collimated reference beams. Reconstruction is performed by read beams that are conjugates of the reference beams.

array configurations in each path. Only one reference beam exposes the crystal at a time to record holograms with all of the desired object beams simultaneously. The angles between neighboring reference beams must be larger than their associated Bragg selectivity. This allows for simultaneous interconnection of more than one read beam without crosstalk.

In our experiment, routing of optical beams to 64 detectors (8×8 array) is achieved with the setup shown in Figure 46 implemented with a 1-cm^3 commercially available bismuth silicon oxide (BSO) crystal. The object beams were generated by diffraction of an argon ion laser beam from a BP-FMLA (8×8), as described in the section on Fresnel microlenses. The focused beams have a spot size of ~ 12 μm and are separated by 1.2 mm. Using a bulk lens, the array was imaged on a 7×7 mm^2 area of the (110) facet of the crystal. The resulting focused beams were ~ 10 μm in diameter within the crystal. The divergence angle of a Gaussian beam with such a waist is $\sim 0.36°$. This beam diverges to a cross section of 70 μm within the 1 cm long crystal. Hence, no significant crosstalk is expected between these beams, which were separated by 900 μm. It is possible to achieve densities higher than 8×8 at the cost of increased crosstalk and fan-out loss. The reference beam is a plane wave with optical power about ~ 0.1–0.3 mW.

A read beam, conjugate of the reference beam, reconstructs the conjugates of the object beams, which are detected by a charge-coupled device (CCD) camera. The photograph in Figure 47 shows part (6×5 elements) of the full reconstructed array. Addition of other reference beams reduces the diffraction efficiency as shown in Figure 48. The recording beams were left on to maintain the gratings. Figure 48a is the baseline with only one reference beam showing an intensity scan across the reconstructed beams. The gradually decreasing intensities of the spots from right to left is due to a slight misalignment of the scan line. We therefore concentrate on the rightmost spot of each photograph. Figure 48b is the same as Figure 48a except that a second reference beam is also turned on, causing a 3.7-dB reduction of the efficiency. When a third reference beam is also turned on, the efficiency is reduced by an additional 2.4 dB (a total of 6.1 dB), as shown in Figure 48c. This reduction is mainly due to the change in the modulation depth of the recorded gratings caused by the addition of new writing beams. Therefore, even in an optimized architecture grating erasure can still be a problem.

B. Interconnect Capacity

The grating angular selectivity is a critical parameter in determining the separation between holograms that would prevent crosstalk. We alluded to angular selectivity in Section II. Equation (7), in conjunction with Eqs. (8)

FREE-SPACE HOLOGRAPHIC GRATING INTERCONNECTS

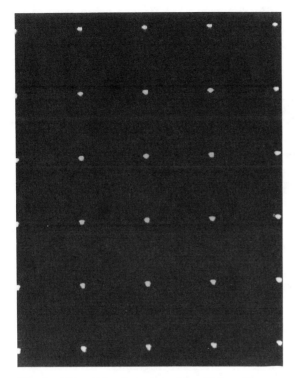

Figure 47 A 6 × 5 section of the 64 (8 × 8) reconstructed beams.

and (9), shows the dependence of the diffraction efficiency on the angular misalignment from the Bragg incidence. Assuming a TE wave, and a lossless transmission hologram, the diffraction efficiency reaches a zero value for $\xi^2 + \nu^2 = \pi^2$. (See Section II for a definition of the different parameters.)

Substituting Eqs. (8) and (9), the full width of the efficiency curve is given by

$$\text{FW} = \frac{1}{n_2 \sin \theta_i} \left(\left(\frac{\lambda_0}{d} \right)^2 - \left(\frac{\Delta n}{\cos \theta_i} \right)^2 \right)^{\frac{1}{2}} \tag{34}$$

In a recording medium where $\Delta n \ll \lambda_0/d$, Eq. (34) can be simplified to

$$\text{FW} \cong \frac{\lambda_0}{n_2 d \sin \theta_i} \tag{35}$$

As an example, let us consider an iron-doped lithium niobate crystal (Fe:LiNbO$_3$). This material has a refractive index $n_2 = 2.28$ and is sensitive

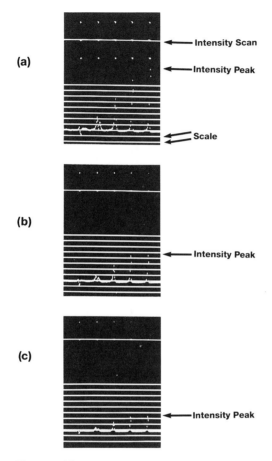

Figure 48 (a) Photographs showing an intensity scan of a few reconstructed beams. The effect of adding (b) one or (c) two reference beams is also illustrated.

at a wavelength $\lambda_0 = 0.5$ μm. Assuming a typical index modulation of $\Delta n = 10^{-4}$, plots of the diffraction efficiency η as a function of the Bragg misalignment $\Delta\theta_B$ are shown in Figure 49a for four different Bragg angles, $\theta_B = 3, 10, 20,$ and $30°$. The peak efficiency at Bragg incidence increases for increasing θ_B for this example. On the average, however, the maximum efficiencies are close to 90%.

In some instances, the sidelobes of the diffraction efficiency plots can be large in magnitude, leading to significant crosstalk. This is shown in Fig.

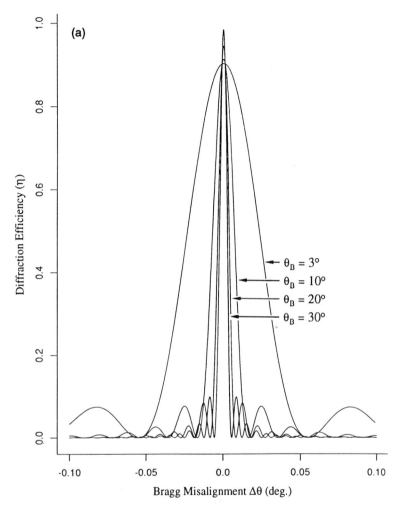

Figure 49 Plot of the diffraction efficiency η as a function of misalignment $\Delta\theta_B$ for an iron-doped LiNbO$_3$ crystal. (b) Same as (a) for a 10° angle and a 1-cm crystal.

49(b) for $\theta_B = 10°$, where a thickness of $d = 1$ cm is used for the preceding Fe:LiNbO$_3$ example. Here, the diffraction efficiency at Bragg incidence is almost zero, and the first sidelobes are at 65% efficiency.

When recording angularly multiplexed holograms, wavevector (or k-vector) degeneracies must be avoided to prevent crosstalk. To understand this better, let us consider the k-space sphere of Figure 50. Two recording beams with wavevectors $\mathbf{k}_1$ and $\mathbf{k}_2$ generate a grating vector $\mathbf{K}_G$. Two paral-

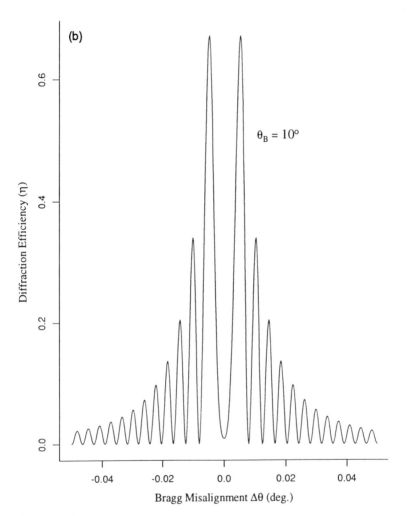

Figure 49 Continued

lel planes orthogonal to $\mathbf{K_G}$ intersect the sphere at two circles, as shown in Figure 50. Any other two recording beams with their wavevectors $\mathbf{k}'_1$ and $\mathbf{k}'_2$ terminating on the upper and lower circles generate $\mathbf{K_G}$ vectors that are indistinguishable, leading to degeneracies. To avoid this problem, and at the same time maximize the number of interconnection nodes, the use of fractal sampling grids has been suggested and experimentally verified [107].

The interconnect capacity can also be affected by the maximum refractive index modulation that can be achieved. Within a recording medium, a

FREE-SPACE HOLOGRAPHIC GRATING INTERCONNECTS

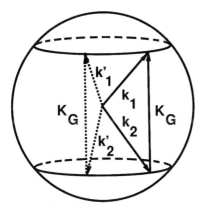

Figure 50 Schematic of the k-space sphere showing k-vector degeneracies.

limited number of charges and traps are available. Usually the space-charge field amplitude is trap limited. Recording multiple holograms reduces the maximum space-charge field of each. If an index modulation for a single hologram is Δn, S holograms would each have $\Delta n/S$, ignoring erasure. Hence, for small Δn, the diffraction efficiency for each hologram is reduced as η/S^2, from an approximation of Eq. (7). Therefore, recording 500 holograms reduces the above efficiency by $\sim 4 \times 10^{-6}$. Such an efficiency could be too small from a signal-to-noise ratio standpoint. One has to record fewer interconnection patterns for improved efficiencies or rely on an optimized detection system.

C. Operation Speed and Power Requirement

In most holographic interconnect architectures, the reconfiguration time is limited by the slower of the photorefractive or the SLM response times. Currently, SLMs with millisecond response times and dynamic ranges of 200:1 are commercially available. The photorefractive response time, on the other hand, is intensity dependent as shown in Section VI. As an example, the time it takes to write a grating to saturation in a BSO crystal is about 5 ms for 100 mW cm^{-2}. This is approximately the time needed to record the appropriate configuration for each beam of the read array in an interconnect application. Hence, 1000 interconnect configurations would require 5 s to complete in a sequential recording process. One can clearly see the speed limitations of such a system. In addition, assuming that the reference beams are generated using diffraction of a single beam from an array of Fresnel microlenses, the intensity of the initial reference beam should be as high as 100 W, which could potentially be a problem.

D. Coherent Beam Fan-In

In this section we describe some of the fan-in issues that relate to a coherent holographic grating interconnect system. The problem of fanning in a number N of coherent input lightwaves into a single output beam has been discussed by Goodman [94]. It arises from the interference effects caused by the diffraction of multiple coherent beams. As mentioned in Section VI.A, there are two general classes of holographic grating interconnects. The problem of fan-in interference is critical only in the shared volume (SV) class, because the multiple beams are diffracted along the same direction, resulting in a detected intensity that depends on the relative phases. When each input beam has its own distinct spatial region, as in the allocated volume (AV) class, the output lightwaves from each region are incident on the photodetector at different angles and hence the intensity will be spatially averaged.

In a fan-in situation, the single output that results from the diffraction of many input beams has an amplitude that depends on the relative phases of the various beams. Keeping tight control over the phase of the readout beams would be quite a challenging task. There are, however, other alternatives that would help minimize the interference effects. These include incoherent readout, temporal sequencing, random phase modulation of the readout beams, and use of the allocated volume configuration. If the main concern is ease of system implementation, then readout of the multiplexed sets of interconnection gratings with arrays of sources that are mutually incoherent is the best alternative. The $1/N$ excess power loss for N gratings, however, is a fundamental limitation that cannot be overcome. In this case, the surface emitting laser diode arrays discussed in Section V.C could be used. If power budget is the primary concern, then it would be preferable to use time sequencing, i.e., readout by turning on sequentially the light sources in the readout array. This could also be an attractive alternative with VC-SELDA, since they can be controlled at speeds much faster than the reconfiguration time of the interconnection gratings (which is limited by the response time of the photorefractive material).

The two alternatives that we have analyzed in more detail are the readout phase modulation and the use of the allocated volume configuration. In the phase modulation technique, the purpose is to introduce a fast random phase variation in the readout beam such that on fan-in and for the observation period the waves do not interfere with each other. If during the observation period the relative phases take on all possible values between 0 and 2π in a random fashion, then the mean value of the phase factor in the interference term is zero. This technique would work only if averaging is

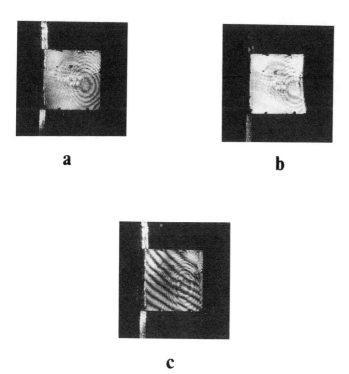

Figure 51 Interference effects caused by the diffraction of two coherent beams along the same direction. (a), (b) The individual diffracted beams just outside the crystal. (c) The resulting fringes when there is a small angle between the diffracted beams.

performed at the detector level, which means that its response time should be large compared with the phase fluctuations.

Figure 51 shows the deleterious effects caused by the interference of the diffracted waves. In this experiment, the recording medium is a single crystal of photorefractive BSO. Two holographic gratings are formed in the bulk of this material by interfering two plane waves from an argon laser operated at a wavelength of 514 nm. The fringe spacings are in the range 1–3 μm. The setup is in a four-wave mixing configuration with two readout beams incident from the back of the crystal, and diffracting along the same direction to form one single output. Figures 51a and 51b show the two diffracted beams separately just after the crystal. One of the readout beams

a

b

Figure 52 Illustration of the elimination of the interference fringes with the phase averaging technique: (a) without phase modulation; (b) with one of the readout beams modulated.

is slightly off Bragg to allow for the visualization in Figure 51c of the fringes caused by coherent interference during fan-in. Figure 52 shows elimination of these fringes when the phase averaging technique is applied. In Figure 52a, both diffracted beams are allowed to interfere. In Figure 52b, one of the readout beams is phase modulated by reflecting it off a piezo-mirror driven with a sine-wave generator. Even though the sine wave is not optimal since it is not a random function, the photographic film acts as the averaging medium and the fringes disappear.

The other alternative way to avoid the fan-in problem is to use the allocated volume configuration. In this case, all of the gratings associated with diffraction from a single input are assigned to a particular region (volume) of the photorefractive material, while gratings associated with diffraction from other inputs are assigned other regions of the material. In this configuration, diffracted outputs directed toward the same detector are incident at different angles, and hence the intensity is spatially averaged and the interference effects are less critical. One limitation that now arises, however, is on spot size of the diffracted beam, which depends on the aperture or area allocated to its corresponding input. The larger the area, the narrower this beam will be, but the fewer the number of input beams that could be accomodated by a given crystal area. One potential advantage of the AV configuration is the gain in the strength of the refractive index modulation. Since only M gratings are written into each region (instead of MN as in the SV method) the strength of each grating can be substantially greater.

VIII. CONCLUSIONS

In this chapter we have shown how one can use the dynamic features of holographic recording in photorefractive materials to implement optical interconnects. Grating diffraction allows one to redirect an incident beam anywhere in a two-dimensional plane. Multiplexing many such gratings allows one to interconnect an array of sources to an array of detectors. Large two-dimensional spatial light modulators with adequate performance characteristics are now available for use in a holographic system. Arrays of surface emitting lasers and photodetectors have also reached a development stage that makes them useful. In this work we have described some recent results on the fabrication and characterization of microlenses, spatial light modulators, surface emitting lasers, and on their integration in an interconnect system.

The dynamics of the photorefractive recording makes the proposed system flexible, since it can be used for both digital and analog interconnects. We have shown that by using phase modulation of the writing beams, the strength of the optical interconnect can be altered without crosstalk with other channels. This feature is essential for the implementation of neural networks.

Our experimental setup can be considered as quite generic. Although its implementation is a modest one with devices that are only 8 × 8 in size, it has raised many critical issues, some of which were analyzed in Section VII. Nevertheless, the potential payoff of an optical interconnect based on holographic gratings in terms of capacity and overall performance compared with an electronic counterpart is enough to justify more effort in developing these photorefractive systems.

REFERENCES

1. J. W. Goodman, F. L. I. Leonberger, S. Kung, and R. A. Athale, Optical interconnections for VLSI systems, *Proc. IEEE*, 72:850 (1984).

2. R. K. Kostuk, J. W. Goodman, and L. Hesselink, Optical imaging applied to microelectronic chip-to-chip interconnects, *Appl. Opt.*, *24*:2851 (1985).
3. D. A. B. Miller, Optics for low-energy communication inside digital processors: Quantum detectors, sources, and modulators as efficient impedance converters, *Opt. Lett.*, *14*:146 (1989).
4. D. H. Hartman, Digital high speed interconnects: a study of the optical alternative, *Opt. Eng.*,*26*:1086 (1986).
5. P. R. Haugen, S. Rychnovsky, and A. Husain, Optical interconnects for high speed computing, *Opt. Eng.*, *25*:1076 (1986).
6. A. Guha, J. Bristow, C. Sulivan, and A. Husain, Optical interconnections for massively parallel architectures, *Appl. Opt.*, *29*:1077 (1990).
7. H. M. Ozaktas and J. W. Goodman, Lower bound for the communication volume required for an optically interconnected array of points, *J. Opt. Soc. Am. A*, *7*:2100 (1990).
8. J. Shamir, H. J. Caulfield, and R. B. Johnson, Massive holographic interconnection networks and their limitations, *Appl. Opt.*, *28*:311 (1989).
9. F. Lin, E. M. Strzelecki, and T. Jannson, Optical multiplanar VLSI interconnects based on multiplexed waveguide holograms, *Appl. Opt.*, *29*:1126 (1990).
10. R. K. Kostuk, Y. Huang, D. Hetherington, and M. Kato, Reducing alignment and chromatic sensitivity of holographic optical interconnects with substrate-mode holograms, *Appl. Opt.*, *28*:4939 (1989).
11. H. Lee, X. Gu, and D. Psaltis, Volume holographic interconnections with maximal capacity and minimal cross talk, *J. Appl. Phys.*, *65*:2191 (1989).
12. D. Z. Anderson and D. M. Lininger, Dynamic optical interconnects: Volume holograms as optical two-port operators, *Appl. Opt.*, *26*:5031 (1987).
13. A. Marrakchi, Continuous coherent erasure of dynamic holographic interconnects in photorefractive crystals, *Opt. Lett.*, *14*:326 (1989).
14. M. Kato, Y. Huang, and R. K. Kostuk, Multiplexed substrate-mode holograms, *J. Opt. Soc. Am. A*, *7*:1441 (1990).
15 R. J. Collier, C. B. Burckhardt, and L. H. Lin, *Optical Holography*, Academic Press, Orlando, Fl. 1971.
16. P. Gunter and J.-P. Huignard (eds.), *Photorefractive Materials and Their Applications*, *I*, *II*, Topics in Applied Physics Springer-Verlag, Berlin and Heidelberg, 1988.
17. E. Hecht and A. Zajac, *Optics*, Addison-Wesley, Menlo Park, CA, 1979, pp. 354–364.
18. M. Born and E. Wolf, *Principles of Optics*, Pergamon Press, New York, NY 1980, pp. 401–407.
19. T. K. Gaylord and M. G. Moharam, Thin and thick gratings: Terminology clarification, *Appl. Opt.*, *20*:3271 (1981)
20. M. G. Moharam, T. K. Gaylord, and R. Magnusson, Criteria for Bragg regime diffraction by phase gratings, *Opt. Comm.*, *32*:14 (1980).
21. M. G. Moharam, T. K. Gaylor, and R. Magnusson, Criteria for Raman–Nath regime diffraction by phase gratings, *Opt. Comm.*, *32*:19 (1980).
22. T. Tamir, H. C. Wang, and A. A. Oliner, Wave propagation in sinusoidally

stratified dielectric media, *IEEE Trans. Microwave Theory and Techniques*, *MTT-12*:323 (1964).
23. R. S. Chu and J. A. Kong, Modal theory of spatially periodic media, *IEEE Trans. Microwave Theory and Techniques*, *MTT-25*:18 (1977).
24. D. Yevick and L. Thylen, Analysis of gratings by the beam propagation method, *J. Opt. Soc. Am.*, *72*:1084 (1989).
25. E. N. Glytsis and T. K. Gaylord, Rigorous 3-D coupled wave diffraction analysis of multiple superposed gratings in anisotropic media, *Appl. Opt.*, *28*:2041 (1989).
26. E. N. Glytsis and T. K. Gaylord, Three-dimensional (vector) rigorous coupled-wave analysis of anisotropic grating diffraction, *J. Opt. Soc. Am. A*,*7*:1399 (1990).
27. H. Kogelnik, Coupled wave theory for thick hologram gratings, *Bell System Techn. J.*, *48*:2909 (1969).
28. A. Marrakchi, Polarization properties of diffraction from elementary gratings in optically active and linearly birefringent materials, *Eletro-Optic and Photorefractive Materials* (P. Gunter, ed.), (Springer Proc. Phys. 18 (1987).
29. A. Marrakchi, J.-P. Huignard, and J.-P. Herriau, Application of phase conjugation in $Bi_{12}SiO_{20}$ crystals to mode pattern visualisation of diffuse vibrating structures, *Opt. Comm.*, *34*:15 (1980).
30. J. Feinberg, Self-pumped, continuous-wave phase-conjugator using internal reflection, *Opt. Lett.*, *7*:486 (1982).
31. R. W. Hellwarth, Generation of time-reversed wave fronts by nonlinear refraction, *J. Opt. Soc. Am.*, *67*:1 (1977).
32. A. Yariv, Phase conjugate optics and real-time holography, *IEEE J. Quantum Electron.*, *QE-14*:650 (1978).
33. A. Ashkin, G. D. Boyd, J. M. Dziedzic, R. G. Smith, A. A. Ballman, J. J. Levinstein, and K. Nassau, Optically-induced refractive index inhomogeneities in $LiNbO_3$ and $LiTaO_3$, *Appl. Phys. Lett.*, *9*:72 (1966).
34. J. Feinberg, D. Heiman, A. R. Tanguay, Jr., and R. W. Hellwarth, Photorefractive effects and light-induced charge migration in barium titanate, *J. Appl. Phys.*, *51*:1297 (1980).
35. N. V. Kukhtarev, V. B. Markov, S. G. Odulov, M. S. Soskin, and V. L. Vinetskii, Holographic storage in electrooptic crystals, *Ferroelectrics*, *22*:949 (1979).
36. V. L. Vinetskii, N. V. Kukhtarev, S. G. Odulov, and M. S. Soskin, Dynamic self-diffraction of coherent light beams, *Sov. Phys. Usp.*, *22*:742 (1979).
37. A. Marrakchi, J.-P. Huignard, and P. Gunter, Diffraction efficiency and energy transfer in two-wave mixing experiments with $Bi_{12}SiO_{20}$ crystals, *Appl. Phys.*, *24*:131 (1981).
38. D. L. Staebler and J. J. Amodei, Coupled-wave analysis of holographic storage in $LiNbO_3$, *J. Appl. Phys.*, *43*:1042 (1972).
39. J.-P. Huignard, J.-P. Herriau, G. Rivet, and P. Gunter, Phase-conjugation and spatial dependence of wave-front reflectivity in $Bi_{12}SiO_{20}$ crystals, *Opt. Lett.*, *5*:102 (1980).

40. J.-P. Huignard and A. Marrakchi, Coherent signal beam amplification in two-wave mixing experiments with photorefractive $Bi_{12}SiO_{20}$ crystals, *Opt. Comm.*, *38*:249 (1981).
41. G. Valley, Two-wave mixing with an applied field and a moving grating, *J. Opt. Soc. Am. B*, *1*:868 (1984).
42. Ph. Refregier, L. Solymar, H. Rajbenbach, and J.-P. Huignard, Two-beam coupling in photorefractive $Bi_{12}SiO_{20}$ crystals with moving grating: Theory and experiment, *J. Appl. Phys.*, *58*:45 (9185).
43. S. I. Stepanov and M. P. Petrov, Efficient unstationary holographic recording in photorefractive crystals under an external alternating electric field, *Opt. Comm.*, *53*:292 (1985).
44. A. Marrakchi, R. V. Johnson, and A. R. Tanguay, Jr., Polarization properties of enhanced self-diffraction in sillenite crystals, *IEEE J. Quantum Electron.*, *QE-23*:2142 (1987).
45. H. Dammann and K. Gortler, High-efficiency in-line multiple imaging by means of multiple phase holograms, *Opt. Comm.*, *3*:312, (1971).
46. H. Dammann and E. Klotz, Coherent optical generation and inspection of two-dimensional periodic structures, *Optica Acta*, *24*:505 (1977).
47. J. Jahns, M. M. Downs, M. E. Prise, N. Streibl, and S. J. Walker, Dammann gratings for laser beam shaping, *Opt. Eng.*, *28*:1267 (1989).
48. M. R. Taghizadeh and J. I. B. Wilson, Optimization and fabrication of grating beamsplitters in silicon nitride, *Appl. Phys. Lett.*, *54*:1492 (1989).
49. L. D'Auria, J. P. Huignard, A. M. Roy, and E. Spitz, Photolithographic fabrication of thin film lenses, *Opt. Comm.*, *5*:232 (1972).
50. G. J. Swanson and W. B. Veldkamp, Binary lenses for use at 10.6 micrometers, *Opt. Eng.*, *24*:791 (1985).
51. T. Shiono, K. Setsune, O. Yamazaki, and K, Wasa, Rectangular-apertured micro-Fresnel lens arrays fabricated by electron-beam lithography, *Appl. Opt.*, *26*:587 (1987).
52. T. Shiono and K. Setsune, Blazed reflection micro-Fresnel lenses fabricated by electron-beam writing and dry development, *Opt. Lett.*, *15*:84 (1990).
53. J. Jahns, and S. J. Walker, Two-dimensional array of diffractive microlenses fabricated by thin film deposition, *Appl. Opt.*, *29*:931 (1990).
54. K. Rastani, S. Habiby, A. Marrakchi, W. M. Hubbard, H. Gilchrist, and R. E. Nahory, Binary phase Fresnel lenses for generation of two-dimensional beam arrays, *Appl. Opt.*, *30*:1347 (1991).
55. M. R. Taghizadeh, I. R. Redmond, A. C. Walker, and S. D. Smith, Design and construction of holographic optical elements for optical photonic switching applications, Topical Meeting on Photonic Switching, Technical Digest Series, Volume 13, Optical Society of America, Washington, D. C., 1987, pp. 104–106.
56. B. Robertson, M. R. Taghizadeh, J. Turunen, and A. Vasara, Fabrication of space invariant fanout components in dichromated gelatin, *Appl. Opt.*, *29*:1134 (1990).
57. A. Yariv and P. Yeh, *Optical Waves in Crystals*, Wiley—Interscience, 1984.

58. I. P. Kaminow, *An Introduction to Electrooptic Devices*, Academic Press, 1974.
59. T. H. Lin, A. Ersen, J. H. Wang, S. Dasgupta, S. Esener, and S. H. Lee, Two-dimensional spatial light modulators fabricated in Si/PLZT, *Appl. Opt.*, 29:1595 (1990).
60. S. Krishnakumar, S. C. Esener, C. Fan, V. H. Ozguz, and S. H. Lee, Thin ferroelectric lead lanthanum zirconate titanate (PLZT) films for spatial light modulators on silicon on sapphire, Technical Digest on Spatial Light Modulators and Applications, Optical Society of America, Washington, D.C., 1990, Vol. 14, pp. 174–177.
61. D. R. Collins, J. B. Sampsell, L. J. Hornbeck, J. M. Florence, P. A. Penz, and M. T. Gately, Deformable mirror device spatial light modulators and their applicability to optical neural networks, *Appl. Opt.*, 28:4900 (1989).
62. A. Larsson and J. Maserjian, Optically induced absorption modulation in a periodically δ-doped InGaAs/GaAs multiple quantum well structure, *Appl. Phys. Lett.*, 58:1946 (1991).
63. L. Buydens, P. De Dobbelaere, P. Demeester, I. Pollentier, and P. Van Daele, GaAs/AlGaAs multiple-quantum-well vertical optical modulators on glass using the epitaxial lift-off technique, *Opt. Lett.*, 16:916 (1991).
64. M. Schadt and W. Helfrich, Voltage-dependent optical activity of a twisted nematic liquid crystal, *Appl. Phys. Lett.*, 18:127 (1971).
65. P. G. DeGennes, *Physics of Liquid Crystals*, Oxford University Press, 1974.
66. J. S. Patel, Liquid crystals for optical modulation, *Spatial Light Modulators and Applications*, Number 3, SPIE Critical Reviews Series, vol. 1150, pp. 14–26, (1989).
67. K. M. Johnson and G. Moddel, Motivations for using ferroelectric liquid crystal spatial light modulators in neurocomputing, *Appl. Opt.*, 28:4888 (1989).
68. A. Marrakchi, S. F. Habiby, and J. R. Wullert II, Generation of programmable coherent source arrays using spatial light modulators, *Appl. Opt.*, 16:931 (1991).
69. J. S. Patel and K. Rastani, Electrically controlled polarization-independent liquid-crystal Fresnel lens arrays, *Opt. Lett.*, 16:532 (1991).
70. H. Soda, K. Iga, C. Kiahara, and Y. Suematsu, GaInAsP/InP surface emitting injection lasers, *Jpn. J. Appl. Phys.*, 18:2329 (1979).
71. J. L. Jewel, S. L. McCall, Y. H. Lee, A. Scherer, A. C. Gossard, and J. H. English, Lasing characteristics of GaAs microresonators, *Appl. Phys. Lett.*, 54:1400 (1989).
72. M. Orenstein, A. C. Von Lehmen, C. Chang-Hasnain, N. G. Stoffel, J. P. Harbison, L. T. Florez, E. Clausen, and J. L. Jewell, Vertical-cavity surface-emitting InGaAs/GaAs lasers with planar lateral definition, *Appl. Phys. Lett.*, 56:2384 (1990).
73. M. Orenstein, A. C. Von Lehmen, C. Chang Hasnain, N. G. Stoffel, J. P. Harbison, L. T. Florez, J. R. Wuller, and A. Scherer, Matrix addressable surface emitting laser array, Conference on Lasers and Electro-Optics, 1990

Technical Digest Series, Vol. 7, Optical Society of America, Washington, D.C., 1990, pp. 88–89.
74. A. C. Von Lehmen, C. Chang Hasnain, J. R. Wullert, L. Carrion, N. G. Stoffel, L. T. Florez, and J. P. Harbison, Independently addressable InGaAs/GaAs vertical-cavity surface-emitting laser arrays, *Electron. Lett.*, *27*:583 (1991).
75. K. Rastani, M. Orenstein, E. Kapon, and A. C. Von Lehmen, Integration of planar Fresnel microlenses with vertical-cavity surface-emitting laser arrays, *Opt. Lett.*, *16*:919 (1991).
76. J. T. LaMacchia and D. L. White, Coded multiple exposure holograms, *Appl. Opt.*, *7*:91 (1968).
77. D. L. Staebler, W. J. Burke, W. Phillips, and J. J. Amodei, Multiple storage and erasure of fixed holograms in Fe-doped $LiNbO_3$, *Appl. Phys. Lett.*, *26*:182 (1975).
78. H. Kurz, Photorefractive recording dynamics and multiple storage of volume holograms in photorefractive $LiNbO_3$, *Optica Acta*, *24*:463 (1977).
79. A. Marrakchi, W. M. Hubbard, S. F. Habiby, and J. S. Patel, Dynamic holographic interconnects with analog weights in photorefractive crystals, *Opt. Eng.*, *29*:215 (1990).
80. D. Z. Anderson and D. M. Lininger, Dynamic optical interconnects: Volume holograms as optical two-port operators, *Appl. Opt.*, *26*:5031 (1987).
81. A. C. Strasser, E. S. Maniloff, K. M. Johnson, and S. D. D. Goggin, Procedure for recording multiple-exposure holograms with equal diffraction efficiency in photorefractive media, *Opt. Lett.*, *14*:6 (1989).
82. D. Psaltis, J. Yu, X. G. Gu, and H. Lee, Optical neural nets implemented with volume holograms, Digest of Topical Meeting on Optical Computing, Optical Society of America, Washington, D.C., 1987, p. 129.
83. H. Lee, Cross-talk effects in multiplexed volume holograms, *Opt. Lett.*, *13*:874 (1988).
84. K. Wagner and D. Psaltis, Multilayer optical learning networks, *Appl. Opt.*, *26*:5061 (1987).
85. P. Yeh, A. E. T. Chiou, and J. Hong, Optical interconnection using photorefractive dynamic holograms, *Appl. Opt.*, *27*:2093 (1988).
86. S. Weiss, M. Segev, S. Sternklar, and B. Fisher, Photorefractive dynamic optical interconnects, *Appl. Opt.*, *27*:3422 (1988).
87. J. Wilde and L. Hesselink, Implementation of dynamic Hopfield-like networks using photorefractive crystals, Digest of Topical Meeting on Optical Computing, Optical Society of America, Washington, D.C., 1989, p. 10.
88. Y. Owechko, Self-pumped optical neural networks, Digest of Topical Meeting on Optical Computing, Optical Society of America, Washington, D.C., 1989, p. 44.
89. W. H. Steier and M. Ziari, Optical nonlinear neurons and dynamic interconnections using the field shielding nonlinearity in CdTe, Digest of Topical Meeting on Optical Computing, Optical Society of America, Washington D.C., 1989, p. 67.

90. D. Gabor, G. W. Stroke, R. Restrick, A. Funkhouser, and D. Brumm, Optical image synthesis (complex amplitude addition and subtraction) by holographic Fourier transformation, *Phys. Lett.*, *18*:116 (1965).
91. J.-P. Huignard, J.-P Herriau, and F. Micheron, Selective erasure and processing in volume holograms superimposed in photosensitive ferroelectrics, *Ferroelectrics*, *11*:393 (1976).
92. A. Marrakchi, Continuous coherent erasure of dynamic holographic interconnects in photorefractive crystals, *Opt. Lett.*, *14*:326 (1989).
93. G. Kavounas and W. H. Steier, Fast hologram erasure in photorefractive materials, Proc. of the Topical Meeting on Photorefractive Materials, Effects, and Devices, 1987, p. 155.
94. J. W. Goodman, Fan-in and fan-out with optical interconnections, *Optica Acta*, *32*:1489 (1985).
95. A. Marrakchi and W. M. Hubbard, Fan-in issues in a holographic grating interconnect system, *Opt. Lett.*, *16*:417 (1991).
96. J.-P. Huignard, J.-P. Herriau, L. Pichon, and A. Marrakchi, Speckle-free imaging in four-wave mixing experiments with $Bi_{12}SiO_{20}$ crystals, *Opt. Lett.*, *5*:436 (1980).
97. J.-P. Huignard and A. Marrakchi, Two-wave mixing and energy transfer in $Bi_{12}SiO_{20}$ crystals: Application to image amplification and vibration analysis, *Opt. Lett.*, *6*:622 (1981).
98. J. O. White and A. Yariv, Real-time image processing via four-wave mixing in a photorefractive medium, *Appl. Phys. Lett.*, *37*:5 (1980).
99. L. Pichon and J.-P. Huignard, Dynamic joint-Fourier-transform correlator by Bragg diffraction in photorefractive $Bi_{12}SiO_{20}$ crystals, *Opt. Comm.*, *36*:277 (1981).
100. J. E. Ford, Y. Fainman, and S. H. Lee, Array interconnection by phase-coded optical correlation, *Opt. Lett.*, *15*:1088 (1990).
101. D. Rak, I. Ledoux, and J.-P. Huignard, Two-wave mixing and energy transfer in $BaTiO_3$: Application to laser beam steering, *Opt. Comm.*, *49*:302 (1984).
102. A. Chiou and P. Yeh, Energy efficiency of optical interconnections using photorefractive holograms, *Appl. Opt.*, *29*:1111 (1990).
103. Y. Owechko, Photorefractive optical neural networks, Proceedings of the Optical Computing Topical Meeting, Kobe, Japan, 1990.
104. F. H. Mok, M. C. Tackitt, and H. M. Stoll, Storage of 500 high resolution holograms in a $LiNbO_3$ crystal, *Opt. Lett.*, *16*:605 (1991).
105. F. Micheron, C. Mayeux, and J. C. Trotier, Electrical control in photoferroelectric materials for optical storage, *Appl. Opt.*, *13*:784 (1974).
106. K. Rastani and W. M. Hubbard, Large interconnects in photorefractives: Grating erasure problem and a proposed solution, *Appl. Opt.*, *31*:598 (1992).
107. D. Psaltis, X. G. Gu, and D. Brady, Fractal sampling grids for holographic interconnects, SPIE Proceedings, *963*:468 (1988).

7
Substrate-Mode Diffractive Optical Elements and Interconnects

Raymond K. Kostuk
The University of Arizona, Tucson, Arizona

Yang-Tung Huang
National Chiao Tung University, Taiwan, Republic of China

Masayuki Kato
Fujitsu Laboratories Ltd., Atsugi, Japan

I. INTRODUCTION

Free-space optical interconnects have been proposed by several researchers as a means of improving the performace of computing and photonic switching systems [1–3]. Free-space configurations have greater interconnect flexibility than guided systems because they allow signal beam crossing and access to three-dimensional spatial regions, information can be transferred in parallel, and individual mechanical point-to-point contacts are not required. In this application, diffractive optical elements have several advantages over conventional refractive and reflective components. Diffractive elements can combine different optical functions such as beam splitting and focusing into a single component. In addition, microoptical diffractive elements can be fabricated by mass production methods that are similar to those used for microelectronic circuit processing. Planar-type diffractive elements can be formed that guide optical signals within a supportive substrate. This form of optical element is referred to as a substrate-mode hologram or diffractive grating. This planar format matches that of microelectronic circuits and makes packaging more tractable than with other types of free-space optics. This chapter examines the characteristics of substrate-mode elements useful in providing interconnects for electronic processing components. The unique polarization properties of substrate-mode gratings will also be discussed, which can be used in reconfigurable interconnect and polarization sensing systems.

II. PROPERTIES OF DIFFRACTIVE OPTICAL ELEMENTS

Two important characteristics of diffractive optical elements are the diffraction angle and efficiency. Understanding these properties is necessary for the design of substrate mode diffractive elements. Methods for calculating these parameters are the topic of this section.

The diffraction angle is determined from the requirement that the frequency of the field remain continuous or phase matched across the interface between the grating and the surrounding medium. Consider the situation for the volume phase hologram in the x–z plane as shown in Figure 1.

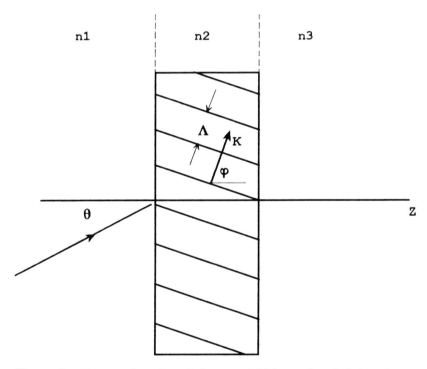

Figure 1 Diagram of a volume hologram of thickness d, period Λ, and average refractive index n_2 surrounded by medium n_1 on the incident side and n_3 at the exit face of the grating. The grating is restricted to the x–z plane.

The grating is formed by modulating the refractive index of medium 2 with a period of Λ between planes of constant index modulation. The grating can be characterized with a grating vector of magnitude $2\pi/\Lambda$, which lies in a direction perpendicular to the planes of constant index modulation. The average refractive index of the incident medium is n_1, n_2 for the grating, and n_3 for the exit medium. The grating is illuminated by an optical field with a propagation vector $\bar{k}_1 = (2\pi n_1/\lambda_0)\hat{k}$, where λ_0 is the free-space wavelength and $\hat{k}$ is a unit vector in the direction of the propagation vector (θ_1 in the x–z plane). Within the grating the field can be considered a summation of plane waves with wavevectors ($\bar{\sigma}_m$) determined from the k-vector closure condition

$$\bar{\sigma}_m = \bar{k}_2 - m\bar{K} \tag{1}$$

The directions of the corresponding diffraction orders outside the grating are again determined by phase matching the components of the propagation vectors along the grating boundary (x axis). In the case of the zero order ($m = 0$), the phase matching condition across the incident boundary results in

$$\bar{k}_{1,0} \cdot \hat{x} = \bar{k}_{2,0} \cdot \hat{x} \tag{2}$$

where $\bar{k}_{2,0}$ has a magnitude $2\pi n_2/\lambda_0$. This is simply a statement of Snell's law, or $n_1 \sin(\theta_1) = n_2 \sin(\theta_2)$. For an arbitrary diffraction order, the phase matching condition is

$$\begin{aligned}\bar{\sigma}_m \cdot \hat{x}_1 &= (\bar{k}_2 - m\bar{K}) \cdot \hat{x} \\ |\sigma_m| \sin(\theta_m) &= k_2 \sin(\theta_o) - mK \sin(\phi) \\ \frac{2\pi n_3}{\lambda} \sin(\theta_m) &= \frac{2\pi n_3}{\lambda} \sin(\theta_o) - m\frac{2\pi}{\Lambda_x}\end{aligned} \tag{3}$$

where the angles θ_o and θ_m are measured in medium n_3. Dividing by 2π and multiplying by λ results in the grating equation

$$\sin(\theta_m) - \sin(\theta_0) = \frac{m\lambda}{\Lambda_x} \tag{4}$$

More complicated holographic components can be analyzed by dividing the interference pattern used to form the hologram into a set of local fringes and then applying the grating equation to the resulting local gratings. If the local fringe is formed by the interference of propagation vectors at angles θ_c (for the reference wave) and θ_o (for the object wave) relative to the z axis with a wavelength λ_1, the construction grating equation becomes

$$\sin(\theta_c) - \sin(\theta_0) = \frac{\lambda_1}{\Lambda_x} \tag{5}$$

After recording and processing the emulsion the resulting grating can be reconstructed with light having a different wavelength (λ_2) at an arbitrary angle of incidence (θ_i) to give a diffraction angle θ_d, where

$$\sin(\theta_i) - \sin(\theta_d) = \frac{\lambda_2}{\Lambda_x} \tag{6}$$

These expressions can be extended to determine all three components of a holographically formed grating vector:

$$K_x = \frac{2\pi}{\lambda_1}(l_c - l_o) = \frac{2\pi}{\lambda_2}(l_r - l_d)$$

$$K_y = \frac{2\pi}{\lambda_1}(m_c - m_o) = \frac{2\pi}{\lambda_2}(m_r - m_d) \tag{7}$$

$$K_z = \frac{2\pi}{\lambda_1}(n_c - n_o) = \frac{2\pi}{\lambda_2}(n_r - n_d)$$

where l, m, and n are direction cosines in the x, y, and z directions, the subscripts c, o, i, and d correspond to the reference, object, reconstruction, and diffracted beam directions, and $K_m = 2\pi/\Lambda_m$ with $m = x, y$, and z. Note that K_z is determined from K_x and K_y since the direction cosines are related by $1 = l^2 + m^2 + n^2$. The surface components K_x and K_y can be used in a general ray tracing expression [4,5] to determine the direction of a diffracted beam. Many commercial optical design programs use these expressions for evaluating diffractive elements.

In addition to the direction of a beam diffracted by a grating, it is also important to determine the efficiency of a grating. Several approaches are possible for evaluating the efficiency of surface relief and volume-type gratings. In some cases scalar diffraction theory can be used to describe the efficiency of multilevel binary-type surface relief gratings [6]. Consider a single binary step grating as illustrated in Figure 2. The transmittance function for this grating can be expressed as

$$t(x) = \sum_{m=-\infty}^{\infty} \delta(x - m\Lambda_x) * \text{rect}(\frac{x}{\Lambda_x}) \exp(j2\pi\phi x) \tag{8}$$

where $\phi = (n - 1)d/(\lambda\Lambda_x)$, and $*$ represents the convolution. Using the relations [7]

$$\delta(ax) = \frac{1}{|a|}\delta(x) \tag{9}$$

OPTICAL ELEMENTS AND INTERCONNECTS

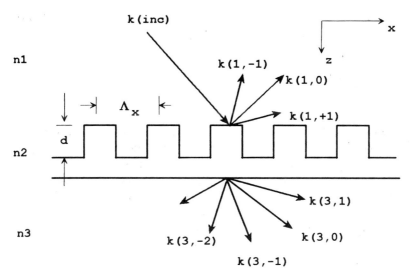

Figure 2 Schematic view of a surface relief grating with period Λ_x and depth d with several diffraction orders formed during reconstruction.

and

$$\text{FT}\left[\text{comb}\left(\frac{x}{\Lambda_x}\right)\right] = \Lambda_x \,\text{comb}(\Lambda_x f)$$

with FT the spatial Fourier transform with frequency f, and $\text{comb}(x) = \sum_{m=-\infty}^{\infty} \delta(x-n)$, the diffracted field becomes

$$A(f) = \sum_{m=-\infty}^{\infty} \delta\!\left(f - \frac{m}{\Lambda_x}\right) \frac{\sin[\pi\Lambda_x(\phi - f)]}{\pi\Lambda_x(\phi - f)} \tag{10}$$

where $f = \sin(\theta)/\lambda$. Notice that the δ function selects frequencies that are equivalent to those determined from the grating equation (Eq. 7) with a normally incident reconstruction beam. The diffraction efficiency is determined by taking the square of the amplitude of a particular diffraction order:

$$\eta_m = \left[\frac{\sin[\pi\Lambda_x(\phi - m/\Lambda_x)]}{\pi\Lambda_x[\phi - m/\Lambda_x]}\right]^2 \tag{11}$$

The maximum diffraction efficiency that can be obtained for the $m = 1$ diffraction order of a single binary step or two-phase level grating is 41%. When more than two phase levels are used to approximate a continuous phase profile, the diffraction efficiency can be significantly improved. For N levels the efficiency follows the relation [6]

$$\eta(N) = \left[\frac{\sin(\pi/N)}{\pi/N} \right]^2 \qquad (12)$$

The high efficiency and lithographic fabrication process make binary optical elements very attractive; for substrate-mode elements, however, relatively high spatial frequencies are required. Scalar theory, which was used to determine the preceding relations, is only accurate when the period to wavelength ratio is about 5 or greater. Below this value, resonance effects in the grating reduce efficiency [8]. In order to accurately determine these limits and characteristics, a more powerful theory is required. One approach suitable for describing both surface relief and volume-type grating efficiencies is rigorous coupled-wave theory [9]. In this method the field within the grating is expanded in terms of harmonic components with propagation vectors determined using the Floquet or K-vector closure condition (Eq. 2). Assuming a volume grating (Fig. 1) with the electric field perpendicular to the plane containing the grating vector and the propagation vector of the reconstruction beam, the electric field within the grating has the form

$$E_2(x, y) = \sum_{m=-\infty}^{\infty} A_m(z) \exp[-j(\bar{k}_2 - m\bar{K}) \cdot \bar{r}] \qquad (13)$$

where $\bar{k}_2$ is again the propagation vector of the reconstruction beam within the grating, and $A_m(z)$ is the amplitude of the mth space harmonic. The field in the incident medium is given by

$$E_1 = \exp(-j\bar{k}_1 \cdot \bar{r}) + \sum_{m=-\infty}^{\infty} A_{R,m} \exp(-j\bar{k}_{1,m} \cdot \bar{r}) \qquad (14)$$

where the first term represents an incoming plane wave with unit amplitude, and the second term represents the sum of the reflected diffracted components with amplitudes $A_{R,m}$. Similarily, the electric field in the grating substrate medium is

$$E_3 = \sum_{m=-\infty}^{\infty} A_{T,m} \exp[-j\bar{k}_{3,m} \cdot (\bar{r} - d\hat{z})] \qquad (15)$$

where $A_{T,m}$ is the amplitude of the mth diffracted order with wavevector $k_{3,m}$. The translation $\bar{r} - d\hat{z}$ allows the same coordinate system to be used for transmitted and reflected fields. The fields in each region must satisfy the Helmholtz equation

OPTICAL ELEMENTS AND INTERCONNECTS

$$\nabla^2 E + k^2 E = 0 \tag{16}$$

where k is the propagation constant of the medium. For a sinusoidal phase grating, $k^2 = (2\pi/\lambda)[\epsilon_{avg} + \epsilon_1 \cos(\vec{K} \cdot \vec{r})]$, with ϵ_{avg} the average permittivity of the grating region, and ϵ_1 the maximum amplitude of the permittivity modulation. Substituting Eq. (13) into Eq. (16) and collecting coefficients of similar exponential terms results in a set of coupled-wave equations

$$\frac{1}{2\pi}\frac{d^2 A_m(z)}{dz^2} - j\frac{2}{\pi}\left[\frac{(\epsilon_{avg})^{1/2}\cos(\theta)}{\lambda} - \frac{m\cos(\phi)}{\Lambda}\right]\frac{dA_m(z)}{dz}$$
$$+ \frac{2m(p-m)}{\Lambda^2}A_m(z) + \frac{\epsilon}{\lambda^2}[A_{m-1}(z) + A_{m+1}(z)] = 0 \tag{17}$$

where p is defined as

$$p = \frac{2\Lambda \epsilon^{1/2}}{\lambda}\cos(\theta - \phi)$$

The system satisfies a Bragg condition whenever $p = m$, and it corresponds to other reconstruction conditions for nonintegral values of p. This set of equations can be solved using the state variable approach developed by Gaylord and Moharam [9]. In their method the state variables are defined as

$$A_{1,m} = A_m(z)$$
$$A_{2,m} = \frac{dA_m(z)}{dz} \tag{18}$$

These relations transform the second-order differential equations into a set of first-order equations:

$$\frac{dA_{1,m}}{dz} = A_{2,m}(z)$$
$$\frac{dA_{2,m}(z)}{dz} = -\frac{2\pi^2\epsilon_1}{\lambda^2}A_{1,m-1}(z) + \frac{4\pi^2(m-p)}{\Lambda^2}A_{1,m}(z) - \frac{2\pi^2\epsilon_1}{\lambda^2}A_{1,m+1}(z) \tag{19}$$
$$+ j4\pi\left(\frac{(\epsilon_{avg})^{1/2}\cos(\theta)}{\lambda} - \frac{m\cos(\phi)}{\Lambda}\right)A_{2,m}(z)$$

This set of equations is of the form $\dot{\mathbf{A}} = \mathbf{C}\mathbf{A}$, where $\dot{\mathbf{A}}$ and $\mathbf{A}$ are column vectors consisting of $2(2m + 1)A_{1,2}$ and $2(2m + 1)\dot{A}_{1,2}$ elements. $\mathbf{C}$ is a $2(2m + 1)$ by $2(2m + 1)$ square coefficient matrix. The solutions to this set of first-order differential equations are of the form

$$A_{l,m}(z) = \sum_{q=-\infty}^{\infty} D_q \omega_{l,mq} \exp(\lambda_q) \qquad (20)$$

where $l = 1, 2$ and the D_q are unknown coefficients, which must be determined by applying the boundary conditions. $\omega_{l,mq}$ is one component of an eigenvector, and λ_q is the corresponding eigenvalue. The eigenvalues are determined by evaluating the determinant of the **C** matrix. The eigenvectors are calculated by inserting the eigenvalues back into the expressions for $A_{l,m}$, reevaluating the state equation, and then using Cramer's rule. In order to evaluate the diffracted amplitude components $A_{R,m}$ and $A_{T,m}$, the tangential components of the electric and magnetic fields corresponding to $E_{q,m}$ and $E_{3,m}$ are matched to the harmonic components within the grating region at $z = 0$ and $z = d$. The magnetic field components are also needed and are related to E_y by the expression $H_x = (-j/\omega\mu) \, \partial E_y/\partial z$, which is derived from Maxwell's equations with time harmonic fields. At the $z = 0$ boundary the tangential E component gives

$$\delta_{m,0} + A_{R,m} = S_m(0) \qquad (21)$$

and the tangential H field component

$$j[k_1^2 - (k_2 \sin \theta - mK \sin \phi)^2]^{1/2} (A_{R,m} - \delta_{m,0})$$
$$= \frac{dA_m(0)}{dz} - j(k_2 \cos \theta - mK \cos \phi) A_m(0) \qquad (22)$$

with $\delta_{m,0}$ the Kronecker delta function. The matching conditions at the $z = d$ boundary are

$$A_{T,m} = A_m(d) \exp[-j(k_2 \cos \theta - mK \cos \phi)d] \qquad (23)$$

for the electric field component and

$$-j[k_3^2 - (k_2 \sin \theta - mK \sin \phi)^2]^{1/2} A_{T,m}$$
$$= \left[\frac{dA_m(d)}{dz} - j(k_2 \cos \theta - mK \cos \phi) A_m(d) \right]$$
$$\exp[-j(k_2 \cos \theta - mK \cos \phi)d] \qquad (24)$$

for the H component at $z = d$. This linear set of equations can then be solved by a method such as Gauss elimination for the $A_{R,m}$ and $A_{T,m}$ amplitudes. Other approaches also exist for solving this system of equations [10]. The resulting diffraction efficiency of a diffracted order in region 1 is

$$\eta_{1,m} = \text{Re}\left(\frac{\bar{k}_{1,m} \cdot \hat{z}}{\bar{k}_{1,0} \cdot \hat{z}} \right) A_{R,m} A^*_{R,m} \qquad (25)$$

OPTICAL ELEMENTS AND INTERCONNECTS

and for region 3,

$$\eta_{3,m} = \text{Re} \frac{\bar{k}_{3,m} \cdot \hat{z}}{\bar{k}_{3,0} \cdot \hat{z}}) A_{T,m} A^*_{T,m}$$

This method for determining the diffraction efficiency for different orders of a volume grating can also be used for evaluating the efficiency of a surface relief grating. This is accomplished by dividing the grating surface into slabs, applying the volume model to each layer, and then boundary matching at each interface [11]. This model is quite general and as mentioned earlier is required in cases where scalar approximations are inappropriate.

III. CONVENTIONAL FREE-SPACE OPTICAL INTERCONNECTS

One important function of an interconnect system is to provide connections in a variety of geometries required by different processing systems. A typical free-space interconnect for connecting different points on a common plane is illustrated in Figure 3. In this case a reflective-type holographic optical element is suspended above a substrate that contains different processing elements. A free-space interconnect system connecting points on different planes is illustrated in Figure 4. This time the optical system acts in transmission transferring data from one plane to another. Each processing element has an optoelectronic interface consisting of a

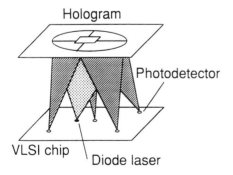

Figure 3 Diagram of a reflective-type holographic interconnect system connecting several processing elements on a plane.

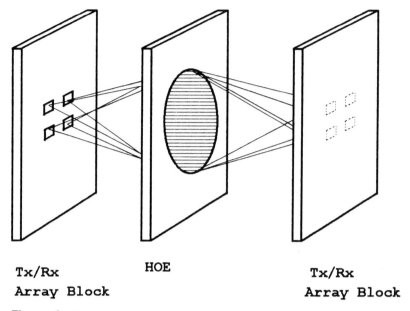

Tx/Rx   HOE   Tx/Rx
Array Block   Array Block

Figure 4 Transmission-type holographic interconnect providing connections between two planes.

transmitting source such as a vertical cavity laser diode and driver [12], along with a detector and receiver circuit.

The region over which interconnects can be made and the separation between planes are affected by off-axis aberrations, diffraction-limited image spot size, beam spreading, and optical crosstalk [13–15]. Misalignment is another important factor that affects the signal propagation distance and density [16]. These performance factors restrict the volume of space over which interconnects can be made. It has been shown that the lateral dimension of the interconnect volume or substrate area is limited to a region approximately equal to the height of the optical element above the substrate. Increasing the suspension height of the optical element above the substrate reduces the alignment tolerance and increases the packaging cost. In many cases these restrictions severely limit the performance of the processing system.

A. Interconnect Design Considerations

An optical interconnect system must be efficient in transmitting and collecting light from source to receiver points. One method to accomplish this is to increase the size of the detector; however, this directly affects the

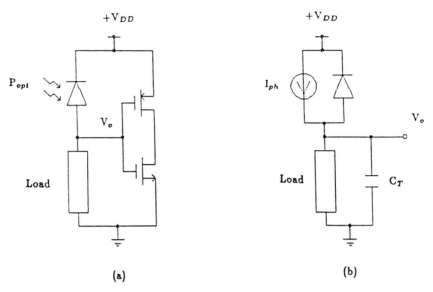

Figure 5 (a) Photodiode with a CMOS inverter as a receiver element. (b) Equivalent circuit for photodiode and CMOS inverter with load.

amount of optical power required to drive the system at a specific operating frequency. To quantify this trade-off, consider the response of a receiver that consists of a CMOS inverter circuit and a photodiode as illustrated in Figure 5a. The corresponding equivalent circuit is also shown in Figure 5b. When an optical pulse is incident on the detector, a constant current I_{ph} is generated and split between the biasing circuit load (Z_B) and a capacitive load (C_T). C_T includes the gate C_G and detector C_D capacitance, or $C_T = C_G + C_D$. If it is assumed that $I_{ph} = I_1 + I_2$ with $I_1 = I_2 = I_{ph}/2$, the charge (Q) required to take C_T from 0 to 90% of the bias potential level (V_{DD}) is given by

$$Q = \int_0^\tau \frac{I_{ph}}{2} dt = c_T 0.9 V_{DD}$$

$$\tau = \frac{1.8 C_T V_{DD}}{I_{ph}}$$

(26)

where τ is the effective time constant of the detector and inverter circuit. If the circuit is biased so that the rise and fall times are equal, the maximum operating frequency is given by $f = 1/2\tau$. Combining this value with the detector responsivity (R_r in A/W), the optical power required to drive this circuit at a given frequency becomes

$$\Phi_{ph} = \frac{I_{ph}}{R_r} = \frac{1.8(C_G + C_D)V_{DD}\,2f}{R_r} \tag{27}$$

The detector capacitance in this relation is given by

$$C_D = \frac{\epsilon_0 \epsilon_r A_D}{d} \tag{28}$$

where ϵ_0 and ϵ_r are respectively the permittivity of free space and relative permittivity of silicon, d is the depletion layer thickness, and A_D is the active area of the detector. The optical power required to drive several different diameter detectors is illustrated in Figure 6. As indicated, the power required to operate the receiver at a specific frequency increases significantly with detector size. This emphasizes the need to confine the focused optical power to the smallest possible region.

The dispersive properties of diffractive elements significantly contributes to the size and displacement of an optical image. This is especially true when laser diodes are used as the reconstruction source, because their emission wavelength can vary by several nanometers from the design wavelength. This results from variations in the bandgap that occur during the fabrication process or when different injection current levels are used during modulation. These chromatic effects must be considered in the design of diffractive optical interconnect systems.

The angular dependence of the image formed by a grating element can be determined from the relations developed in the previous section (Eqs. 4 and 5). For interferometric gratings, the wavelengths used during recording (λ_1) set up the grating period (Λ_x):

$$\frac{\sin \alpha_c}{\lambda_1} - \frac{\sin \alpha_o}{\lambda_1} = \frac{1}{\Lambda_x}$$

where the subscripts c and o refer to the construction and object beam angles, respectively. If λ_1 represents the nominal laser diode wavelength, and $\lambda_2 = \lambda_1 \pm \Delta\lambda$ the actual reconstruction wavelength, the angle of the diffracted beam will become

$$\alpha_{o'} = \sin^{-1}\left[\lambda_2 \left(\frac{\sin \alpha_r}{\lambda_2} + \frac{1}{\Lambda_x}\right)\right] \tag{29}$$

where the angle of incidence for the reconstruction beam α_r may be different from that used during construction (α_c). In order to examine the effect of changing the reconstruction wavelength on image position and quality,

OPTICAL ELEMENTS AND INTERCONNECTS

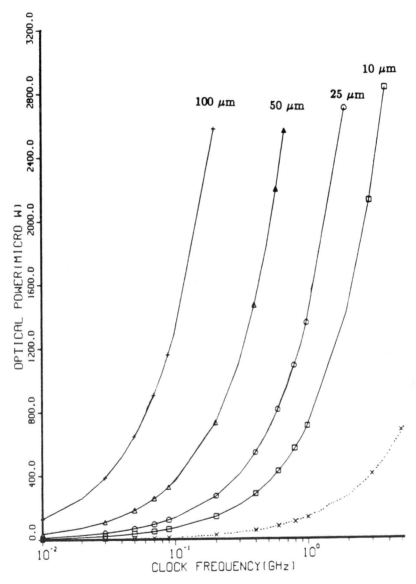

Figure 6 Optical power required to drive different diameter detectors at increasing clock frequencies. Also shown is the power required to drive the inverter without the detector (dotted line with X's).

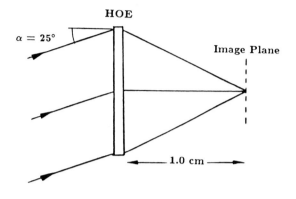

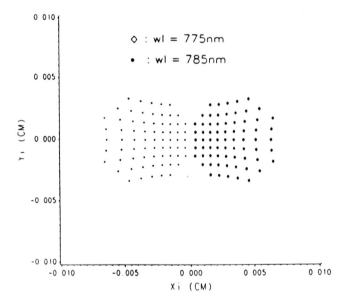

Figure 7 Effect of chromatic dispersion on the focused spot location formed by a diffractive lens.

consider the hologram shown in Figure 7. This hologram is formed with a 780-nm wavelength collimated beam incident at an angle of 25° to the hologram normal and the field from a point source located 1.0 cm from the hologram plane. If the wavelength varies by ±5 nm from the construction wavelength, the reconstructed image shifts considerably and spherical

OPTICAL ELEMENTS AND INTERCONNECTS

aberration is introduced. In this case, the diameter of the detector must be about 100 μm to encompass the aberrated images. The receiver analysis and results shown in Figure 6 indicate that a significant amount of optical power is required to drive a detector of this size.

One method to reduce the shift in the diffracted beam location and the corresponding detector diameter is to use a compensating grating in conjunction with the focusing element [17]. Complete compensation can be achieved if the corrective grating diffracts light in the opposite direction from the focusing element at each point in the aperture of the focusing element [18]. This requires accurate positioning of the two gratings, however, which makes packaging difficult. A less demanding method is to use a planar grating designed to compensate light diffracted at the center of the focusing element. This method does not correct for spherical chromatic aberration, but it does eliminate the wavelength-dependent lateral shift. Figure 8 shows the effect of using a compensating grating with the focusing element described above. As can be seen, the lateral shift is eliminated, and although the spherical chromatic aberration still exists, the light is confined within a 50-μm radius. The corresponding detector radius and the optical power required to drive it can be considerably reduced.

An experimental verification of this corrective effect was performed by constructing a 15-cm focal length focusing element in a silver halide emulsion using a temperature-stabilized laser diode operating at 780 nm. After construction, the emulsion is developed and bleached and reconstructed with the same laser diode system. During reconstruction the temperature of the diode's heat sink was changed by 5°C, which varied the emission wavelength by about 2 nm. As indicated in the time-superimposed photograph (Fig. 9A), the image shifts by 280 μm. When the compensating plate is used (Fig. 9B), the image remains at a fixed location even when the laser diode heat sink temperature is changed by 10°C. This verifies that the technique can significantly reduce the detector size and the optical power required for the interconnect system.

B. Substrate-Mode Grating Components

The previous discussion indicates that several properties should be incorporated into the design of diffractive optical interconnect systems. First, the distance from the focusing element to the image should be kept short in order to reduce the effects of aberrations and beam spreading. On-axis imaging components also minimize these effects. In addition, grating pairs should be used to reduce image displacement due to source emission wavelength variation. Finally, the system must connect points separated by considerable distances on a substrate. This allows a larger number of processing elements to be connected.

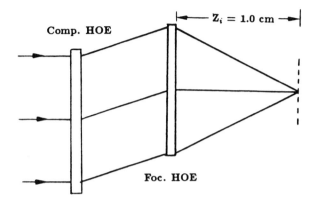

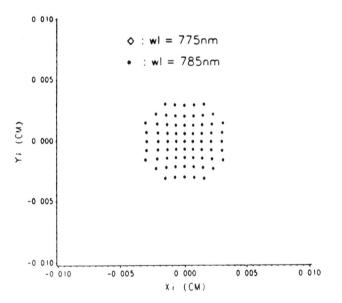

Figure 8 Focused spot formed by the same diffractive lens shown in Fig. 7, but with compensation grating.

Many of these features can be realized using a substrate-mode hologram (Fig. 10). This element diffracts an incident beam beyond the critical angle of the substrate material, and the total internal reflected beam is guided by the substrate to a receiver location. A second grating at the receiver location couples light out of the substrate and focuses it onto a detector. A prime advantage of this component is that it separates the functions of

OPTICAL ELEMENTS AND INTERCONNECTS 339

TEMPERATURE RANGE: 25° C - 30° C

|←— 280 μm —→|

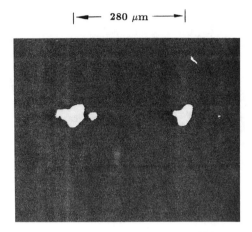

A

TEMPERATURE RANGE: 20° C - 30° C

40 μm

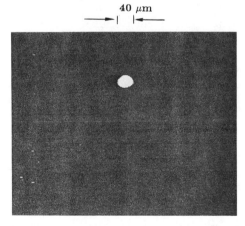

B

Figure 9 Experimental results showing the elimination of lateral shift due to changes in laser diode wavelength. (A) Image formed with uncompensated focusing element and an effective wavelength change of 2 nm. (B) Image formed by same focusing element with compensating grating and wavelength variation greater than 2 nm.

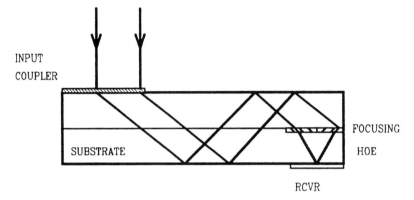

Figure 10 Substrate-mode hologram device consisting of an input coupler, which diffracts a beam beyond the critical angle of the substrate, and an output coupler, which focuses light onto a receiver.

beam translation and imaging. This allows signals to propagate over a much larger region of the substrate. On-axis imaging components can also be used with relatively short focal lengths, which reduce the degrading effects of aberrations and off-axis image spreading. In addition, the input and output coupler act as a compensating grating pair to correct for wavelength variations of the reconstruction source. Finally, another important benefit of the substrate-mode arrangement is that it is more suitable for packaging with planar electronic substrates.

C. Construction Techniques for Substrate-Mode Diffractive Elements

Two basic methods have been used to form substrate-mode optical elements. The first is by volume holography and the second by binary step lithography [6].

Most available high-efficiency holographic recording materials are sensitive to blue–green (488–514.5 nm) wavelengths. In most interconnect applications, however, it is desirable to use laser diode reconstruction sources with emission wavelengths ranging from 670 to 1300 nm. One design method for these elements is to start with the desired reconstruction parameters and then work backward to determine the construction angles required with the formation wavelength (Fig. 11). The physical properties of the recording material and changes that occur during processing must also be included in the design to obtain accurate results. The first step, therefore, is to characterize the properties of the recording material. This includes the average refractive index and thickness of the emulsion before

OPTICAL ELEMENTS AND INTERCONNECTS 341

Desired Reconstruction Conditions

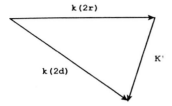

Grating Vector Shift Due to Thickness Change

Construction Conditions with Shorter Wavelength

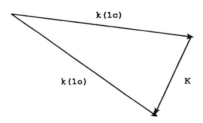

Figure 11 Holographic construction for forming a substrate-mode element at a short wavelength (488 nm) and reconstructing at a longer wavelength (780 nm). The design starts by using the desired reconstruction (k_{2r}) and diffracted (k_{2d}) propagation vectors to determine the final grating vector **K′**. The grating vector rotation due to thickness change is then included to determine the grating vector during construction (**K**). This grating vector is then used to determine the directions for the construction propagation vectors (k_{1c} and k_{1o}).

and after processing, and the exposure conditions that maximize the diffraction efficiency at the reconstruction wavelength.

Starting with the desired angle in the substrate (θ_s), the diffraction angle inside the emulsion is calculated using Snell's law:

$$\theta_e = \sin^{-1}\left(\frac{n_s \sin(\theta_s)}{n_{e,2}}\right) \tag{30}$$

where $n_{e,2}$ is the average refractive index of the emulsion after processing, and θ_e is the angle in the emulsion. The magnitude of the reconstruction propagation vectors in the processed emulsion are

$$|k_2| = \frac{2\pi n_{e,2}}{\lambda_2} \qquad (31)$$

where λ_2 is the reconstruction wavelength. The grating vector and period are obtained from the propagation vectors and the Bragg condition:

$$K' = k_{2,r} - k_{2,d} \qquad (32)$$

where K' is the grating vector after processing, $k_{2,r}$ is the reconstruction propagation vector, and $k_{2,d}$ is the diffracted propagation vector.

If the emulsion swells or shrinks during processing, the grating period in the z direction changes in proportion to the effective thickness change. This material factor can be incorporated into the design by subtracting the predetermined fractional thickness change from the z component of the grating vector:

$$K_z = \frac{2\pi}{\Lambda'} \frac{1}{(1 - \Delta d/d)} = \frac{K'_z}{1 - \Delta d/d} \qquad (33)$$

where K_z and K'_z are the z components of the grating vector before and after processing, respectively, d is the emulsion thickness, and Δd is the thickness change.

Lithographic formation of diffractive optical elements (binary optics) offers several advantages for substrate-mode grating design. Since these gratings are defined mathematically, a large variety of complex components can be realized for use at any desired wavelength. In addition, since integrated circuit fabrication techniques are used, elements that perform different optical functions can be accurately aligned on the same substrate, and these components can be readily replicated.

The construction process for binary optical elements is illustrated in Figure 12. In the first step, a mask is designed by specifying the phase transition points for the grating. For a single rectangular groove grating design, the grating period and duty cycle determine the transition points. The grating periods are determined in the x–y plane using Eq. (7) and the desired incident and diffraction angles. The duty cycle and grating depth are then optimized to obtain high diffraction efficiency either using rigorous coupled-wave theory or scalar theory as appropriate. The resulting coordinate map is then translated into a pattern generation code, which controls the movement of an e-beam mask writer. Next, the substrate is coated with photoresist and exposed to light passing through the mask. After exposure, the photoresist is

OPTICAL ELEMENTS AND INTERCONNECTS

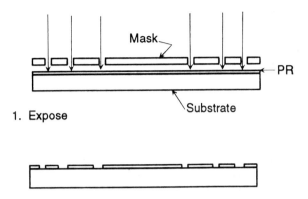

1. Expose

2. Develop Photoresist

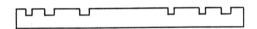

3. Etch Substrate/Repeat Procedure

Figure 12 Binary phase grating fabrication steps. First, a mask is formed and used to pattern an exposure in photoresist. The photoresist is then developed, and the substrate is etched to the required phase depth. The process is repeated for a multiple-phase step element.

developed and acts as a protective coating of the substrate. The uncoated substrate areas are etched to a depth that gives maximum efficiency. The remaining photoresist is then stripped from the substrate. If a blazed or nonrectangular grating structure is desired, this process is repeated using several masks with smaller feature sizes. In the scalar diffraction regime ($\Lambda > 5\lambda$), the grating efficiency varies with the number of etch levels (N) according to Eq. (12):

$$\eta = \left| \frac{\sin(\pi/N)}{\pi/N} \right|^2$$

In this expression the number of phase levels is related to the number of masks (M) by $N = 2^M$. The optimum etch depth (d_M) for the Mth mask pattern is

$$d_M = \frac{\lambda_0}{2^M(n-1)} \tag{34}$$

where λ_0 is the free-space design wavelength, and n is the refractive index of the substrate.

Although binary optic techniques are attractive, they do have some limitations in substrate-mode designs. Perhaps the most serious is the resolution limit of the lithographic process. Typically the smallest feature size that can be aligned using a high-resolution mask aligner is 0.2–0.5 μm. At visible wavelengths, gratings with these dimensions are not sufficient to diffract light beyond the critical angle of glass substrates. Direct e-beam writing can give greater spatial resolution, but this reduces the manufacturing capacity and increases cost. Substrate-mode elements can be formed with smaller diffraction angles using a reflective coating on one of the surfaces. In order to propagate signals over significant substrate distances, however, thicker substrates must be used or the beam must be reflected many times [19]. This increases loss and delay in the interconnect system.

One approach to resolving this problem is to combine volume grating components with lithographic diffractive elements. High spatial frequency gratings can readily be formed in volume holographic materials and used as input and output couplers, while binary optics is used for applications requiring accurate alignment. In this manner the best properties of each fabrication technique are used; several issues must be resolved, however, before this approach becomes viable. In particular, the problem of packaging and aligning volume and surface relief gratings must be resolved.

IV. MULTIPLEXED SUBSTRATE-MODE HOLOGRAPHIC ELEMENTS

Many interconnect systems must split a signal and simultaneously send information to several receiving locations. These broadcasting operations are important for system functions such as synchronous clock distribution and data exchange in memory. Optical interconnect components must also perform these splitting operations. One method to accomplish this is by holographic multiplexing of several gratings in the same recording material. In the following section, the design and performance of a multiplexed substrate-mode optical element is described [20].

A schematic view of a multiplexed grating that splits a normally incident beam into four total internal reflected beams is illustrated in Figure 13. If the incident light is polarized in the y direction, the fields diffracted by gratings K_1 and K_3 are effectively s-polarized, while those diffracted by gratings K_2 and K_4 are p-polarized. The efficiency dependence on refractive index modulation differs for the two polarization states. Therefore, each pair of gratings must be recorded with a different exposure value to maxi-

OPTICAL ELEMENTS AND INTERCONNECTS

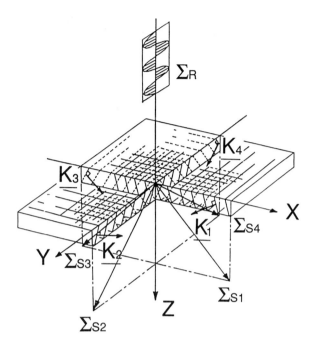

Figure 13 Schematic view of multiplexed substrate-mode hologram that diffracts a normally incident beam polarized in the y direction into four orthogonal substrate-mode beams.

mize the efficiency. In addition, saturation effects also occur and must be included in calculations for determining the optimum grating exposure.

The multiplexed grating efficiency can be analyzed using coupled-wave theory [21,22]. Since high-efficiency dichromated gelatin and photopolymers are dielectric phase gratings, it is assumed that the dielectric permitivity is modulated according to

$$\epsilon = \epsilon_0 + \sum_{m=1}^{4} \epsilon_m \cos(\mathbf{K}_m \cdot \bar{\mathbf{r}}) \tag{35}$$

where ϵ_0 is the average dielectric constant of the recording material, the ϵ_m ($m = 1, 2, 3, 4$) are the amplitudes of the dielectric modulation for gratings $\mathbf{K}_m$, and $\bar{\mathbf{r}}$ is the postion vector. If the multiplexed grating is illuminated by a wave Σ_R, the electric field $\mathbf{E}$ in the medium is given by

$$\mathbf{E} = \mathbf{R}(z) \exp[-j(\bar{\rho} \cdot \bar{\mathbf{r}})] + \sum_{m=1}^{4} \mathbf{S}_m(z) \exp[-j(\bar{\sigma}_m \cdot \bar{\mathbf{r}})] \tag{36}$$

where the first term represents the undiffracted beam and the second term the diffracted beams from each grating $\mathbf{K}_m$. $\bar{\rho}$ and $\bar{\sigma}_m$ are the propagation vectors of Σ_R and Σ_{Sm}, respectively, and satisfy the following Bragg conditions:

$$\bar{\sigma}_m = \bar{\rho} - \mathbf{K}_m \tag{37}$$

The amplitudes $\mathbf{R}(z)$, $\mathbf{S}_m(z)$ are

$$\begin{aligned}\mathbf{R}(z) &= R_y(z)\hat{j} \\ \mathbf{S}_m(z) &= S_{my}(z)\hat{j} & (m = 1, 3) \\ \mathbf{S}_m(z) &= S_{my}(z)\hat{j} + S_{mz}(z)\hat{k} & (m = 2, 4)\end{aligned} \tag{38}$$

where $\hat{j}$ and $\hat{k}$ are unit vectors along the y and z axes, respectively.

Substituting into the expression for $\mathbf{E}$, the x, y, z components of the electric field are

$$E_x = 0$$

$$E_y = R_y(z) \exp[-j(\rho \cdot \mathbf{x})] + \sum_{m=1}^{4} S_{my}(z) \exp[-j(\sigma_m \cdot \mathbf{x})] \tag{39}$$

$$E_z = S_{2z}(z) \exp[-j(\sigma_2 \cdot \mathbf{x})] + S_{4z}(z) \exp[-j(\sigma_4 \cdot \mathbf{x})]$$

Substituting these field values into the wave equation

$$\nabla^2 \mathbf{E} - \nabla(\nabla \cdot \mathbf{E}) + \frac{\epsilon \omega^2}{c^2} \mathbf{E} = 0 \tag{40}$$

assuming Bragg angle reconstruction, and neglecting second-order derivatives and diffraction orders other than -1 [23], the resulting coupled-wave equations become

$$\begin{aligned}0 &= c_{Sm}S'_m(z) + j\kappa_m R(z) & (m = 1, 3) \\ 0 &= c_{Sm}S'_m(z) + j\kappa_m R(z)(\hat{s}_m \cdot \hat{r}) & (m = 2, 4) \\ 0 &= c_R R'(z) + j\kappa_1 S_1(z) + j\kappa_2 S_2(\hat{s}_2 \cdot \hat{r}) + k\kappa_3 S_3(z) + j\kappa_4 S_4(\hat{s}_4 \cdot \hat{r})\end{aligned} \tag{41}$$

where $\hat{s}_2$, $\hat{s}_4$, and $\hat{r}$ are unit vectors in directions $\mathbf{S}_2(z)$, $\mathbf{S}_4(z)$, and $\mathbf{R}(z)$, respectively, and

$$c_{Sm} = \frac{\sigma_{mz}}{\beta}, \quad c_R = \frac{\rho_z}{\beta}, \quad \beta = \frac{2\pi n_0}{\lambda}$$

and

$$\kappa_m = \frac{\pi n_m}{\lambda} \quad (m = 1, 2, 3, 4)$$

OPTICAL ELEMENTS AND INTERCONNECTS

n_m is the index modulation of the mth grating, n_0 is the average refractive index, and λ is the free-space wavelength of the reconstruction beam. After solving these expressions using the method described by Kogelnik [21] and Case [22], the diffraction efficiencies are

$$\eta_{S_i} = \frac{\nu_i^2 \sin^2\left(\sum_{m=1}^{4} \nu_m^2\right)^{1/2}}{\sum_{m=1}^{4} \nu_m^2} \quad (i = 1, 2, 3, 4)$$

$$\eta_r = \cos^2\left(\sum_{m=1}^{4} \nu_m^2\right)^{1/2} \tag{42}$$

where

$$\nu_m = \frac{\pi n_m d}{\lambda (c_{S_m} c_R)^{1/2}} \quad (m = 1, 3)$$

$$\nu_m = \frac{\pi n_m d}{\lambda (c_{S_m} c_R)^{1/2}} (\hat{s}_m \cdot \hat{r}) \quad (m = 2, 3)$$

and d is the thickness. Note that for a lossless medium such as a phase grating the sum of η_{S_i} and η_R is unity. Also if n_2, n_3, and n_4 are set to zero, the efficiencies of η_{S_2}, η_{S_3}, and η_{S_4} are zero, and the situation reduces to the single grating case with the familiar $\eta_{S_1} = \sin^2(\nu_1)$ relation.

Figure 14 shows the diffraction efficiency versus refractive index modulation for a multiplexed SMH using Eqs. (42). In this case, equal refractive index is assumed for each grating. The diffraction angle θ in the emulsion is 45° ($c_{S_m} = \cos 45°$), and the thickness $d = 15$ μm. The ratio of diffraction efficiencies for s- and p-grating sets is

$$\frac{\eta_{2,4}^{(P)}}{\eta_{1,3}^{(S)}} = \left(\frac{\nu_{2,4}}{\nu_{1,3}}\right)^2 = \left(\frac{\kappa'_{2,4}}{\kappa_{1,3}}\right)^2 = \frac{\eta_{2,4}}{\eta_{1,3}} \cos^2\theta = \cos^2\theta = 0.5 \tag{43}$$

This ratio is constant with respect to refractive index modulation; however, the p-polarized beam efficiency is about half that of the s-light.

One way to increase the efficiency of the p-polarized beam is to increase the refractive index modulation of the gratings that diffract into this polarization state so that

$$\frac{\eta_{2,4}^{(P)}}{\eta_{2,4}^{(S)}} = \left(\frac{n_2}{n_1}\right)^2 \cos^2\theta = 1 \tag{44}$$

In this case $n_{1,3}$ and $n_{2,4}$ must be related by

$$n_{1,3} = n_{2,4} \cos\theta \quad (n_3 = n_1, n_2 = n_4) \tag{45}$$

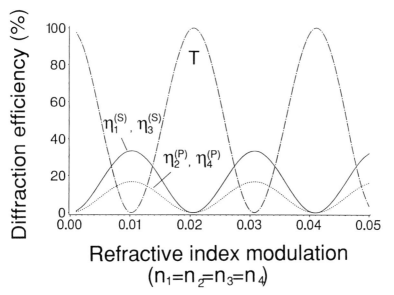

Figure 14 Diffraction efficiency versus refractive index modulation for four gratings with equal index modulation.

The resulting diffraction efficiencies for this condition are shown in Figure 15, and they provide equal efficiency for s- and p-polarized components.

This model is suitable for designing holographic elements in volume recording materials such as dichromated gelatin (DCG). This material has a large dynamic exposure range, which allows multiple gratings to be recorded sequentially without complete saturation of the medium. Sequentially recorded gratings do not produce intermodulation gratings and allow individual grating exposures to be optimized.

In order to illustrate these factors, an experimental multiplexed element was designed with four gratings formed at 488 nm and reconstructed at 632.8 nm with a normally incident beam (Fig. 13). The construction angle for the total internal reflected beam within the glass substrate ($n = 1.517$) is 47.5°, and 45.8° in the emulsion ($n = 1.56$). This forms a grating with a spatial frequency of approximately 1765 lines/mm. In order to form large slant angles, the construction beams illuminate the emulsion through a prism (Fig. 16). Multiple reflections at the emulsion–prism interface are suppressed by using an index matching oil to fill the gap at the interface. The orthogonal grating orientation is achieved by rotating the grating by 90° between subsequent exposures. The incident angles of the two construction beams are optimized to satisfy the Bragg condition when reconstructed

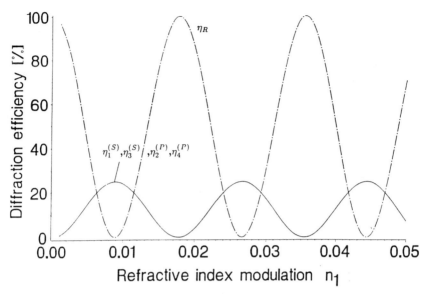

Figure 15 Diffraction efficiency versus index modulation with individually optimized index modulation for each grating.

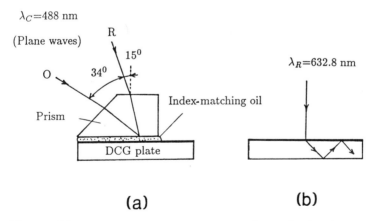

Figure 16 Construction method for experimentally forming a multiplexed substrate-mode quadrant beam divider in DCG at 488 nm (a) with reconstruction at 632.8 nm (b). A prism is used to couple into the emulsion and substrate at large angles of incidence.

with a common, normally incident beam of 632.8 nm light, and the beam intensities are adjusted for a 1:1 ratio. The DCG emulsions were prepared by extracting silver halide crystals from Kodak 649F photographic plates, and then they were sensitized and processed according to the procedure outlined by Georgekutty and Liu [23].

In order to check the DCG recording process, single exposure gratings are formed using the substrate-mode geometry. Typical single grating diffraction efficiencies of 94% were obtained with sensitized Kodak 649F plates for s- and p-polarized beams with exposures of 45 and 150 mJ/cm^2, respectively. The diffraction efficiency is defined as the ratio of diffracted light coupled out of the substrate with a prism relative to the incident light intensity after adjusting for boundary reflection losses.

After optimizing the exposure conditions for single gratings, the exposure conditions for multiple grating recording were optimized. From the analysis it was determined that the rates of change in diffraction efficiency with index modulation for s- and p-polarized light are not the same. In addition, it has been found that the order in which the exposure is made will affect the index modulation and diffraction efficiency of the resulting grating. Higher exposure values must be used to record gratings recorded later in the sequence to achieve the same index modulation. These results are combined in the plot shown in Figure 17, which shows the efficiency versus exposure with different exposure ratios ($E_2/E_1 = E_4/E_3$). From this curve, if $E_2/E_1 = E_4/E_3 = 1.8$ the four gratings will have equal diffraction efficiency, and the total efficiency is approximately 80%. In one experimental evaluation, individual efficiencies of 24.2, 21.2, 18.6, and 17.0% were obtained for $\eta_1^{(S)}$, $\eta_3^{(S)}$, $\eta_2^{(P)}$, $\eta_4^{(P)}$ with a total efficiency of 81%. Figure 18 shows the fields reconstructed from this grating when illuminated with a normally incident beam at 632.8 nm. Scattering from the glass substrate was enhanced to show the totally internally reflected beam more clearly.

V. POLARIZATION-SELECTIVE ELEMENTS

Several parameters affect the sensitivity of volume holograms to the polarization state of the reconstruction beam. These include the grating period, the amplitude of the index modulation, and the angle between the reconstruction and diffracted beams. These relationships can be described in general using rigorous coupled-wave theory [10]. If the reconstruction and diffracted beams are restricted to either s- or p-polarization states, however, the fields are uncoupled, and simpler two-wave theory provides a suitable description of the diffraction process [21,24]. This discussion is restricted to the simpler uncoupled condition.

OPTICAL ELEMENTS AND INTERCONNECTS

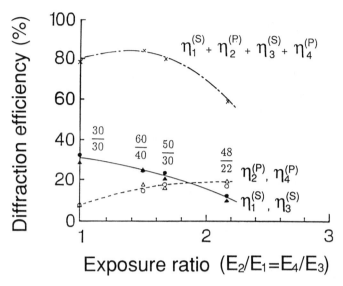

Figure 17 Diffraction efficiency versus exposure ratio for different pairs of gratings in the multiplexed quadrant beam divider.

For this approximation it is assumed that the grating vector lies within the x–z plane and the polarization is either perpendicular (s) or parallel (p) to this plane (Fig. 19). For the case of a sinusoidal phase grating reconstructed at the Bragg angle, the coupled-wave equations describing the amplitudes of the diffracted $S(z)$ and reconstruction $R(z)$ fields are

$$c_R R' = -j\kappa_{s,p} S$$
$$c_S S' = -j\kappa_{s,p} R \tag{46}$$

where c_R and c_S are the components of the reconstruction and diffracted propagation vectors along the z axis. When the polarization of the reconstruction field is perpendicular to the plane of incidence (s-polarized) the coupling coefficient is

$$\kappa_s = \frac{\pi n_1}{\lambda} \tag{47}$$

and for a p-polarized reconstruction beam,

$$\kappa_p = -\kappa_s \cos 2(\theta_0 - \phi) \tag{48}$$

where n_1 is the amplitude of the refractive index modulation, θ_0 is the angle of the reconstruction beam within the grating, and ϕ is the grating slant

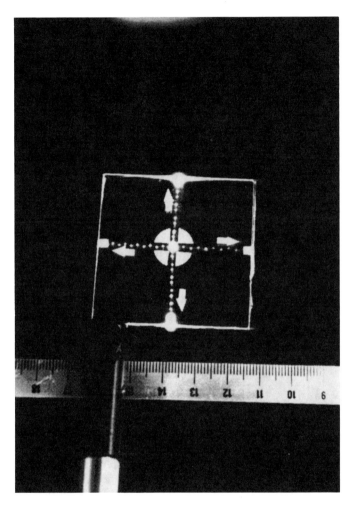

Figure 18 Photograph of the experimental multiplexed substrate-mode quadrant beam divider. Scattering is enhanced by placing oil on the surface of the glass in order to show the guided beam.

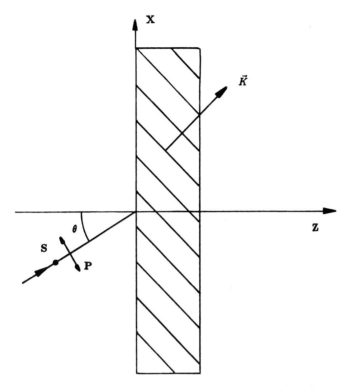

Figure 19 Geometry for a volume grating showing the s- and p-polarization directions.

angle. The diffraction efficiency can then be determined by solving these expressions for S and R at the boundary of interest and applying the relation

$$\eta_{s,p} = \frac{|c_S|}{c_R} SS^*_{s,p} \tag{49}$$

One requirement for a substrate-mode hologram is to diffract an incident beam at an angle that exceeds the total internal reflection angle of the substrate supporting the recording material. Consider the input coupler illustrated in Fig. 20, with the reconstruction beam illuminating the grating at normal incidence, an average emulsion thickness of 11 μm and refractive index of 1.54, and a glass substrate with index 1.51. The beam is diffracted at 45° to the normal within the glass substrate. Figure 21 shows the resulting efficiencies for the s- and p-polarized diffracted fields as a function of

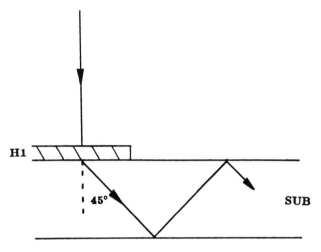

Figure 20 Substrate-mode input coupler that diffracts a beam at a 45° angle within the substrate.

increasing index modulation. The first maximum for the s-field occurs with $n_1 = 0.025$, while the efficiency for a p-polarized beam increases at a slower rate, with the first maximum occuring at $n_1 = 0.035$. Since these curves overlap, a condition for high and equal s- and p-efficiencies exists. In this case this condition occurs when the index modulation is near 0.028. An element formed with these construction parameters can serve as an input coupler that is not selective to the state of polarization of the reconstruction beam and diffracts different polarized beams in the same direction within the substrate.

Examination of the diffraction efficiency properties of substrate-mode elements shows that a polarization selective grating can also be formed. If the interbeam angle between the reconstruction and diffracted beams is 90°, κ_p and the p-polarized efficiency are zero. Therefore, the grating functions as a polarization beam splitter that is essentially transparent to p-polarized light and diffracts s-polarized light as determined from the grating parameters.

The diffraction efficiency versus index modulation relation plotted in Figure 21 shows a second way of making a polarization-selective element. In this case the interbeam angle is less than 90°; the different rate of change in efficiency with index modulation for s- and p-polarized light, however, displaces their maxima and minima values. Using a suitable grating design

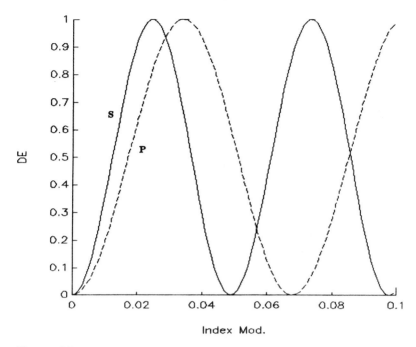

Figure 21 Diffraction efficiency versus index modulation for s- and p-polarized light diffracted by the input coupler of Fig. 20. Note that the rate of increase in diffraction efficiency for s- and p-polarized light is different.

and a recording material with sufficient dynamic range, high s- and low p-efficiency or the converse can be realized.

The nonselective substrate-mode input coupler can be combined with the selective grating to form a compact polarization sensing element. An experimental test of this device concept was realized by forming two separate gratings in DCG and then assembling and sealing them with a UV curable cement. The efficiency and transmittance of the two components as a function of polarization angle of the reconstruction beam is shown in Figure 22. At 0° (s-polarization) the efficiency is high and transmitance is low. This condition is reversed as the polarization angle is rotated to a p-polarized state. This illustrates that two substrate-mode grating elements with properly controlled grating parameters can function as a polarization sensing switch that reroutes the reconstruction beam in a different direction depending on its state of polarization.

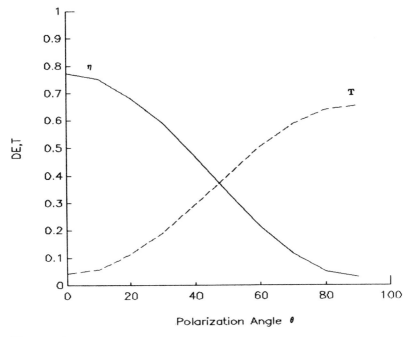

Figure 22 Diffraction efficiency and transmittance as a function of polarization angle of the reconstruction beam illuminating a non-polarization-selective input coupler and a polarization-selective element.

VI. APPLICATIONS OF SUBSTRATE-MODE ELEMENTS

It was previously mentioned that one advantage of substrate-mode elements is their planar format. This is important for interconnect applications, because it improves the packaging compatibility with planarized electronic components. Systems have been proposed that use substrate-mode or planar optics at the chip-to-chip [19], multichip module [25], and board level [26] of electronic packaging. The polarization-selective routing element has also been proposed for use in a reconfigurable interconnect system [27].

In order to illustrate how substrate-mode elements might be used, consider the problem of connecting devices on multichip modules that in turn are mounted on several boards. Signals are exchanged laterally between devices on individual modules and between modules, as well as longitudinally between boards. In synchronous signal operations such as clock distribution, the skew or delay between received signals must be minimized.

OPTICAL ELEMENTS AND INTERCONNECTS

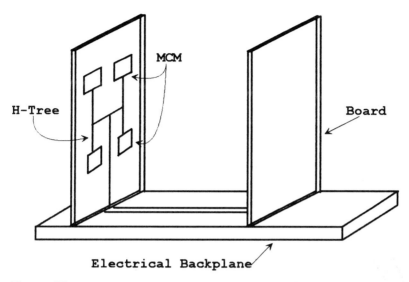

Figure 23 Electrical board–board connection through a backplane delivering a synchronous signal to four multichip modules on a board.

Some electrical interconnect solutions to these problems use equalized electrical lines in an H-tree form (Fig. 23) to distribute signals and an electrical backplane for longitudinal distribution. A similar approach using substrate-mode elements can also be realized with the configuration illustrated in Figure 24. Each circuit has an electrooptic interface unit consisting of surface emitting laser and receiver arrays. Substrate-mode gratings are fabricated that exchange data between module circuits, and to a set of gratings at the center that transfer data to adjacent boards. The multiplexed grating element described previously can be used as the central grating components and replace the electrical backplane. Relocating the backplane to the center of the board also helps to reduce timing skew.

Lohmann [28] and DeBiase [29] have suggested optical bus networks that use one or more polarization-selective elements. The systems they described used polarizing beam splitters and Wollaston and Rochon prisms, in conjunction with electrooptic half-wave plates. These components can be replaced with the substrate-mode device illustrated in Figure 25. It consists of an input coupler and polarization-selective element discussed in the previous section, a ferroelectric liquid crystal (FLC) at the input to control the state of polarization of the reconstruction beam, and an additional grating to couple light out of the substrate. The two separated beams would then serve as an input to two similar devices, which form the second stage of the bus. This arrange-

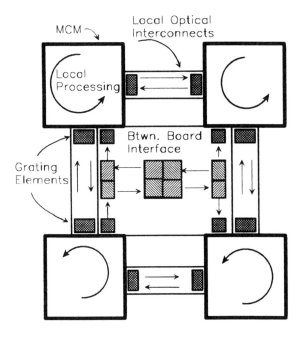

Figure 24 Optical technique for lateral distribution between multichip modules on a processing board and longitudinal distribution between boards.

ment expands in pyramid fashion. Once the path is established by setting the electrooptic half-wave plates, a high-data-rate signal can be transmitted through the network to a set of receiving nodes.

Another application for the polarization-selective device is as a sensor for magnetooptic data storage systems. A typical magnetooptical data storage read–write head assembly must detect a small rotation in the polarization of a beam reflected from the magnetooptic media. This is accomplished by passing the return signal through a "leaky beam splitter," which diffracts 100% of the s-polarized light and 10–30% of p-polarized light. This light is then passed through a polarization analyzer to detect the data. The use of conventional optical components results in a relatively large, bulky device that requires careful alignment of the individual components. The design of this device can be simplified using the arrangement shown in Figure 26. In this configuration the leaky beam splitter is formed as an

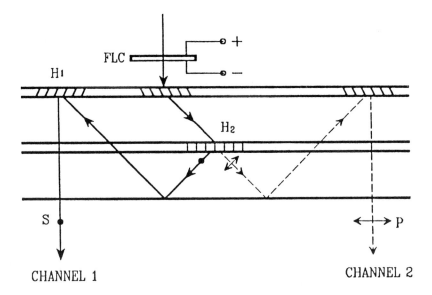

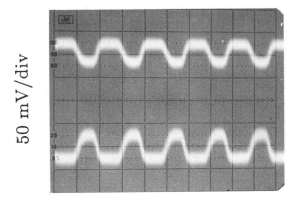

200 μs/div

Figure 25 Experimental demonstration of a polarization routing element consisting of a non-polarization-selective input coupler and a polarization-selective element. The state of polarization is changed by the ferroelectric liquid crystal (FLC). The complementary signals from the s- and p-channels monitored by two detectors are shown in the lower photograph.

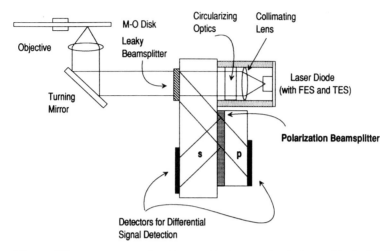

Figure 26 System for using the polarization sensing holographic device for monitoring magnetooptic data storage signals.

input coupler substrate-mode hologram, and a polarization-selective grating separates s- and p-polarization states. Since both elements are planar-type gratings, alignment is less critical, and they can be cemented together to form a compact device.

VII. CONCLUDING REMARKS

Substrate-mode diffractive elements can be used to form compact, free-space optical systems. These elements separate the tasks of beam translation and imaging, which allows interconnection over much greater distances than can be obtained with a single imaging element. The components are also configured on a planar substrate, which is more compatible with planar electronic substrates; techniques for packaging, however, are still in an early stage of development. These gratings can be multiplexed and used for transmitting information laterally and longitudinally in a processing system. The large interbeam angles that can be realized in substrate-guided conditions can be used to enhance the polarization properties of the grating. This feature combined with proper control of the grating parameters can be used to form elements that are either polarization selective or not selective. This flexibility can be used to design polarization-controlled reconfigurable interconnects or sensor systems.

Several aspects of this technology require further development. First, better techniques of grating fabrication must be developed. These should include improved volume hologram fabrication and investigation of hybrid-type lithographic and volume hologram components. This would make many different types of components more readily available and improve fabrication repeatability. Techniques for packaging planar optics with electronic systems must also be examined in order to make optics easier to incorporate into electronic system designs. Finally, improved electrooptic interface component performance is necessary in order to make optical interconnects more competitive with electronic counterparts. These investigations should increase the application of the substrate-mode optical element concept to many different types of practical electrooptic systems.

REFERENCES

1. J. W. Goodman, F. I. Leonberger, S. Y. Kung, and R. A. Athale, Optical interconnection for VLSI systems, *Proc. IEEE, 72*:850 (1984).
2. P. B. Berra, A. Ghafoor, M. Guizani, S. J. Marcinkowski, and P. A. Mitkas, Optics and supercomputing, *Proc. IEEE, 77*:1797 (1989).
3. N. Streibl, K.-H. Brenner, A. Huang, J. Jahns, J. Jewell, A. W. Lohmann, D. A. B. Miller, M. Murdocca, M. E. Prise, and T. Sizer, Digital optics, *Proc. IEEE, 77*:1954 (1989).
4. H. W. Hollaway and R. A. Ferrante, Computer analysis of holographic systems by means of vector ray tracing, *Appl. Opt., 20*:2081 (1981).
5. W.T. Welford, A vector raytracing equation for hologram lenses of arbitrary shape, *Opt. Comm., 14*:322 (1975).
6. G. J. Swanson, *Binary Optics Technology: The Theory and Design of Multi-Level Diffractive Optical Elements*, MIT Lincoln Laboratory Tech. Report 854, 14, Aug. 1989.
7. J. W. Goodman, *Introduction to Fourier Optics,* McGraw-Hill, San Francisco, CA, 1968, Chapter 2, Appendix A.
8. T. Shiono, M. Kitagawa, K. Setsune, and T. Mitsuyu, Reflection micro-Fresnel lenses and their use in an integrated focus sensor, *Appl. Opt., 28*:3434 (1989).
9. T. K. Gaylord and M. G. Moharam, Analysis and applications of optical diffraction gratings, *IEEE Proc., 73*:894 (1985).
10. M. G. Moharam and T. K. Gaylord, Three-dimensional vector coupled-wave analysis of planar-grating diffraction, *J. Opt. Soc. Am., 73*:1105 (1983).
11. M. G. Moharam and T. K. Gaylord, Diffraction analysis of dielectric surface relief gratings, *J. Opt. Soc. Am., 72*:1385 (1982).
12. A. Von Lehman, T.C. Banwell, R. Cordell, C. Chang-Hasnain, J. Harbison, and L. Florez, High speed operation of hybrid CMOS-VCSE laser array, *Electron. Lett., 27*:1189 (1991).
13. J. Shamir, H. J. Caufield, and R. B. Johnson, Massive holographic interconnection networks and their limitations, *Appl. Opt., 28*:311 (1989).

14. T. Sakano, K. Noguchi, and T. Matsumoto, Optical limits for spatial interconnection networks using 2-D optical array devices, *Appl. Opt., 29:*1094 (1990).
15. R. K. Kostuk, J. W. Goodman, and L. Hesselink, Design considerations for holographic optical interconnects, *Appl. Opt., 26:*3947 (1987).
16. R. K. Kostuk, Simulation of board-level free-space optical interconnects for electronic processing, *Appl. Opt., 31:*2438 (1992).
17. D. J. De Bitetto, White-light viewing of surface holograms by simple dispersion compensation, *Appl. Phys. Lett., 9:*417 (1966).
18. M. Kato, S. Maeda, F. Yamagishi, H. Ikeda, and T. Inagaki, Wavelength independent grating lens system, *Appl. Opt., 28:*682 (1989).
19. J. Jahns, Y. H. Lee, C. A. Burrus, and J. L. Jewell, Optical interconnects using top-surface-emitting microlasers and planar optics, *Appl. Opt., 31:*592 (1992).
20. M. Kato, Y.-T. Huang, and R. K. Kostuk, Multiplexed substrate-mode holograms, *J. Opt. Soc. Am., 7:*1441 (1990).
21. H. Kogelnik, Coupled wave theory for thick hologram gratings, *Bell Syst. Techn. J., 48:*2909 (1969).
22. S. K. Case,. Coupled wave theory for multiply exposed thick holographic gratings, *J. Opt. Soc. Am., 65:*730 (1975).
23. T. G. Georgekutty and H. K. Liu, Simplified dichromated gelatin hologram recording process, *Appl. Opt., 26:*372 (1987).
24. R. R. A. Syms, Vector effects in holographic optical elements, *Optica Acta, 32:*1413 (1985).
25. R. K. Kostuk, Y.-T. Huang, and M. Kato, Multiprocessor optical bus, *SPIE Proceedings, 1389:*515 (1990).
26. C. Sebillotte, Holographic optical backplane for board interconnections, *SPIE Proceedings, 1389:*600 (1990).
27. R. K. Kostuk, M. Kato, and Y.-T. Huang, Polarization properties of substrate mode holographic interconnects, *Appl. Opt., 29:*3848 (1990).
28. A. Lohmann, Optical bus network, *Optik, 74:*30 (1986).
29. G. A. DeBiase, Optical multistage interconnection networks for large-scale multiprocessor systems, *Appl. Opt., 27:*2017 (1988).

8
Optical Interconnects in Japan

Takakiyo Nakagami
Fujitsu Laboratories Ltd., Kawasaki, Japan
Satoshi Ishihara
Electrotechnical Laboratory, Tsukuba Science City, Japan

I. INTRODUCTION

The ability to transmit huge amounts of information from one point to another in a short period of time is the fundamental and essential feature of optics. Optical interconnects are expected not only to solve the present data transfer bottleneck in high-speed computers but also to provide a key technology in the creation of a new architecture for the next generation of information processing systems [1]. In Japan, much research and development on optical interconnects is underway, focusing on interprocessor optical communications, short-distance data transfer at the board-to-board, chip-to-chip, and gate-to-gate levels, and the dense parallel interconnects between processing units. In this chapter, we review recent research activities in optical interconnects in Japan.

Before describing the individual technologies, we provide a general discussion in Section II on the features and categories of optical interconnects. In Section III, optical interconnects research for short-distance applications such as those between circuit boards and large scale integrated (LSI) circuit chips are reviewed. Sections IV and V deal with such novel systems as reconfigurable multichannel interconnects and optical computing systems, respectively. In Section VI, parallel optical fiber links are introduced as an example of near-term applications. Finally, research on optical devices essential to optical interconnect implementation are introduced in Section VII.

II. GENERAL DISCUSSION

A. Optical Versus Electrical Interconnects

A comparison of optical and electrical interconnects and the various technologies available are shown in Figure 1 [1,2]. Important features of optical interconnects include high-speed and long-distance data transfer capability, dense parallel transmission without conflict, large fan-out broadcasting, and reconfigurability of network arrangement. In addition, the problems associated with electrical lines such as signal delays due to parasitic reactances, pulse reflection due to terminal impedance mismatching, and difference in ground potential between equipment can be solved by optical means.

B. Categories of Optical Interconnects

Figure 2 shows the categories and effective ranges of optical interconnects indicated in a three-dimensional space with respect to distance, flexibility, and dimension of data multiplexing [2]. The shaded area indicates the range where optical interconnects can play a major role. Some of the key devices reported recently are also indicated in the figure.

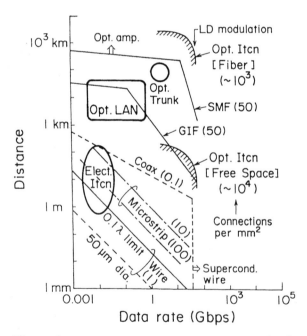

Figure 1 Optical versus electrical interconnects [1,2].

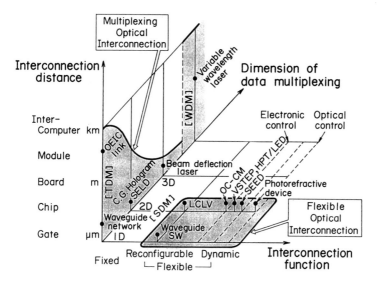

Figure 2 Categories of optical interconnects [2].

The distances of optical interconnects range from intercomputer communication at distances on the order of kilometers to gate-level distances on an LSI chip of a few micrometers. Since the signal delay time increases in proportion to the transmission distance, a reduction of the interconnection distance is essential to increase system speed. The break-even distance beyond which the optical interconnect has advantages over the electrical interconnect is largely dependent on the performance and cost of the transmitter and receiver. Owing to the recent progress in optoelectronic technology, this distance has been reduced to a few tens of micrometers. For intercomputer and board-level interconnects, optical fiber links are becoming increasingly popular. Parallel fiber links using array transmitters and receivers comprising such elements as laser and detector arrays and optoelectronic integrated circuits (OEICs) are of increasing importance in terms of performance, compactness, and reliability. For short-distance interconnects, that is, from board to gate levels, waveguides are often used because accurate optical coupling and alignment between waveguides and elements becomes critical.

Three parameters, space, time, and wavelength, are possible for data multiplexing in multichannel interconnects. The axis of flexibility includes point-to-point fixed interconnects and flexible interconnects in which the connection pattern is reconfigured dynamically depending on the nature of the data transmitted. Spatial optical devices such as surface emitting laser arrays and liquid crystal light valves (LCLVs) play important roles in space-multiplexed flexible interconnects. The research on new architectures is seriously dependent on the development of such new devices.

III. OPTICAL INTERCONNECTS FOR CIRCUIT BOARDS AND LSI CHIPS

The recent advance in semiconductor technology has increased the scale and speed of LSIs enormously, enabling computer engineers to design a compact high-performance computer with a parallel architecture. As a large number of high-speed gates are densely packed in circuit boards and LSI chips, however, data communication channels between boards and chips have become a serious bottleneck that limits performance and scaling of such a computer. Because the break-even distance of optical interconnects has been reduced to several tens of micrometers recently, there is a great possibility of solving the problem by optical means. Therefore, the application of optical technology to I/O channels of circuit boards and interchip wirings is a crucial issue in the development of future high-performance computers.

Two ways of optical implementation, one using free-space optics and the other using optical waveguides, are possible for short-distance applications in board-level and chip-level interconnects. Figure 3 shows a conceptual illustration of an optically interconnected massively parallel computer using both schemes [3]. The free-space optical interconnects have several advantages: a large number of optical paths can be densely packed into a small space; broadcast data transmission with a large fan-out is easily realized; and fast path reconfiguration is possible. On the other hand, optical waveguide interconnects have such features as optical waves being confined in a guiding medium such as a glass plate so that precise control of optical path and secure coupling of optical beams to many optical elements can be achieved.

Various ideas have been proposed and demonstrated for such short-distance optical interconnect implementations. In this section, we review some typical examples.

A. Board-Level and Chip-Level Interconnects Using Free-Space Optics

The free-space optical bus for a multiprocessor system shown in Figure 4 has been proposed [4]. It consists of a cylindrical mirror and arrays of optical transmitters and receivers attached to the surrounding processor units. In this configuration, each beam from the transmitter is reflected by

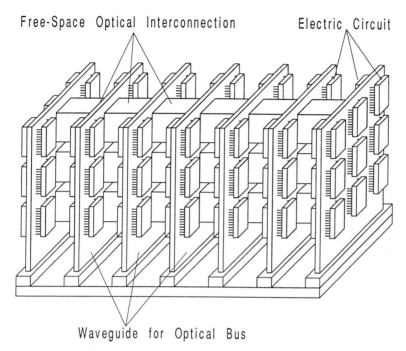

Figure 3 Optical interconnects for massively parallel processing [3].

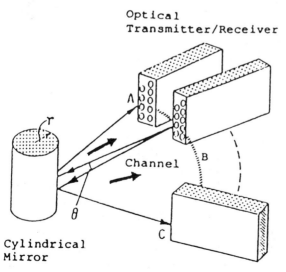

Figure 4 A configuration of an optical bus [4].

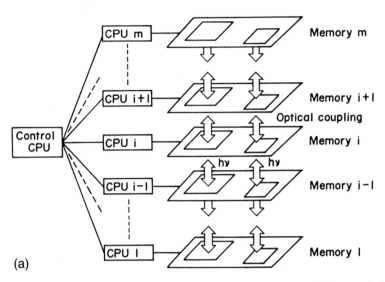

Figure 5 (a) Parallel processor system with the 3-D optically coupled common memory [6]. (b) Cross-sectional view of the memory.

the mirror and expanded to reach other receivers, forming many bus channels for global communication among processors. Skew in the data transmission can be made negligible by adjusting every interconnection path length among processors to be almost equal. In addition to broadcast communication, dedicated fixed local interconnects using holographic mirrors have been also reported [5], and a prototype bus system interconnecting processor units at a clock rate of 100 MHz has been successfully demonstrated.

As for the free-space interconnects between LSI chips, a concept of an optically coupled multiprocessor common memory system has been proposed (Fig. 5a) [6]. In this configuration, memory chips on stacked processor boards are optically interconnected to form a three-dimensional (3-D) common memory. The memory in each board layer is composed of two-dimensional (2-D) memories having light emitting diodes (LEDs) and photoconductors. Each memory is connected to its respective CPU in each layer and is optically coupled to the memories in the other board layers. Figure 5b is a cross section showing the structure of the layered 2-D memories. Since the data can be transferred through parallel optical paths from layer to layer without passing through lengthy electrical lines and backplanes, high-speed data communication is achieved without conflict. A prototype circuit for a $4k$-bit × 4-layer memory system was designed, and a data transfer speed of 32 Gb/s per layer was predicted by a computer simulation.

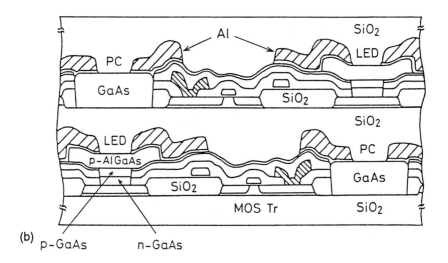

(b)

B. Chip-to-Chip Interconnects Using Integrated Optical Waveguides

Integrated optical waveguides are advantageous for chip-to-chip interconnects on the circuit board, where precise and secure connection between LSI chips is essential. Figure 6 shows an example of this scheme [7]. To enable inter-LSI chip data transmission, a silica waveguide is formed on a silicon substrate, and the chips are connected by a laser diode and a photodiode through a multiport optical mixer. To minimize the space for the waveguide, a bidirectional transmission is done by using mirrors at waveguide ports and at the end of the coupler. The mirror at the waveguide port directs the received signal to the detector. In an experiment using a four-port reflective coupler, the measured average loss was 13.7 dB including the theoretical distribution loss of 6 dB.

Optical surface mount technology (SMT) has also been proposed for chip-level interconnects [8]. The cross section in Figure 7 shows how the optical and electrical components are connected on the same circuit board. The board includes optical waveguides as well as electrical printed circuits. Optical elements can be precisely surface mounted with the aid of a guide-pin. The feature of this scheme is that it enables precise assembly and alignment of optical components while mounting them on the board. In this way, optical and electrical interconnection on the board can be done simultaneously, avoiding complicated mounting and wiring of optoelectronic elements. Prototype optical isolators assembled by this technology had an average loss of 2 dB with a variation of 0.16 dB.

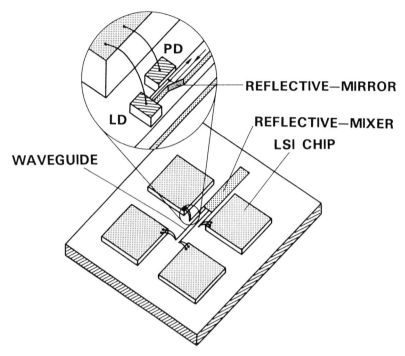

Figure 6 Guided-wave optical chip-to-chip interconnects [7].

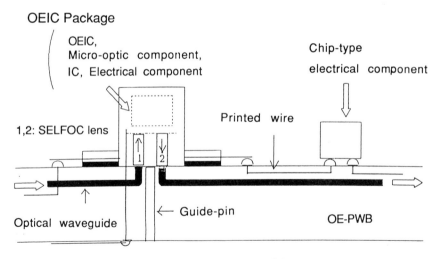

Figure 7 Structure of surface mount technology [8].

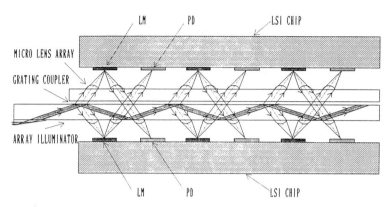

Figure 8 Concept of vertical optical interchip interconnects using an integrated optical array illuminator [9].

C. Interchip Interconnects Using Optical Waveguides

Highly accurate positioning and high optical beam packing densities are required for gate-level interchip connections. Optical waveguides fabricated by photolithography or other precise patterning methods provide quite suitable solutions.

Figure 8 shows a new idea using an integrated optical array illuminator [9]. Islands of grating couplers made by surface relief holograms are fabricated regularly at beam positions on the surface of the waveguide plate. The illuminator is sandwiched between two LSI chips. When the light is incident to the waveguide plate, the grating couplers generate upward and downward light beams that are then focused by the microlens array onto an array of reflection-type light modulators (LM) on the LSI chip. The beams are modulated by the LMs and reflected back toward the photodiodes on the other LSI chip across the array illuminator. By expanding the array illuminator for a two-dimensional configuration, a large number of optical paths can be established. Theoretically, more than 30,000 paths are achievable under practical design conditions at the 0.63-μm wavelength.

D. Optical Interconnects Using Microoptics

For highly accurate control of light beam paths, a two-dimensionally arranged microlens array is effective. For this, a concept of stacked planar optics, shown in Figure 9, has been proposed [10]. A variety of arrayed optical components such as an optical tap array, fiber coupler array, and focusing beam splitter array can be realized in an integrated form. In addition, the microlens array is also essential for the optical coupling sys-

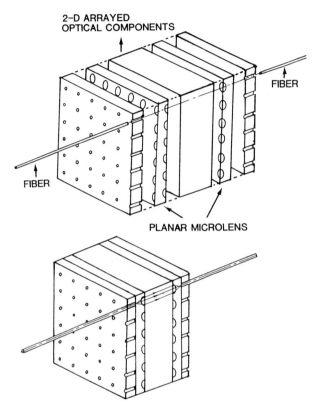

Figure 9 Concept of stacked planar optics [10].

tem for 1-D and 2-D arrayed light sources, and there has been remarkable progress made recently using this technology. Surface emitting laser arrays [11] are the most suitable elements for multibeam transmitters incorporating stacked planar optics in parallel free-space interconnects.

Using this technique, a new board-to-board optical interconnecting scheme has been proposed (Figs. 10a and b) [12] in which self-focusing glass (SELFOC®) rods with perpendicular gaps formed by slicing are aligned and fixed on a mother board. Gaps are determined so that conjugate image planes are formed at the same positions (I_1–I_5 in Fig. 10b) in all gaps. Placing a light source array at one end and a mirror at the other and inserting boards with transparent photodetectors and spatial light modulators (SLMs) in the gaps produces an optical bus that interconnects arbitrary boards. It is also conceivable that an optical neural network can be con-

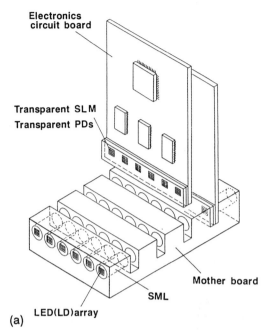

Figure 10 (a) Optical bus interconnection system [12]. (b) Cross section of the optical bus.

structed in this way by inserting functional elements such as a weighting mask and a transparent detector array with a thresholding function between the microlens arrays.

Random multiple reflection of optical beams within a glass plate can be applied to the optical bus. Figure 11a shows one such example in which a 2-D optical bus with microlenses consisting of small concave holes at the surface serve as optical beam inputs and outputs [3]. ICs mounted on the waveguide plate have light sources and detectors positioned near the lenses to enable them to communicate with each other. Light beams from the light source are incident on the waveguide from the concave lens, and those whose incident angles are greater than the critical angle can propagate in it. They then emerge from the waveguide at the lens and impinge on the photodetector. Figure 11b shows a microlens variation made by changing its refractive index and fabricating a wedge-shaped ditch at the surface of the waveguide. This structure is easily fabricated by photolithography and sufficiently accurate to be implemented with ICs. Theoretical loss calculations for 250 μm thick glass plates and 1×1-mm^2 detectors show that a

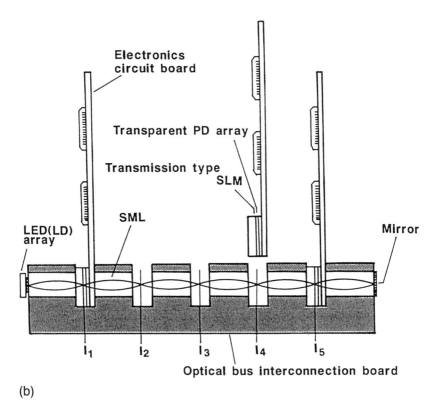

(b)

Figure 10 Continued

propagation distance of more than 100 mm with a loss of less than 30 dB can be achieved. The feasibility of this concept has been demonstrated using an experimental optical bus fabricated with a 5 mm thick glass plate having 2-mm diameter concave lenses.

IV. OPTICAL INTERCONNECTS FOR A RECONFIGURABLE NETWORK

A high degree of flexibility and reconfigurability in the interconnection networks is often required for parallel processor systems. It is also important to route information from one port to another in digital communication systems, such as asynchronous transfer mode (ATM) switching systems and cross-connect systems. A number of such reconfigurable optical interconnects have been proposed, for example, using a switching element or providing switching functionality in the network itself.

Figure 12 shows an addressable optical switch named LISA (lightwave interconnections employing spatial addressing) [13]. The LISA architec-

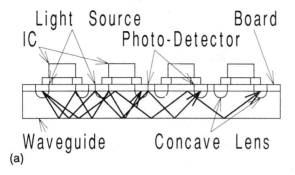
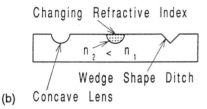

Figure 11 (a) Optical bus using waveguide [3]. (b) Waveguide variation.

ture is based on a combination of free-space optics and integrated electrical circuits in which flexible connection paths are arranged by the free-space optics, while switching operations are carried out electronically. As the figure shows, a 2 × 2 optical source array, four lenses, and four detector arrays perform 1 × 4 addressing. A 2-bit binary address is applied to the input processing section that drives the light source array, making an optical pattern of the input address. All elements in the detector array receive an optical signal only when the arrangement of the detectors agrees with the optical pattern transmitted from the optical source array. The output processing section that follows the detector array performs the AND operation for the detector output, and outputs "1" only when the assigned address pattern agrees with the transmitted one. Extending this basic principle, a large fan-out optical switch to interconnect one processor to many can be constructed. The principle of operation was confirmed experimentally using a 1 × 16 interconnect and a 3 × 3 laser diode array as the light source. The experimental system demonstrated a switching operation of 200 Mb/s data with a switch repetition rate of 25 MHz.

Figure 13 is another example of a free-space optical switch. In this configuration, $N \times N$ switching is done by controlling the optical beam deflection angle by phase-only holograms generated by a liquid crystal display (LCD) panel [14]. An 4 × 4 switching experiment was conducted using an

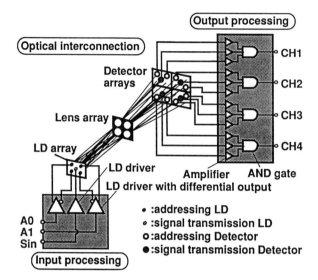

Figure 12 Schematic diagram of 1 × 4 LISA [13].

LCD panel having a 640 × 300 matrix of 0.33 mm × 0.33 mm dots. The voltage across each cell of the LCD is controlled by a computer to reconfigure the hologram pattern, and the outgoing direction of the incident optical beam is changed. Successful switching was achieved by inputting four 633-

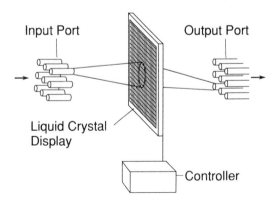

Figure 13 Free-space optical switch using a phase-only hologram generated by an LCD [14].

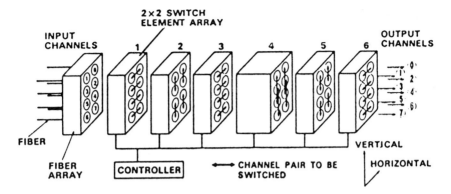

Figure 14 8 × 8 reconfigurable interconnection network using polarization switch arrays [15].

nm wavelength beams having a beam diameter of 20 mm. The optical loss between input and output ports was 13 dB, and crosstalk was 12 dB.

Polarization-controlled free-space optical switches in which beam paths are changed by controlling the polarization states of beams incident on a birefringent optical system have such features that a large number of channels can be input, and they can be constructed in an integrated structure. Figure 14 shows an 8 × 8 switch that consists of six cascaded stages, each having an array of four 2 × 2 element switches [15]. In this configuration, an 8 × 8 bitonic sorting network is established in which element switches shift the beam positions of two channel beams according to a bypass or exchange configuration determined by a bitonic sorting argorithm. The element switch structures are shown in Figure 15. The structure in Figure 15a shifts beams equal to the channel interval, and that in Figure 15b shifts them twice at the channel interval used only in the fourth stage. An optical switch experiment conducted at the He–Ne wavelength using liquid crystal polarization controller arrays having 330 × 330-mm^2 pixels demonstrated successful optical beam switching with an insertion loss of 22 dB and crosstalk of about −9 dB.

Figure 16 shows another polarization-controlled optical switch having an 8 × 8 Banyan network configuration [16]. Figure 17 shows the element switch configuration and its principle of operation. When a single-polarized beam (p-polarized in the figure) is incident on this switch, its polarization is rotated by 90° or left unchanged according to the applied voltage across the polarization control device (PCD), and the position of the output beam is shifted as shown in the figure. This optical system therefore operates as a 2 × 2 element switch having a bypass and exchange mode switch configuration. A multichannel optical switch can be constructed by cascading and multiplexing these element switches. The arrangement shown in Figure 16

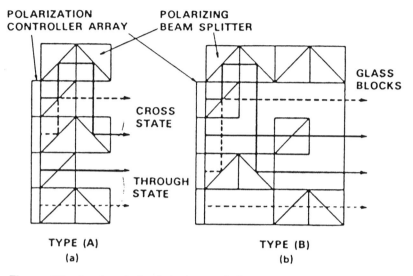

Figure 15 2 × 2 optical switch element [15].

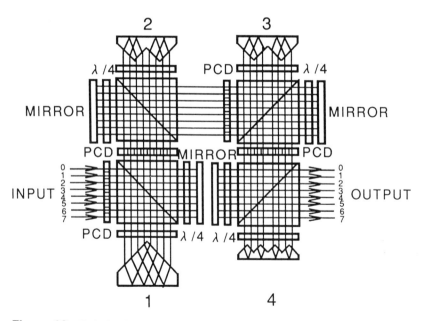

Figure 16 Polarization-controlled 8 × 8 optical switch [16].

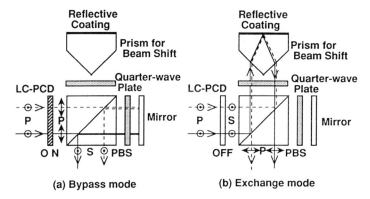

Figure 17 2 × 2 element switch configuration and operation principle [16].

is designed so that nonblocking characteristics can be achieved by inserting a redundant third section. An experimental 4 × 4 switch has been developed by using LCLVs for the PCDs and switching operation for four channel inputs has been successfully demonstrated.

In these three examples, the possibility of realizing a large-scale optical switch using liquid crystal devices have been demonstrated. As a large array scale of 2-D integrated spatial devices can be easily achieved by liquid crystal (LC) technology, it will be possible to develop optical switches having a large number of channels. By applying an advanced LCD technology including hybrid integration with thin film transistor (TFT) circuits to spatial devices for the switch, such functions as address decoding, data routing, and data storing will be realized in the future.

Two-dimensionally arrayed optoelectronic devices have advantages of realizing various processing functions. Using a 2-D optoelectronic memory array, a self-addressing crossbar switch as shown in Figure 18 has been developed [17]. In this switch, parallel streams of electrical signal containing reset, address, and data bits are input to the row of the optical memory array, and the timing address data is input to the column array simultaneously. The optical memory used in the system is a 2-D bistable optical array called a VSTEP (vertical to surface transmission electrophotonic device) [18]. As explained in Section VII.C, each element of the VSTEP can be addressed individually and switched to emit light according to the voltages applied between its cathode and anode. Parallel streams of negative and positive addressing pulses are applied simultaneously to the cathode array and anode array, and only the elements to which voltages are applied to both the cathode and anode are switched on and addressed. Input signals are then applied to the anode array, and only the addressed

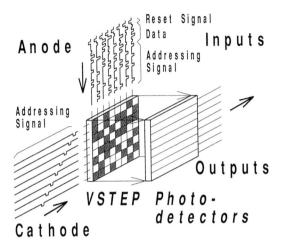

Figure 18 Optical crossbar interconnection switch using VSTEP [17].

elements emit light. The light from the VSTEP element is incident to the corresponding element of the photodetector array, achieving the crossbar connection. With large-scale integrated VSTEP devices, we can expect to interconnect several tens or more input and output channels.

V. OPTICAL INTERCONNECTS FOR OPTICAL COMPUTING AND INFORMATION PROCESSING

The concept of optical interconnects is essential in optical computing and information processing. It is the trunk of the "tree" of optical information processing technology (Fig. 19) and supports all processing techniques [19].

There are two major roles of the optical interconnects in information processing and computing. One is its near-term application, as described in Section VI, to high-speed transfer of information in electronic computers to cope with the need for increasing information processing power in electronic computers. This is because the factors that limit computing speed are not only the switching speed of logic and memory devices but also the data transfer rate between them. Optics is preferable to electronics for interconnection lengths longer than the critical length, l_c, and recent research has shown that l_c is a few tens of micrometers in terms of energy dissipation [20]. Optical interconnects look promising, mainly because optical detectors, sources, and modulators are effective impedance transformers [21]. In the future, optical interconnection networks will be promising

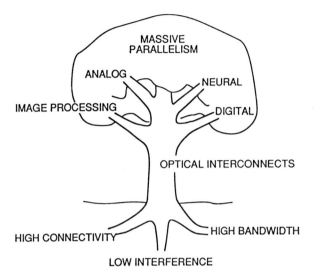

Figure 19 Tree of optical information processing systems [19].

processors in a parallel computer having 10^6 processors and a computing performance on the order of 10^{15} flops (floating operations per second). Optical data transmission in the GByte/s range has already been attained. One of the most challenging problems confronting high-speed interconnects will be not in optics, but in the interface between the optics and electronics, and research on optoelectronic interface devices (light sources, detectors, and OEICs) and fundamental research in materials will be very important.

The other role of the optical interconnects is to provide a fundamental technology for constructing almost all types of optical computing and information processing systems. One of the advantages of optics over electronics in general is its ability to transfer information at large bandwidths in free space with less crosstalk, thus enabling parallel processing as well as interconnection of spatially distributed information. A simple image projection with a single lens can be regarded as multichannel point-to-point interconnects between pixels in an object plane and an image plane, while a two-dimensional Fourier transform by a single lens can be regarded as interconnects from a single point to many. The anamorphic optical system, shown in Figure 20, used for a vector-matrix multiplier or crossbar switching is another typical example. At any point in the space between input/output planes in both cases, light beams from various

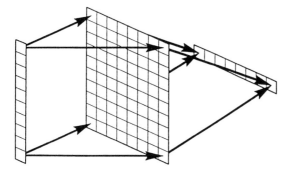

Figure 20 Spatial light modulator that can be used for optical interconnection or information processing [19].

points in the object plane intersect but do not affect each other. This interconnection characteristic can be used in such applications as optical analog/digital conversion and neural computing. Most optical computing and information processing systems take advantage of this characteristic, and optics will play a significant role in supercomputing.

To meet the potential demands for ultrahigh performance information processing systems in the coming information-based society, it will be necessary to develop massively parallel and massively distributed computers. The parallelism and interconnection ability of light makes it suitable for adoption in future computer systems. For next-generation massively parallel, massively distributed information processing systems [22] in which "distribution and coordination" or "parallelism and flexibility" will be important, any architecture will utilize the interconnection ability of optics [23]. Much effort should be directed at exploiting new architectures for such optical computing and information processing systems. Although the technology is not yet fully mature, a variety of pioneering research is ongoing in Japan, some of which are collected in [23].

The following discussion reviews some of the results of optical computing and information processing studies in Japan. Optical interconnects play important roles in all instances.

A. Optical Parallel Array Logic System (OPALS)

An optical parallel logic gate technique for two binary images with a shadow casting system has been developed [24]. In Figure 21 two binary discrete images are converted into a coded input image, which is set in the shadow casting system and illuminated by an array of point sources. Paral-

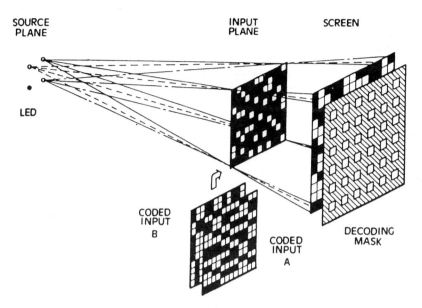

Figure 21 Optical logic array processor [24].

lel logic operation can be obtained through the decoding mask on the screen. Various kinds of operations are implemented merely by selecting the switching pattern of the point source array. Parallel digital image processing, parallel digital filtering, template matching, and processing of gray-level images have been discussed.

B. Cellular Logic Architecture for Optical Computers

Optical implementations of cellular logic computers (CLCs) for two-dimensional logical neighborhood functions were investigated with conventional space-invariant optical filtering techniques for cellular logic operations [25]. Its basic configuration is shown in Figure 22. For a 3 × 3 pixel matrix operation, a lookup table with 512 elements is required, and patterns corresponding to the 512 elements are searched by 512 different matched filters. To enable parallel processing, multiple imaging systems are used, and parallel matched filtering is performed using a microlens array. A binary look-up table is positioned behind the matched filter array. The function of the system can be changed by replacing the optical mask corresponding to the look-up table. The superimposed image of the look-up table and the output of the matched filter array is demultiplex-imaged on the output plane.

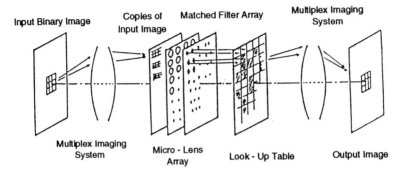

Figure 22 Optical cellular logic using a matched filer array [25].

C. Computing with Guided Wave Devices Controlled by Spatial Light

It is important to develop new architectural concepts suitable for optical information processing. The concept and applications of optical array processing with new optical guided wave devices, in which device functions are controlled by external "optical" signals, have been proposed [26]. Figure 23 shows its concept. The external light can take a spatially distributed form to control many arrayed devices simultaneously. The drawbacks of electrical control can be eliminated, and some additional advantages of optical computing, such as simplified processor architecture, can be obtained. For example, the permutation of the cyclic residue adder with a generalized crossbar switch can be dynamically programmed simply by changing the controlling spatial light pattern, as shown in Figure 24.

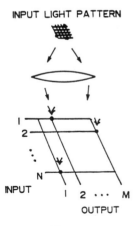

Figure 23 Optically controlled array processing [26].

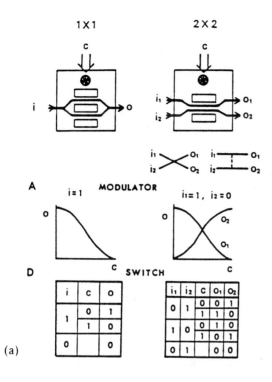

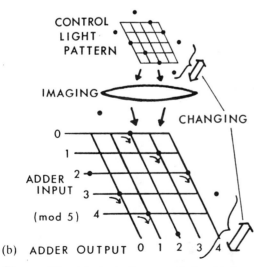

Figure 24 (a) Optically controlled guided wave device for switching [26]. (b) Residue adder (mod 5) using the devices shown in (a).

D. Optical Neural Networks

Recently, neural networks have received attention in optical computing. Optoelectronic networks are attractive because their parallel processing and a dense interconnection capability make it possible to implement a great number of neurons and the connections between them.

In Japan, a variety of approaches can be found. Among these are the following:

1. GaAs optical neurochips consisting of an LED array, an interconnection synaptic mask, and a photodiode array with a 3-D layered structure as shown in Figure 25. Operation of this chip is described in Section VII.C [27].

2. A new architecture for an optical associative memory with learning capability, the optical associatron [28]. In Figure 26 a recalling process of the associative memory is shown schematically. After the Hadamard product is optically executed between the input pattern and the memory matrix M calculated by the computer, its output is converted to an electric signal. It is then accumulated locally in sequence processed for a threshold operation by an image processor, and finally it is displayed by the LED monitor.

3. Investigation of the effects of optical device imperfection on the performance of the optoelectronic neural network [29]. The optical backpropagation learning operation was demonstrated under the condition that there is the nonuniformity of ±35% in synaptic connection coefficients and 20% in the contrast of neuron outputs. It was shown that optical devices

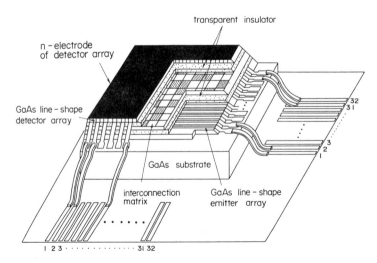

Figure 25 Optical neurochip [27].

OPTICAL INTERCONNECTS IN JAPAN 387

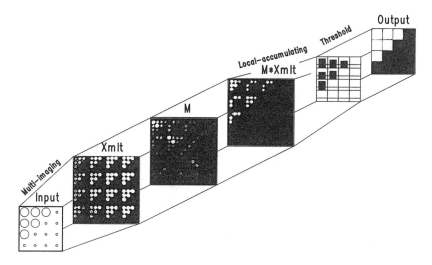

Figure 26 Optical associative memory [28].

that interconnect neurons can work effectively as a whole in an optoelectronic neural network system even if the performance of each component is imperfect and differs from the theoretical one.

VI. HIGH-SPEED OPTICAL FIBER LINKS

The use of optical fiber links is the most practical technology for near-term applications. Their large bandwidth, low crosstalk, and small cable diameter can effectively solve the present problem of data transfer bottleneck in computers and digital equipment. Especially important, multifiber parallel links have the potential to replace the current bulky electrical bus cables in high-speed I/O channels of large-scale computers and digital communication systems. As the propagation time difference in arrayed fibers is typically less than one-tenth that of electrical cables, data and clock skew problems in parallel electrical cables are solved by parallel fibers. When developing parallel links, it is important to minimize their size and cost so that conventional electrical connectors on the board can be replaced. Due to the serious consequences of data errors in computer systems, an extremely low bit error rate of several orders of magnitude less than that of conventional optical transmission systems is needed, requiring the introduction of special data coding and error correction circuits. To minimize the size and complexity of these circuits, the arraying of electrical circuits as well as optical elements is essential. An OEIC in which optical elements and electrical circuits are

Table 1 Typical Examples of High-Speed Optical Links for Parallel I/O Interface

Transmission (fiber count)	Total throughput	Fiber type and length	Wavelength	Light source	Detector	Transmitter and receiver circuit	Organization	Reference
Parallel (12)	1.8 Gb/s (150 Mb/s × 12)	62.5 μm GI 100 m	1.3 μm	Zn-doped InGaAsP LED array	InGaAs PIN-PD array	Transistor driver and Si receiver ICs	NEC	[33]
Parallel (8)	1.6 Gb/s (200 Mb/s × 8)	single-mode 100 m	1.3 μm	InGaAsP LD array	InGaAs PIN-PD array	Driver and receiver ICs	Hitachi	[35]
Parallel (6)	6 Gb/s (1 Gb/s × 6)	50 μm GI	0.83 μm	LD array	6-channel GaAs MSM-receiver array OEIC and GaAs driver array IC		NTT	[30]
Parallel (4)	8 Gb/s (2 Gb/s × 4)	50μmGI	1.3 μm	InGaAsP LD array	InGaAs PIN-PD array	Driver and receiver ICs	Toshiba	[34]

OPTICAL INTERCONNECTS IN JAPAN

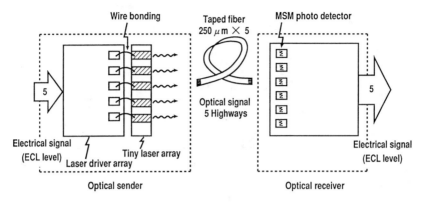

Figure 27 Schematic block diagram of a parallel optical link [30].

integrated on the same substrate is the ideal solution. Table 1 shows some typical parallel links recently reported [30–35].

As for the OEIC link, a five-highway optical link having a block diagram as shown in Figure 27 has been developed using a GaAs optoelectronic LSI chip set [30]. Figure 28 shows a microphotograph of the receiver chip. On this chip, six metal–semiconductor–metal (MSM) photodiodes, one clock receiver circuit, and five data receiver circuits are integrated. The transmitter is composed of a 0.85-μm laser diode array and a driver circuit array. A five-highway link experiment showed successful simultaneous operation at 1 Gb/s with a satisfactory signal-to-noise ratio and jitter of 50 ps. Another parallel-link OEIC, an 8-channel PIN/HEMT receiver, has been reported [31]. Eight PIN-PDs, 40 high-electron-mobility transistor (HEMT) elements, and 32 resistors are integrated on a semiinsulating InP substrate. An average bandwidth of 1.2 GHz for eight receivers, a sensitivity of −24 dBm at 1 Gb/s, and crosstalk of less than −30 dB were obtained. In the development of OEIC transmitters, the integration of laser diodes and driver circuits on a single chip is an important technical issue. A development effort of integrating four channels of laser–transmitter pair on a GaAs substrate has been reported [32].

For an increased number of channels, a 12-channel transmitter and receiver pair has been developed using a special LED array and a photodiode (PD) array at 1.3-μm wavelength [33]. Figure 29 shows the structure of the transmitter module. The LED array is composed of 12 high-speed InGaAsP LEDs, and the PD array of 12 InGaAs PIN-PDs. These devices are coupled to tape fibers and packaged together with the driver and receiver ICs. The developed link showed the ability of data transmission at 150 Mb/s per channel over 100 m of graded-index (GI) fiber with a loss

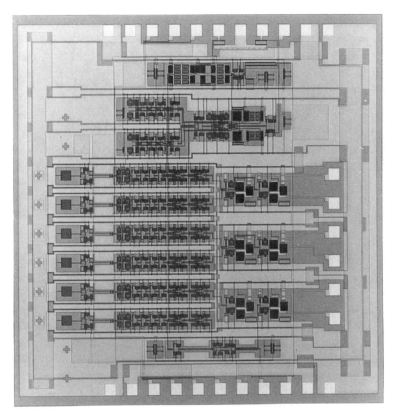

Figure 28 Microphotograph of the receiver array LSI [30].

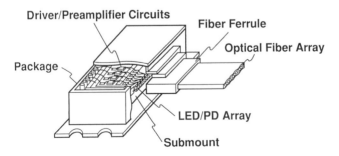

Figure 29 The structure of the array module [33].

margin of more than 7 dB and a maximum skew of 1.6 ns among all 12 channels. Another example is an 8-channel link developed for internal data transmission in an ATM switching system [35]. Its eight transmitters with 1.3-μm InGaAsP edge emitting LEDs and eight receivers with InGaAs PIN-PDs were packaged in an optical multiconnector housing. Using eight optical fiber pairs, 8-channel two-way transmission was demonstrated with each channel carrying 155 Mb/s data over 200 m of GI fiber.

VII. DEVICES FOR OPTICAL INTERCONNECTS

One of the most challenging problems in the development of optical interconnects is the development of devices. Since conventional devices for optical fiber communications such as discrete laser diodes and photodiodes do not fully satisfy the requirements unique to optical interconnects in terms of functionality and performance, a variety of devices with novel structures are being developed. Spatial light devices that generate, control, and receive optical signals in three-dimensional space have special importance for taking full advantage of the parallelism and flexibility of optics.

A. One-Dimensional Light Source and Detector Arrays

One-dimensionally arrayed light sources and detectors are essential to parallel links and many other optical interconnection systems. A monolithic integration of light sources solves the problems of highly accurate alignment of many beams in a limited space and reliability. The integration of PIN-FETs (field effect transistors) and PIN-preamplifiers as well as OEICs in an arrayed configuration have special importance (Section VI).

To obtain high power levels, narrow spectra, and sharply focused parallel light beams, an integrated laser diode array is important. Various fabrication methods have been applied to get highly uniform characteristics from integrated lasers. For example, an 8-channel array of 0.83-μm lasers has been developed by metal-organic vapor phase epitaxy (MOVPE) that has an output power level larger than 60 mW per channel, uniform characteristics with threshold currents from 59 to 62 mA, and slope efficiencies ranging from 0.87 to 0.95 W/A [36]. A similar 4-channel array of GaAlAs window diffusion stripe lasers has been fabricated by metal–organic chemical vapor deposition (MOCVD) that features an output power of more than 100 mW per channel and a low threshold current of 28 mA [37].

For an increased number of laser integrations, a very dense 102-laser array with a 4.5-μm separation has been developed for the 0.86-μm wavelength by MOCVD [38], and successful simultaneous operation of all 102 lasers at room temperature was reported. The total CW output power of the array was 850 mW at a total current of 2 A, and the slope efficiency was 87%. These laser arrays were originally developed for the high-power light source in optical disk memories, but it is conceivable that the technology is extended to optical interconnection applications.

A multiwavelength integrated laser array is suitable for a wavelength-division-multiplexing architecture for optical interconnection and switching. Since it is necessary to adjust each laser on the chip to a different lasing wavelength, fabrication becomes very complex in terms of structure and reproducibility. One example of such a laser array that has been developed integrates five distributed feedback (DFB) lasers having a 50 ± 5 Å wavelength separation at the 1.3-μm wavelength [39]. Its output power exceeds 6 mW/facet for each laser at room temperature. More recently, a 20-wavelength DFB laser array with the separation of 10 Å was developed at 1.3-μm wavelength [40]. X-ray lithography was used to achieve precise patterning of all different gratings of the individual lasers in one exposure. Each laser in the array yields a CW output power of more than 20 mW with adequate sidemode suppression and operates at over 400 MHz by intensity modulation.

B. Two-Dimensionally Integrated Laser Diode Arrays

Surface emitting laser diodes (SELDs) are advantageous for realizing light sources with 2-D arrayed sharp parallel beams. A flat SELD matrix makes an excellent compact multibeam transmitter for optical interconnects, photonic switching, and optical computing. Figure 30 shows an example of a SELD incorporating a AlGaAs/GaAs heterostructure and a dielectric multilayer mirror [11]. These SELDs are the result of continuous research since first proposed in 1979, and remarkable progress has been made in recent years. CW operation at room temperature with GaAs SELDs was reported recently for the first time. Recent devices with improved cavity design have shown submilliampere threshold currents and improved efficiency at both the 1.3- and 1.55-μm wavelengths, and integrated SELD arrays with various scales of integration have been reported [11,41].

C. Two-Dimensional Optoelectronic Functional Devices

A variety of 2-D functional optoelectronic devices based on quantum mechanical phenomena have recently been developed using III-V semiconductor materials [18,42–44].

The vertical-to-surface transmission electrophotonic (VSTEP) device having a double heterostructure configuration as shown in Figure 31 performs memory and switching operations [18]. The inner GaAs layers of the

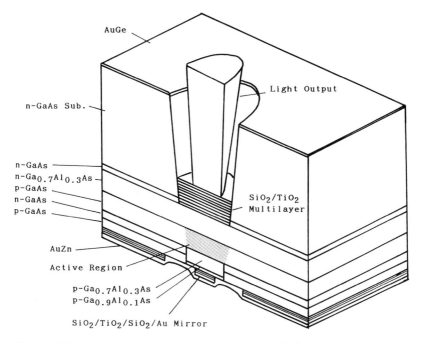

Figure 30 A surface emitting vertical cavity laser [11].

pnpn multilayer structure with dual gate electrodes exhibits a thyristor-like electronic nonlinearity that provides a latching function. With an applied bias voltage below the turn-on voltage, the device can be switched on by an optical input. It remains in the on state until switched off by a negative voltage pulse applied to its anode. It is structured so that its optical input and output transmit vertically to the device surface, thereby enabling integration in a 2-D matrix form. Low power consumption and fast switching operation have been reported for an integrated 32 × 32 array of VSTEP chips having a 1 mm × 1 mm size [42].

Another heterostructure 2-D optoelectronic device, a 32 × 32 array of optical memories consisting of an LED and a heterostructure phototransistor (HPT), has been fabricated with InGaAsP material [43]. The memory element structure is shown in Figure 32. Each element works as an optical gate switch in which the LED can be switched on and off with a small incident optical power and bias voltage across the device. A 5-ns light pulse is sufficient to turn the memory on, and the product of the switch-on power and pulse width is as small as 1.2 pJ. Successful write and read operation with a 1-Kb parallel optical signal pattern has been demonstrated.

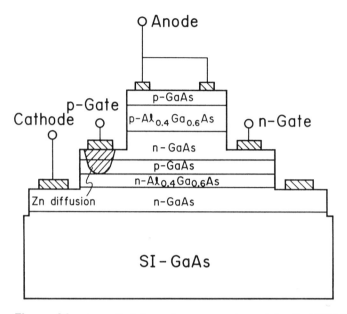

Figure 31 An optical dynamic memory element for the VSTEP [18].

In an attempt to develop the novel optical neural computer described in Section V, a GaAs-based optoelectronic neural chip having the structure shown in Figure 25 has been developed [27]. It has a hybrid structure consisting of an array of 32 stripe LEDs, a 32×32 interconnection matrix mask, and an array of 32 stripe PDs. In this chip, optical vector-matrix multiplication is performed in parallel between the LED and PD arrays with the matrix weighting parameters determined by the mask pattern. By connecting external electrical nonlinear thresholding elements and feedback circuits to this chip, an optical associative memory with a feedback-type neural network can be constructed. Since the chip has a fixed mask pattern, it has been difficult to perform a learning function by dynamically rearranging connections through a learning process. To solve this problem, an optical learning chip consisting of fast operating and variable sensitivity photodiodes to achieve dynamic interconnections in the chip has been developed [44].

D. Liquid Crystal Spatial Light Modulators

Spatial light modulators (SLMs), including liquid crystal light valves, microchannel spatial light modulators (MSLMs), and Pockels readout optical

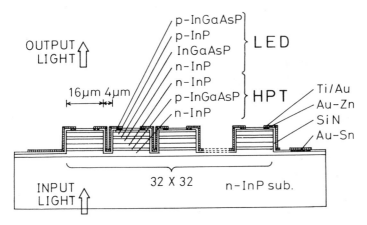

Figure 32 Cross section of a photonic parallel memory [43].

modulators (PROMs), are essential functional elements for the free-space flexible interconnects in optical computers and switching systems. SLMs using liquid crystal materials are receiving considerable attention recently for their high resolution, high sensitivity, low operating voltage, and logic and memory functions that cannot be realized with other materials [45–48]. Owing to the recent progress in LCD technology, a great deal of improvement in performance of liquid crystal SLMs have been achieved.

There are two major types of electrically addressable SLMs: those using electron beam addressing and those using addressing with an electrical circuit integrated with liquid crystal elements. As an example of the electron beam addressing type, a liquid crystal MSLM, in which a nematic liquid crystal material is used in the active layers, has been reported that achieved a minimum writing intensity of 6 mW/cm^2, contrast ratios from 30:1 to 100:1, and response times of 150 to 200 ms [45].

To improve the resolution, response time, and contrast ratio of LCLVs fabricated from conventional nematic crystal and CdS, a combination of ferroelectric materials and such photoconductors as BSO, Si, and GaAs have been studied. A bistable SLM using liquid crystal materials having a nonlinear output power versus input power characteristic is used for such image processing functions as memory, logic, thresholding, and amplification. Recently, an attempt has been made to use ferroelectric liquid crystal and an amorphous Si photoconductor to achieve fast response times [46]. Figure 33 shows the device structure and its write/read characteristics. An experimental device showed a response time of 70 ms, a contrast ratio of 60:1, a resolution of 60 lines/mm, and an image holding time of 2 h.

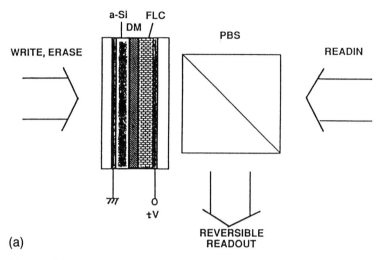

Figure 33 (a) Device structure of the bistable spatial light modulator (BSLM) [46]. (b) Reversible operation mode of BSLM.

A gray-scale controllable LCLV based on the gray-scale effect in the memory state of surface-stabilized ferroelectric liquid crystal has also been reported [47]. It has potential applications in analog optical computing such as optical neural computers having a graded interconnection.

Another optically addressable LCLV having a new structure consisted of a 120 × 120 LC active cell matrix with each cell driven by a TFT circuit and a photoconductor [48]. This device operates at 125 Hz and has a delay time of 740 ms and a 10–90% rise time of 290 ms.

VIII. CONCLUSION

Optical interconnect research in Japan was reviewed. The concept of optical interconnects covers a wide technical area, from simple optical fiber links, to dense parallel free-space communications, to flexible interconnects between 2-D image planes. The potential advantages of the optical interconnect are clear, and we can expect that novel information systems using this technology such as optical computing and information processing systems will be realized using this technology. Research is still in an early stage, however, and much more work remains to be done.

Due to space limitations, we were able to introduce only a few examples of current research activity in Japan. Since optical interconnect re-

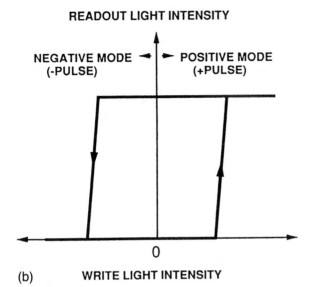

(b)

search is one of the major targets of Japanese researchers and engineers in the fields of optics, optoelectronics, and information processing, advancement in system architectures as well as devices can be expected in the near future.

We should emphasize that the development of spatial optical devices that can handle two-dimensionally arranged light is essential for making the system concept a reality. Such optical spatial devices include not only active devices such as 2-D surface emitting laser arrays, 2-D functional optoelectronic device arrays, spatial light modulators, and 2-D OEICs, but also passive devices such as microlens arrays, integrated waveguide devices, and computer-generated holograms.

ACKNOWLEDGMENTS

We are grateful to the authors of the papers cited in this chapter who kindly agreed to include their results. Special thanks are also due to the many members of the Research Committees of the New Information Processing Technology, Optoelectronic Division of the Electrotechnical Laboratory of MITI, and the Optical Interconnect Division of Fujitsu Laboratories for valuable information and discussions on optical interconnects and related subjects.

REFERENCES

1. S. Ishihara and O. Wada, Optical interconnects and optical computing, *Trans. IEICE,* J75-C-I:235(1992).
2. O. Wada, T. Kamijoh, and M. Nakamura, Integrated optoelectronics for optical interconnection, *IEEE Circuits and Devices Magazine,* Nov. 1992:37.
3. S. Kawai and M. Mizoguchi, Two-dimensional optical buses for massively parallel processing, 1991 Topical Meeting on Optical Computing, Salt Lake City, Utah, paper ME20.
4. H. Tajima, M. Suzuki, Y. Hamazaki, N. Sanechika, Y. Okada, and K. Tamura, A highspeed optical bus and its design, *Trans. IEICE, J70-D:*1718 (1987).
5. M. Suzuki, H. Tajima, Y. Hamazaki, Y. Okada, and K. Tamura, Experiments on an optical bus using holography, IEICE Technical Report OQE85-175, Tokyo, Japan, 1986, pp. 31–35.
6. M. Koyanagi, H. Takata, H. Mori, and J. Iba, Design of 4-kbit × 4-layer optically coupled three-dimensional common memory for parallel processor system, *IEEE J. Solid State Circuits 25:*109 (1990).
7. M. Kobayashi, M. Yamada, Y. Yamada, A. Himeno, and H. Terui, Guided-wave optical chip-to-chip interconnections, *Electron. Lett., 23:*143 (1987).
8. T. Uchida and Y. Masuda, Optical surface mount technology, Proc. of the 1990 IEICE Fall Conf., Hiroshima, Japan, 1990, paper C-189.
9. M. Takeda and T. Kubota, Integrated optic array illuminator: A design for efficient and uniform power distribution, *Appl. Opt., 30:*1090 (1991).
10. K. Iga and S. Misawa, Distributed-index planar microlens and stacked planar optics: A review of progress, *Appl. Opt., 25:*3388 (1986).
11. F. Koyama, S. Kinoshita, and K. Iga, Room-temperature continuous wave lasing characteristics of a GaAs vertical cavity surface-emitting laser, *Appl. Phys. Lett., 55:*221 (1989).
12. K. Hamanaka, Optical bus interconnection system by using SELFOC lenses and planar microlens arrays, 1991 Topical Meeting on Optical Computing, Salt Lake City, Utah, paper MB4.
13. T. Sakano, K. Noguchi, and T. Matsumoto, Lightwave interconnection employing spatial addressing: LISA, 1991 Topical Meeting on Optical Computing, Salt Lake City, Utah, paper ThD1.
14. H. Yamazaki and M. Yamaguchi, 4 × 4 free-space optical switching using real-time binary phase-only holograms generated by a liquid-crystal display, *Opt. Lett.,* 16:1415 (1991).
15. K. Hogari, K. Noguchi, and T. Matsumoto, Two-dimensional multichannel optical switch employing polarization control technique, *Trans. IEICE, E73:* 1863 (1990).
16. T. Nakagami, T. Yamamoto, and H. N. Itoh, Multistage reconfigurable optical interconnection network using polarization switch arrays, *OSA Proc. on Photonic Switching,* 8:67 (1991).
17. S. Kawai, Y. Tashiro, I. Ogura, K. Yamada, and K. Kasahara, Optical crossbar interconnection with self-addressing, 1990 Topical Meeting on Optical Computing, Kobe, Japan paper 10E7.

18. K. Kasahara, Y. Tashiro, N. Hamao, M. Sugimoto, and T. Yanase, Double heterostructure optoelectronic switch as a dynamic memory with low-power consumption. *Appl. Phys. Lett., 52:*679 (1988).
19. W. T. Cathey, S. Ishihara, S. Y. Lee, and J. Chrostowski, Optical information processing systems, *IEICE Trans. Fundamentals, E75-A:*28 (1992).
20. M. R. Feldman, S. C. Esener, C. C. Guest, and S. H. Lee, Comparison between optical and electrical interconnects based on power and speed considerations, *Appl. Opt., 27:*1742 (1988).
21. D. A. B. Miller, Optics for low-energy communication inside digital processors: Quantum detectors, sources, and modulators, as efficient impedance converters, *Opt. Lett., 14:*146 (1989).
22. Report of the Research Committee on New Information Processing Technology, Industrial Electronics Division, Tokyo, MITI, March, 1991.
23. S. Ishihara (ed.), *Optical Computing in Japan,* NOVA Science Publishers, Commack, NY, 1990, p. 525.
24. Y. Ichioka and J. Tanida, Optical parallel logic gates using a shadow-casting system for optical digital computing, *Proc. IEEE, 72:*787 (1984).
25. T. Yatagai, Cellular logic architectures for optical computing, *Appl. Opt., 25:*1571 (1986).
26. S. Ishihara and H. Yajima, Optical array processing with optically controlled guided wave devices, *Optical Computing in Japan* (S. Ishihara, ed.), NOVA Science Publishers, Commack, NY, 1990, p. 271.
27. Y. Nitta, J. Ohta, K. Mitsunaga, and K. Kyuma, GaAs/AlGaAs optical neurochip (second report), 1990 Topical Meeting on Optical Computing, Kobe, Japan, paper 10I13.
28. M. Ishikawa, N. Mukohzaka, H. Toyoda, and Y. Suzuki, Optical associatron: A simple model for optical associative memory, *Appl. Opt., 28:*291 (1989).
29. S. Ishihara, N. Kasama, M. Mori, Y. Hayasaki, and T. Yatagai, Effects of imperfection in spatial optical devices on backpropagation learning capability of optoelectronic neural network, Postdeadline Papers of 1991 Topical Meeting on Optical Computing, Salt Lake City, Utah, paper PD2, pp. 7–10.
30. N. Yamanaka, M. Sasaki, S. Kikuchi, T. Takada, and M. Idda, A gigabit-rate five-highway GaAs OE-LSI chipset for high-speed optical interconnections between modules or VLSI's, *J. Sel. Areas Comm. SAC-9:*689 (1991).
31. H. Yano, M. Murata, G. Sasaki, and H. Hayashi, Fabrication of 8-channel pin/HEMT Receiver OEIC Array, 1991 IEICE Spring Conf., Tokushima, Japan, 1991, paper C-179.
32. O. Wada, H. Nobuhara, T. Sanada, M. Kuno, M. Makiuchi, T. Fujii, and T. Sakurai, Optoelectronic integrated four-channel transmitter array incorporating AlGaAs/GaAs quantum-well lasers, *IEEE J. Lightwave Technol., JLT-7:*186 (1989).
33. T. Nagahori, H. Kohashi, M. Itoh, I. Watanabe, T. Uji, H. Honmou, H. Suzuki, H. Kaneko, and M. Fujiwara, 150-Mbit/s/ch 12-channel optical parallel transmission experiment using an LED and a PD array module, IEICE Technical Meeting Rec., 1990, Tokyo, Japan, OCS-90, pp. 25–30.
34. F. Shimizu, H. Furuyama, H. Hamasaki, F. Kuroda, M. Nakamura, and T. Tamura, Optical parallel interconnection characteristics of 4-channel 2-Gbit/s

bit synchronous data transmission modules, Proc. of 42nd Electronic Components and Technology Conf., San Diego, California, 1992, pp. 77–82.
35. A. Takai, H. Abe, and T. Kato, Subsystem optical interconnections using long-wavelength LD and single-mode fiber arrays, ibid, pp. 115–119.
36. M. Tsunekane, K. Endo, M. Nido, I. Komazaki, R. Katayama, K. Yoshihara, Y. Yamanaka, and T. Yuasa, High-power individually addressable monolithic laser diode array, *Electron. Lett.*, 25:1091 (1989).
37. K. Isshiki, T. Kadowaki, A. Takami, S. Karakida, T. Kamizato, Y. Kokubo, and M. Aiga, High-power, low threshold current, individually addressable monolithic four-beam array of GaAlAs window diffusion stripe lasers, 1991 Conf. on Lasers and Electro-Optics (CLEO '91), Baltimore, Maryland, paper CTuQ1.
38. S. Hirata, H. Narui, and Y. Mori, Very dense 102-laser arrays with extremely low threshold current, 1990 Laser Diode Conf., Davos, Switzerland, paper PD-7.
39. H. Okuda, Y. Hirayama, H. Furuyama, J. Kinoshita, and M. Nakamura, Five-wavelength integrated DFB laser arrays with quarter-wave-shift structures, *IEEE J. Quantum Electron.*, QE-23:843 (1987).
40. M. Nakao, M. Fukuda, K. Sato, Y. Kondo, T. Nishida, and T. Tamamura, 20-DFB laser arrays fabricated by SOR lithography, 1989 Int'l Conf. on Integrated Optics and Optical Fiber Commun. (IOOC '89), Kobe, Japan, paper 21D3.
41. A. Ibaraki, K. Furusawa, T. Ishikawa, K. Yodoshi, T. Yamguchi, and T. Niina, GaAs buried heterostructure vertical cavity top-surface emitting lasers, *IEEE J., Quantum Electron,* QE-27:1386 (1991).
42. K. Kasahara, Y. Tashiro, I. Ogura, M. Sugimoto, S. Kawai, and K. Kubota, Drastically reduced power consumption in vertical to surface transmission electro-photonic device for large scale integration, 1989 Int'l Conf. on Integrated Optics and Optical Fiber Commun. (IOOC '89), Kobe, Japan, paper 20C3.
43. K. Matsuda, K. Takimoto, J. Shibata, and T. Kajiwara, Integration of 1024 InGaAsP/InP optoelectronic bistable switches, *IEEE Trans. Electron. Devices, ED-37:*1630 (1990).
44. K. Kyuma, Y. Nitta, J. Ohta, S. Tai, and M. Takahashi, The first demonstration of an optical learning chip, 1991 Topical Meeting on Optical Computing, Salt Lake City, Utah, paper WB2.
45. N. Mukohzaka, T. Hara, and Y. Suzuki, Microchannel spatial light modulator using liquid crystal for modulating material, 1990 Topical Meeting on Optical Computing, Kobe, Japan, paper 9C5.
46. T. Kurokawa and S. Fukushima, "Bipolar-operational spatial light modulator for optical processing, ibid., paper 9B3.
47. C. M. Gomes, S. Tsujikawa, H. Maeda, H. Sekine, T. Yamazaki, M. Sakamoto, F. Okumura, and S. Kobayashi, Gray-scale controllable ferroelectric liquid crystal spatial light modulator, 1991 Topical Meeting on Optical Computing, Salt Lake City, Utah, paper ME4.
48. S. Tsujikawa, F. Okumara, M. Sakamato, H. Ichinose, M. Imai, K. Sera, H. Asada, and K. Kubota, Optically-addressed TFT liquid crystal light valve, 1990 Topical Meeting on Optical Computing, Kobe, Japan, paper 9C3.

9
Photonic Interconnect Applications in High-Performance Systems

Sarry Fouad Habiby and Gail R. Lalk

Bellcore, Red Bank, New Jersey

I. INTRODUCTION

One of the trends that is likely to affect society as we move from the twentieth to the twenty-first century is the advent of a global information infrastructure. The presence of such an infrastructure will have an impact on the speed of communication as well as the quality and amount of information that can be exchanged among physically distant locations. Applications for networks capable of instantaneously processing large amounts of information affect many facets of society including education, medicine, business, entertainment, and defense [1]. Realization of these applications requires significant advances in the hardware that supports these networks. For some of these applications, computers capable of more than 1×10^{12} operations per second (teraflops) are envisioned. The highest performance computers available in 1992 are capable of about 10 gigaflops [2,3], implying a need for at least a factor of 100 improvement in performance. Similar advances are required of the telecommunications switches that will be at the center of these high-speed networks. Photonic interconnections are likely to play a major role in enabling these advances in computing and telecommunications.

The evolution from low- to high-bandwidth systems has been achieved by using optical fibers to connect network elements. This is true of both computer and telecommunications systems. The key technology change at the network element interface is the conversion from electrical to optoelectronic terminations. Availability of high-bandwidth transmission links among switches and network elements implies the need for commensurate high-speed processing electronics within those elements and consequently increased demand on hardware performance [4].

In this chapter, we will discuss the potential roles of photonic technologies in future-generation computers and telecommunications systems, specifically for intramodule high-speed processing and interconnect applications. Models of systems that might require the use of photonics to achieve greatly enhanced performance over present systems will be outlined, as will technologies being considered for application in these systems.

Among the most significant challenges in implementing future-generation computers and telecommunications networks are the requirements for them to handle data links that are operating at increasingly higher speeds with more sophisticated control protocols. Computing and communications are examples of areas in which performance of transmission and switching technologies is advancing so rapidly that without the introduction of new interconnection technology or methodology intramodule interconnections are likely to reach performance limits. Historically, innovations in technology and system architecture have been used to enhance the information carrying capacity and processing capability of computing and telecommunication systems. Examples include

- VLSI circuit design and implementation
- Multichip module (MCM) and other advanced packaging methods (with increased circuit density)
- Parallel, distributed, and pipelined computer system architectures
- Computing technologies (cache memory, RISC microprocessors)
- Coding and compression algorithms

Yet for all the promises each of these advances possess, their net effect on system performance fails to overcome many constraints imposed by electrical interconnect technology. Examples include susceptibility to noise, large power consumption and thermal dissipation, relatively small chip and connector signal pin-out density (the ratio of ground and bias pins to signal pins is significantly large in many of the higher density pin-outs), and limited frequency response. Many of these limitations affect the practical interconnect density of electronic components because of capacitive loading, attenuation, and mismatched impedance. Other more subtle limitations are manifested in the combination of high-performance components. For example, a system's frequency response is determined by the ensemble response of all the components; therefore, individual component responses must be superior to that of the system. To ensure the sharp clock and signal edges necessary to maintain synchronization and low jitter operation, the 3-dB frequency often must extend beyond the fundamental to include third and fifth harmonics. Photonic interconnects may offer an alternative platform to overcome these limitations, improving overall system performance.

Parallel and distributed computing architectures represent an innovative way of using electrical transmission and switching technologies to improve the processing times for large amounts of information; since these systems rely so heavily on massive interconnection densities, however, their performance will eventually reach a plateau because of limitations in electrical interconnections [5]. Photonic interconnections have the potential to overcome many of the electrical interconnection limitations, and they are likely to be a key factor for achieving future advances in high-speed computing.

In telecommunications the emergence of broadband systems for transmission of voice, data, and video over the public communications network implies major changes in switching technologies and architectures. In one scenario for a future-generation broadband network, every subscriber would have access to an information capacity of 155.52 Mb/s carried over at least one optical fiber to a central switching point. In the model that will be discussed later in this chapter, a switch in this type of network may be required to handle an information throughput of about one terabit per second. An examination of the physical issues associated with the implementation of such a switch reveals several areas where alternatives to conventional electrical transmission and switching may be required to achieve the desired switch performance [6]. The broadband telecommunications switch is an example of a system in which optical interconnections may play a significant role in increasing the processing speed and information carrying capacity.

There are significant physical design issues that need to be addressed in pursuit of scalable technologies that enable both broadband telecommunication networks and advanced computing networks. Many issues that will be outlined in this chapter are relevant in both environments. Among the critical issues is the need for high density interframe, interboard, and intraboard interconnections. In addition, optical interconnections may play a role in the management of the fiber-terminated interfaces in these applications and in the distribution of clock signals to the individual components of these systems. The requirements for these interconnects will be reviewed here, as will the attributes of optical interconnect technology and their potential impact as solutions to physical design limitations.

Since our work has concentrated on telecommunications technology, the emphasis in this chapter will be on a telecommunications example. We have participated in a detailed study of the potential critical physical design issues of a broadband switch that is scalable to a terabit per second capacity.

Included in Section II is a discussion of the model used in this study. We will draw on the results from that study to show potential applications of optical interconnect technologies. We have tried to indicate, however, the duality of many of these issues (with relevance to both telecommunications and computer industries) throughout the chapter. In Section III we briefly outline some analogies that can be drawn between the broadband switch and high-speed computing. We have attempted to assess the impact of optical interconnect technologies as they apply to high-speed computing, and to assess how advanced computing, switching, transmission, and data storage techniques can merge to form a global broadband information network.

II. LARGE-SCALE BROADBAND SWITCHING SYSTEMS

The telecommunication services that are anticipated as the cornerstone of future networks entail delivering a diverse set of capabilities to subscribers. In addition to regular telephone service (POTS—plain old telephone service), consumers can expect to have access to a variety of multimedia services including data, video, and large information databases. The widespread deployment of optical fibers enables the provision of these services, collectively referred to as broadband services, in an integrated form. International and U.S. standards committees have arrived at numerous standards for the transmission of broadband services and protocols for the operation of broadband networks. Two such standards are SONET (synchronous optical network) [7] (synchronous digital hierarchy, SDH, internationally*), and ATM—asynchronous transfer mode.**

To provide broadband services to many subscribers may require central office switching systems that are much more complex and have much higher

---

*SONET, or equivalently, SDH, is a transmission standard for multiplexed data streams composed of integer multiples of the basic rate of 51.84 Mb/s known as STS-1, or 155.52 Mb/s known as STM-1 (STM denotes synchronous transfer mode). The notation STS-N denotes synchronous transport signal, level N. The STS-3 rate is equivalent to the STM-1 rate. These data streams are byte multiplexed and can be combined and separated using add–drop multiplexing nodes. Data is organized in header and payload sections.
**Asynchronous transfer mode (ATM) is a technique for switching variable rate traffic. The ATM layer of the broadband ISDN transfer protocol provides for the transparent transfer of fixed size data units between a source and destination with an agreed grade of service. An ATM cell consists of an information field (48 bytes) and a header (54 bytes). The header contains a label that uniquely identifies a logical channel and is used for multiplexing, routing, and switching. The ATM technique can be used to support both connection-oriented and connectionless services. Since the service mix and information transfer rates in ATM are decoupled from the characteristics of the physical interface, ATM can be used to support voice, data, and video services.

PHOTONIC INTERCONNECT APPLICATIONS

throughput than today's local digital switches. One can assume a central office (CO) switch capable of supporting 50,000–100,000 subscriber lines, a size typical of today's large switches. If each line is served by a SONET OC-3 (155.52 Mb/s) or OC-12 (622.08 Mb/s) access channel, the total bandwidth at the subscriber interface to a large switch would be greater than 10 Tb/s. With an average 10% utilization of the access channel bandwidth during the busy hour, the required switching throughput is 1 Tb/s or more, with 8–16 Tb/s of optical termination capacity. This is two to three orders of magnitude greater throughput than the capacity of existing commercial end-office switches of the same scale. Today's POTS COs have 64-kb/s access channels, for a total switching throughput on the order of 1 Gb/s.

Many of the optical and electrical components necessary to realize a large ATM switch have been demonstrated, either individually or integrated into relatively small (<100 lines) switch prototypes [8,9]. The issue of whether this hardware can be scaled to provide switches with one terabit per second or more of throughout will be decided, in large part, by the capabilities of interconnection and packaging technology. This is true both because these technologies will determine whether the performance of individual components can be maintained when many of them are integrated into a complex system and also because these technologies will be important elements in determining the economics of large ATM switches. While it is likely that switch manufacturers will initially build small-scale broadband switches that can be attached as extensions of their current switching products, the issues addressed in the aforementioned study pertain to large switches that would be needed to provide universal broadband access to subscribers.

Traditionally the evolution from low- to high-bandwidth systems has been achieved by using optical fibers to interconnect network elements. This is true of both computer and telecommunications systems. The key technology change at the network element interface is from electrical to optoelectronic terminations. Availability of high-bandwidth interconnections between network elements implies the need for high-speed processing electronics within the network elements and consequently improved system performance (e.g., enhanced thermal dissipation techniques).

In the broadband switch, convective cooling techniques may not be sufficient to meet established environmental standards. This is one example of a physical design challenge that is introduced by a high-performance system and needs to be addressed through innovative design and methodology. In a more general case, one might consider alternative interconnection and packaging techniques within the network elements to improve system performance. Photonic interconnections represent an alternative technology that may be used to address such challenges.

An additional consideration is that physical design choices are often key factors in determining the cost, power dissipation, power supply (and its cost), space, and environmental requirements of these systems. Therefore, it is necessary to identify physical design alternatives that do not dramatically increase these factors.

A goal of the broadband switch physical design study was to identify and quantify the interconnection and packaging technology requirements that, if addressed now, would lead to future realization of practical large broadband switches. Additional research in these areas may lead to system advantages [6,10] compared with the current state of the art in conventional physical design choices (e.g., standard backplane topologies, convective cooling, etc.). Thus, even if component technology in connectors, switching chips, memory, etc., continues to make the significant gains that we have come to expect, these gains may not be adequate to build such a switching system.

A critical element in the scalability of the switch designs is the ability to provide dense yet efficient interconnections among boards and modules of a large switch. Our investigation of physical design issues led to the identification of several areas where optical interconnections may play a role in the practical realization of required system performance. The model functional architecture used will be described next, highlighting areas where speed and interconnection densities as well as requirements for ease of access and efficient power utilization challenge conventional partitioning and packaging strategies. We will indicate where optical interconnections may relieve some of the physical design bottlenecks including fiber management at the subscriber interface to the switch, routing and distribution of high-density interconnections within the fabric of the switch, backplane interconnections (allowing increased system throughput), and in the processing necessary to perform contention resolution in self-routing switches.

A. Broadband Switch System Architecture

To analyze the potential critical physical design issues in a broadband switching system, a modular framework architecture that takes into account the fundamental functional properties of this system was selected. Figure 1 is a block diagram of the architecture assumed in this study [6]. (The architecture shown is a representative one that includes the major functional elements necessary in a broadband switch. The chosen arrangement of the functional entities is one of many viable choices and is not intended to be a recommendation.) The architecture consists of a set of interconnected modules that are functionally specific. This includes interface modules (IMs) that contain most of the user-specific circuitry, control modules that are the

PHOTONIC INTERCONNECT APPLICATIONS 407

primary interface between the switch hardware and software, service modules that provide customized special services that are too application-specific to be included in every switching system, and fabric modules (FMs) that are responsible for most of the switching and routing. We have chosen to concentrate on the modules that are hardware intensive (and therefore bear the bulk of cost, power, and space requirements of such a switch), the IMs and the FMs. By their nature these network modules are required to handle high-speed data. They typically have optoelectronic terminations, thereby offering potential photonic interconnects applications.

The target subscriber base for a single switching office assumed in this analysis is between 50,000 and 100,000 lines, each capable of accessing the network at an STS-3c or higher rate via an interface module. Several IMs collectively provide access for the total number of subscribers. Scaling to a larger subscriber base may require the addition of new IMs. For consistency we will assume a 64k (65,536) subscriber base model in the remainder of this chapter. The IM is responsible for terminating subscriber lines,

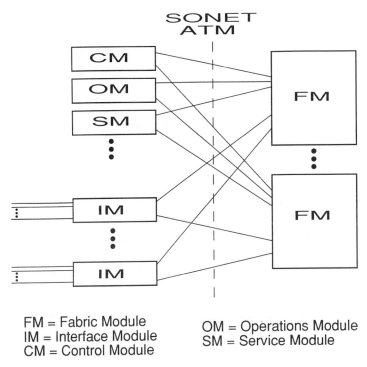

Figure 1 Broadband switch architecture model.

interoffice and intermodule facilities, and traffic concentration. A fabric module within the switch is responsible for routing all the intermodule interconnections (including signaling) within the broadband switch. The FM responds to messages from control and operations modules. Switch modules are connected using open interfaces based on the current SONET/ATM standards. The bandwidths at these interfaces range from STS-3c (155.52 Mb/s) and STS-12 (622.08 Mb/s) at the user–network interface (UNI) to STS-12 and STS-48 rates (2.488 Gb/s) for internal feeders and at the network–node interface (NNI). The availability of standard interfaces within a broadband switch may simplify interoperability among various modules provided by different suppliers.

Placing a typical call in the proposed architecture would require interaction among several modules. A call setup message is routed through an IM and an FM to the control module, which in turn notifies the appropriate destination fabric and interface modules involved in the call, to arrange for the appropriate paths through the system. The call proceeds with packets flowing into one IM where they are routed to the appropriate FM. The FM switches the packets to another link that terminates either on the originating or any other IM; the IM then routes the outgoing packets to the appropriate output channel to another subscriber (through a UNI) or to another switching center in the network (through an NNI). To accomplish this, several functions are performed at each node in this process. Concentrating on the IM and FM functionality, we will describe in more detail the relevant physical design issues that may have optical interconnect applications.

B. Physical Design Issues in the Interface and Fabric Modules

1. Interface Module Issues

The interface module (IM) is responsible for terminating optical fiber–based subscriber lines and interoffice facilities, and for traffic concentration and routing. IMs are expected to handle a variety of narrowband and broadband services including ISDN, POTS, data, and video. The external IM interfaces are the UNI and the NNI, depending on whether the call is to be routed to another subscriber or switching center, respectively. SONET format interfaces were also assumed for intermodule interfaces that provide connectivity to the switch FMs. These interfaces consist of high-speed signals composed of byte-interleaved STS-3c signals referred to as feeders and may be transported over optical links (OC-3 to OC-48). Groups of STS-3c (or STS-12c) signals between the same two modules are called feeder groups.

PHOTONIC INTERCONNECT APPLICATIONS

In the model we used, each IM was dedicated to serving 512 subscriber lines; two such IMs can be placed in a single 7-foot frame. Thus, serving 65,536 lines would require about 128 IMs (64 frames). This is based on an aggressive estimate of technology capabilities. Scaling to a larger subscriber base can be done by adding ports per IM or adding IMs. Even more aggressive assumptions about advances in technology may indicate an improvement in overall system performance, reducing the size of the IM footprint. There is much room for progress in integration, termination, and interconnection technologies.

The IM functional architecture is constrained by the features necessary to provide broadband services. In the following discussion, IMs terminating only broadband subscriber signals are considered. Broadband interfaces make extensive use of optoelectronic termination technology and represent a hardware-intensive portion of the cost, power dissipation, and size of the IM. The functional model described here is not intended to represent an actual implementation but rather represents the functions that have to be performed. Though external interfaces to an IM are based on the SONET/ATM standard, internal interfaces within each module may be proprietary.

There are three basic categories of functions performed in the IM. We will refer to these as the per-line, interline and perfeeder functions. A detailed description of these functions and an estimate of chip count and power dissipation are available in previously published journals [6]. The relationships among these functions in the context of the IM are shown in Figure 2. The per-line functions are those that are performed on a per-subscriber basis, such as upstream and downstream optoelectronic conversion, SONET framing, cell header translation and address processing, and bandwidth management. Dedicated broadband subscriber interface cards are used to implement these functions. The interline functions consist of upstream multiplexing and downstream demultiplexing, sorting, concentrating, and routing of subscriber traffic. The per-feeder functions consist of upstream and downstream optoelectronic conversion, buffering, additional multiplexing or demultiplexing, cell formatting, and cell header processing. The per-line and per-feeder functions may also require additional local memory chips for efficient operation.

Per-line functions do not require shared resources; we have assumed that they are placed on independent boards. It is the optical interfaces of these boards that offer the potential for optical interconnect applications. The interline functions span many per-line and per-feeder ports, thus requiring significant resource sharing; they act as the shared links between the per-line and per-feeder circuit boards. Optical interconnects may play a

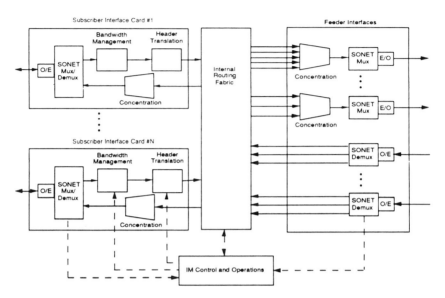

Figure 2 Interface module functional architecture.

significant role in acting as an active high-speed, high-capacity backplane that accomplishes the interline functionality.

Conventional interconnect technologies partitioning strategies may not be able to support the bandwidth and capacity required in a typical IM described here. Limitations in both connector pin-out numbers and bandwidth and backplane capacity may be severely taxed. Alternative electronic approaches have been described elsewhere [6,10], but optical interconnects (free space or guided wave) may be practical using recent advances in large one- and two-dimensional source arrays (surface emitting laser arrays, or SELA [11]) and detector arrays. For a more detailed description of one type of free-space optical interconnect technology, see Chapter 6 on holographic interconnects in this book. Potential applications of this technology are described in Section II.C.

2. Fabric Module Issues

The FM is responsible for routing all the intermodule interconnections (including control and signaling) in response to messages from control and operations processors. One difference from the UNI, however, is that the FM interface will probably be at the STS-12 (622 Mb/s) or STS-48 (2.488 Gb/s) rates. Several technologies and architectures for electronic, self-routing, packet switching fabrics have been proposed. Many rely on multi-

PHOTONIC INTERCONNECT APPLICATIONS

ple interconnected stages. These switch architectures typically require a stage of input buffers and/or a stage of output buffers. To serve 65,536 subscribers through 128 IMs, with an 8:1 concentration ratio, switch fabrics that switch at the 155 Mb/s rate would have to scale to a 8192 × 8192 size. Interconnections among stages for a switch of this size have densities that scale as integer multiples of the switch size. This burdens the electronic interconnection capacity of the intrafabric boards. We will discuss the potential role for optical interconnects in this application in Section II.C. Thus, the roles of optics in the FM go beyond the optoelectronic interfaces, and include high-density interconnects between two adjacent stages of a multistage fabric, e.g., optical contention resolution [12], or within the switching fabric [13,14].

The following section contains a description of potential optical interconnect applications within the modules of a broadband switching architecture. We will concentrate on the broadband subscriber interface cards, fiber management issues, backplane and high-density frame-to-frame interconnects in the FM, and contention resolution in a self-routing switch.

C. Optical Interconnect Opportunities in a Broadband Switch

Based on the physical design study and the assumptions we presented earlier in this chapter, we now describe a set of optical interconnect applications in a specific high-performance system—a broadband switch. Many of these applications have parallels in other high-performance systems.

1. Broadband Subscriber Interface Card

The broadband subscriber interface card is responsible for the per-line functions of the IM. It handles subscriber-specific functions such as optical fiber termination, optical-to-electrical (O/E) and electrical-to-optical (E/O) conversion, clock recovery, SONET frame header interpretation, and SONET payload definition. In addition, ATM-layer functions such as ATM cell header and payload identification and some error checking are performed on this card. The SONET frame header and ATM cell headers are used to identify routing, service type, and error conditions.

The specific role of optics on this card is the O/E and E/O conversion; it is critical, however, that these elements are designed for implementation in this context. This includes consideration of issues such as maximum allowable signal jitter, delays, and minimum required data run lengths necessary to maintain clock recovery capability consistent with SONET specifications. Since the functions on these cards are subscriber specific, their contribution to total system power consumption, heat dissipation, space, and cost is in direct proportion to the number of subscribers. Therefore, the conservation of these resources should be a priority in the implementation of the components on these cards, providing an incentive to integrate as many of the functions as possible to reduce total power, space, and cost.

Prototype versions of the interface card are partitioned into roughly four separate interconnected functional components including the O/E and E/O conversion (detectors and lasers), the receivers and transmitters, the clock recovery, and VLSI circuits for the SONET and ATM layer functions [15]. In the near future, this partitioning might be reduced to three levels, combining the detectors, receivers, and clock recovery functions on a single chip, integrating the lasers and transmitters, and combining the SONET and ATM functions on one VLSI. Ultimately, the complete integration of these functions would revolutionize the approaches now proposed to accomplish the broadband switching system architecture.

Another path in the evolution of this type of interface might be through increased user bandwidth. Although this discussion has centered around OC-3 (155 Mb/s) rate applications, in some proposed applications it is desirable to carry information at the OC-12 (622 Mb/s) and OC-48 (2.448 Gb/s) rates. In these cases, the optical fiber might be carrying multiplexed information from a series of hosts such as computer workstations. It would be advantageous to be able to maintain the same interface board designs throughout the evolution of these systems and simply replace existing optoelectronic components with pin-compatible, higher-speed components.

2. Fiber Management

The large number of fibers that are terminated on an IM could potentially limit craft access to and maintenance of this portion of the switch. Innovative packaging and partitioning of the functions in the IM, particularly the subscriber line card, will be required to prevent fiber management from being a system bottleneck. An issue that affects the design of the subscriber line card is the need to readily access individual subscriber optoelectronics and electronics. This requirement is driven by the need to evolve the capacity and performance of the switch over time and to provide means for quick replacement of components on failure. A first-order reliability analysis indicates that the provision of standby modules may improve the availability of the system [16]. This would require a method to switch to a standby, however, which in turn affects the board design and the fiber management strategy.

As discussed above, it is desirable to be able to individually access the subscriber-specific functions to maintain and upgrade the system. One objective is that unaffected subscribers do not lose service when system maintenance is being performed. Thus, it is envisioned that an N-subscriber

PHOTONIC INTERCONNECT APPLICATIONS 413

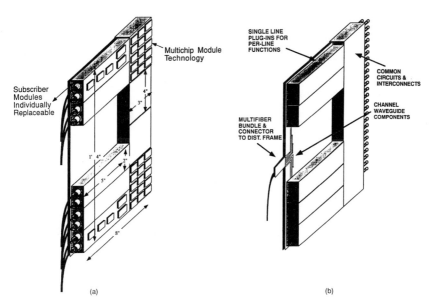

Figure 3 (a) Interface module line card with conventional optical fiber access. (b) Interface module line card with multifiber ribbon access and on-board waveguide interconnects.

board would contain N separate, removable subscriber modules. These modules contain the functions of the broadband interface cards described above. If each module terminates a subscriber's fibers using conventional optical connector (e.g., ST-type) technology, a line card might look like the illustration shown in Figure 3a. This type of connector can accommodate only one fiber. Because of the size of these connectors, the option of supplying two or more fibers to each subscriber places a restriction on the number of subscribers that can be served per frame. Smaller multifiber single-mode fiber connectors that are now commercially available can support up to 12 fibers in a fiber ribbon [17].* Using these connectors, it is possible to extend the number of subscriber lines terminating on each card and the overall number served per IM. The multifiber connector allows more effective management of the fiber terminations but introduces a new issue, the distribution of signals to the individual subscriber modules. This distribution might be accomplished using guided wave technologies as illustrated in Figure 3b.

---

*AT&T 9630 Multifiber Array Connector (MAC) can be used to interconnect 2–18 multimode fibers.

An additional concern is the constraint that the partitioning strategy of Figure 3a imposes on the overall system availability. Putting each set of subscriber-specific functions on removable modules allows for uninterrupted service to other subscribers while one subscriber module is being repaired. It would also be desirable, however, to be able to reroute subscriber traffic to spare modules in cases of failure or required maintenance. The conclusion of a first-order reliability analysis in [14] is that over widely varying failure rate distributions, the addition of one standby module for every eight subscribers substantially improves the time to first failure of the system. The availability of uninterrupted service to every subscriber is important, and provision of standby protection increases the chance of satisfying this criterion.

One advantage to the approach shown in Figure 3b is the amount of space that is opened up at the front of the IM for access and fiber management. Other functions could be implemented on a passive routing board (not shown), such as optical protection switching to a standby module on the same board and providing broadcast signals to the subscribers by integrating splitters into the optical paths.

There are several technological challenges to realizing a board like that depicted in Figure 3b. These include finding material systems that will be amenable to routing, switching, splitting, and coupling light within a limited power budget. A typical interconnection length between the multifiber connector and the transmit/receive locations on the subscriber modules is about 25 cm. Therefore, for the practical realization of such a board, it is imperative to find a relatively low-loss material. There are several materials that are candidates for an optical routing board. Optical fibers are appealing because of their low intrinsic loss, but they are not easily integrable on a board. The amount of space required for a fiber splitter has decreased enormously, but the mechanical difficulty of routing the fibers on the board remains. Technologies such as lithium niobate can be used to accomplish integrable splits, but the size of the substrates on which these technologies are based limits their use to relatively small area applications. Polymer waveguides are integrable and can be used on a very large substrate [18]. Splitting is accomplished by patterning on the board. The main disadvantage of polymers is the relatively high intrinsic loss (0.1 dB/cm at 820 nm and 0.3–0.5 dB/cm at 1300 nm) in comparison with optical fiber. Perhaps a realistic solution will be an optical routing board that takes advantage of a mixture of technologies. For example, one might use integrated optics to accomplish splits, optical fibers for long, straight runs, and polymers to pattern smooth interconnections among the pieces. An important research topic is the feasibility of using a mixture of technologies to achieve reliable and efficient interconnections.

PHOTONIC INTERCONNECT APPLICATIONS

Other technological challenges include achieving reliable (and potentially tight tolerance) mechanical partitioning of the system. The mechanical partitioning occurs in three locations: at the multifiber connector, the removable subscriber module, and the connection of the board into the backplane. The partitioning shown in Figure 3b suggests a clear separation of the N-subscriber board into an optical routing portion for optical signal distribution and an electrical processing portion for each subscriber. This concept represents the next step in the migration of optical interconnections into lower levels of system interconnection. ("Levels" refer to a hierarchy of system interconnections ranging from full systems—highest level—to subsystems, and to components.) An analogy can be drawn to the effects that the introduction of optical fiber between system elements has had on the demands placed on the system electronics. The introduction of optical interconnections tends to alleviate bandwidth and distribution limitations at one level while pushing the burden of system performance to the next lower level of interconnection. Thus, in our example the management of optical signals and the availability of service has been improved by the introduction of board-level optical interconnections; this implies a performance capability in the subscriber processing electronics that may be difficult to achieve using conventional packaging and interconnection techniques.

The next step might be to use an advanced packaging technology such as multichip-module packaging to enhance the system performance. (The multichip module here could be the removable subscriber module referred to earlier.) Optical interconnections are likely to impact multichip-module technology at both the intermodule and intramodule interconnection levels. One of the challenges in realizing the partitioning strategy suggested by Figure 3b is the optical interface at the subscriber module. The mechanical alignment at this interface is difficult because of the tight tolerance (submicron) and because the module must be removable so the optical interface cannot be permanently aligned. The orthogonal interface geometry suggested in the figure creates a nontrivial coupling problem. If the impact of optical interconnections into progressively lower interconnection levels continues, one can imagine that intramodule optical interconnects are a likely future application. Research into intra- and intermodule optical interconnections is gaining increased support as the migration of optics into systems continues [19].

A configuration that accommodates many of the design constraints discussed in this chapter is illustrated in Figure 4. In this configuration, solutions to the fiber management issues include the use of a multifiber optical connector, an optical routing board, and demountable subscriber modules in a mechanical arrangement that facilitates front access for maintenance. The multifiber connector reduces the space required for fiber termination

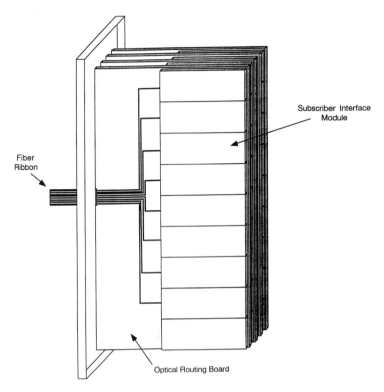

Figure 4 Example of multiple line card terminations using fiber ribbon cable and optical routing board.

by bringing groups of subscriber fibers to a common point on the N-subscriber board. This board provides a means of distributing subscriber traffic to the dedicated subscriber modules with a method that allows some flexibility in routing. One way that this flexibility might be utilized is to provide a standby module on each board to improve the system availability. In this configuration, the plane for optical routing can be elevated with respect to the base of the subscriber modules, allowing access to the modules through the sidewalls. Terminating on the line cards from the backplane facilitates optical coupling into the modules and improves access.

3. Backplane Applications

There are ways to enhance electrical backplane performance including the use of higher-performance insulating materials and/or additional signal layers (in a multilayer board), the use of alternative packaging and partitioning approaches such as three-dimensional structures [20], and the use of alternative transmission media such as optical interconnects.

Backplane technology strongly affects the expected performance of our model switching system. Conventional interconnection and partitioning of the IM functions and systems is shown in Figure 5a, where line, interline, and feeder circuit boards are interconnected through a single backplane. Fiber access to the line and feeder cards is also shown. Assuming conservative reduction in line card size and increased fiber access, and using 40 cards in a 2-foot shelf, a 7-foot equipment bay could accommodate about 1000 subscribers [10]. There are, however, significant physical design challenges in implementing conventional backplane interconnect technology. Specifically, requirements for connectivity among line, interline, and feeder cards cannot be met by current and near-future connector I/O density, and estimates for power needed to drive the backplane interconnects are high. At high-speed operation (greater than 155 Mb/s), high-density interconnects are susceptible to severe crosstalk problems.

An alternative that provides significant advantages in meeting the needed interconnection requirements is shown in Figure 5b—a three-dimensional (3-D) approach for coupling the line cards and feeder cards through the interline cards. This also allows the use of an active backplane, the stack of parallel horizontal interline cards, to replace the passive backplane used to interconnect active line and feeder cards in the conventional approach. In this structure, the point-to-point interconnections among the cards are shortened since the interline cards span all the line and feeder cards. An estimate of the size of the boards in this structure, extending the assumptions made in estimating the conventional structure size, also indicates that about 1000 subscribers can be served from a single 7-foot frame. The advantage is that the structure shown in Figure 5b allows for increased spacing between signal traces and between connector pins, reducing potential crosstalk. This new structure, however, requires advances in packaging and high-density connectors, and in reduced power dissipation and improved thermal management [10].

Optical interconnections have the potential to enhance backplane performance and can be applied to both the conventional and 3-D topologies. For long interconnection lengths (kilometers), optical fiber interconnects have advantages over electrical interconnections because of the large bandwidth and noise immunity of the media. The limitations to electrical technology arise largely because transmission lines can be subject to electrical interference from signals on adjacent lines. Exploitation of optical interconnects over shorter distances (less than 1 meter) has been slow, in part because of uncertainty about the power efficiency and reliability of optical technology. The advent of low-threshold lasers, however, may significantly decrease

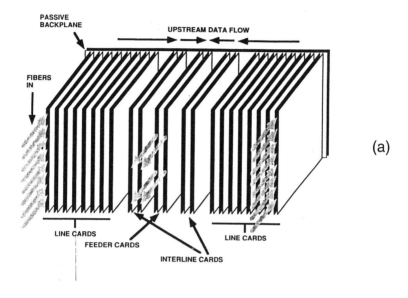

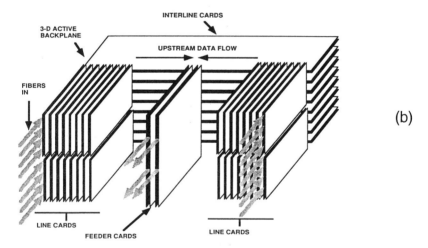

Figure 5 (a) Conventional backplane architecture for an interface module. (b) Three-dimensional backplane architecture for an interface module.

PHOTONIC INTERCONNECT APPLICATIONS

the power consumed in the optical transmitter, making relatively short distance optical transmission competitive with metallic transmission from a power consumption standpoint. At the backplane, alternate passive optical technologies such as polymer waveguides, free-space interconnections, or integrated waveguides may provide many of the advantages of optical fiber (e.g., increased bandwidth and noise immunity) while avoiding some of the disadvantages (e.g., mechanical bending stress).

In the near term, it appears that the interconnect density of electrical backplanes needs to be enhanced, but the electrical backplanes do not have to be fully replaced. A hybrid mix of optical and electrical interconnections may meet the performance requirements of this system application. For example, one might select a set of signal lines on a particularly congested area of the backplane and replace them with optical interconnections. Fewer electrical interconnections may result in a reduction in the required number of board layers and thus an overall cost savings for the system.

4. Frame-to-Frame Optical Interconnects

The FM in the high-speed switching system described is required to switch ATM cells within STS-3 level signals for over 8000 input and output lines. The switch fabrics studied as potential technologies to implement the functionality of the FM typically involve multiple stages of switching boards and buffer (input and output buffering) boards. The cross section of interconnections between any two stages is 8000 lines or more (8000 in the most conservative fabric studied), with some fabrics requiring interconnect densities of 128,000 to 256,000 connections [21]. The switching system that contains these boards is estimated to be large enough that it will have to be housed in several frames. The number of equipment bays and their physical size indicates that connectivity between two frames may span a distance of several feet (10–30 feet, likely). Furthermore, a significant portion of the power dissipation within the fabric module (25–30%) can be attributed to the termination drivers that are needed to drive any coaxial lines used to connect these frames (in an electrical interconnect).

The interconnect application described above has more constraints than a conventional high-speed transmission problem because of three factors. First, the bit error ratio (BER) requirements of these interconnects may be difficult to meet because error detection and correction protocols may not be in place in the application described, where the link is one of many (8000 or more) simple interconnects between two boards or frames within a single module. Adding error detection circuitry on each link would increase the complexity and cost of the modules. While there is no agreement on just what the BER targets should be, it is safe to assume that they will be between 10^{-12} and 10^{-17}. Second, the thermal and electrical environments will probably be hostile. Large power dissipation will raise the ambient temperature to perhaps 85°C, and large current surges may be caused by nearby switching elements. And third, the space requirements will dictate that both the transmission medium and the circuits themselves must be physically small.

Frame-to-frame optical interconnects already have proven to be a viable alternative. Examples of simple fiber interconnects already exist in current networks in telecommunications (the 5ESS switch has modules that are interconnected by fibers), and between high-speed terminals in advanced computer networks. What has yet to be proven is their benefit in large-scale interconnection structures. Optical interconnect technology promises the potential of low-power terminations, as evident in recent advances in low-threshold lasers and low-power receiver circuit design [22,23].

Because the power dissipation savings over electronic interconnects alone may not be significant enough to warrant using optical interconnects, it may turn out that the choice of photonic or electronic interconnection will be decided by other factors. In a one-to-one comparison of high-speed electrical and optical interconnects one must consider the power dissipation in driver and receiver circuits, as well as losses and crosstalk attributed to the transmission medium. A typical 100-Mb/s electronic interconnect may involve 100-mW power dissipation at each end and may be limited to 10s of feet (10–30 feet) before crosstalk effects are significant. A typical optical interconnect dissipates 100 mW at the transmitter and 50 mW at the receiver, but the crosstalk and bandwidth limitations are significantly different. Multiplexing of several STS-3 signals onto a single fiber for lengths that span the distances required in the CO provides more bandwidth and reduced crosstalk than using several coaxial connections (e.g., coaxial ribbon cables). In addition, photonic interconnects offer the potential of a major reduction in electromagnetic interference (EMI) problems associated with the transmission medium itself. The elimination of EMI from the photonic receiver, however, will be a challenge in circuit design and packaging. Another advantage of photonic interconnects lies in the size of the transmission medium. Fibers could be arranged in ribbon cables that would be much smaller, lighter, and more practical than the twisted pairs (probably shielded) or coaxial cables that would be required with the electronic alternative. Loss may limit the length of dense electronic interconnects to less than 10 meters if twisted pairs or ribbon coaxial cables are used. In addition, ongoing research on fabricating arrays of surface emitting lasers may also lead to miniature packages containing many lasers that serve as the transmitters for several SONET STS-3 sources from one board [22]. The source arrays can be interconnected to receivers (or detectors) through guided-wave or free-space interconnects.

While commercially available photonic components are consistently too power hungry to look attractive compared with their electronic counterparts, experimental transmitters reported in the literature [24] appear to be attractive as far as power consumption is concerned. In this respect, photonic receivers show promise, and this makes it reasonable to anticipate that soon power consumption of photonic interconnects will compete favorably with that of their electronic counterparts. One important question is whether the relatively low photocurrent from these receivers is enough to overcome noise and power supply transients in the switching environment.

On the other hand, line termination receivers all have output stages that need to be driven, for example, by an ECL-level output stage. If the receiver were integrated with the electronic circuitry that followed it, then the output stage could be eliminated and the termination drive power could be substantially reduced. The same low-level signal problems that are present in the photonic case, however, would be encountered in the electronic case as well. To make matters worse, the electronic circuit would have the long transmission line to induce pickup and current surges that in addition to inducing crosstalk could also burn out sensitive electronics in adjacent circuitry. Optical fibers, of course, would be immune to this problem.

5. Contention Resolution

In the applications described so far, the role of optical interconnects is to provide transmission media (optical fiber, waveguides, free space) for preassigned point-to-point, high-density, and high-bandwidth connectivity among electronic components or modules with enhanced performance such as improved transmission and reduced EMI. We now describe an example of another role for optical interconnects, one where the connection performed is dependent on the outcome of a signal processor within a system or subsystem. The use of optics in such an application is driven by the need for improved system throughput, generally achieved by the parallelism optics has to offer. With recent advances in laser and detector array technology, these types of interconnects will become more practical.

The major component in the FM is a self-routing packet switch. This type of switch typically has either input or output buffers, or both. Output port contention resolution is needed when two or more input ports request transmission of a cell to the same output port; only a single request can be satisfied to that output port during a cell cycle time. A contention resolution device (CRD) must be able to handle all requests within that cycle time, implying the need for high-speed processing.

The potential for using optical self-routing switching technology to further advance high-speed switching capability in an FM also suggests the need for optical contention resolution devices (OCRDs). While several methods for performing electrical contention resolution have been proposed, they have limited size and speed scalability. The higher the signal rate that can be switched, for a given number of ports, the shorter the cell cycle time, requiring processing of electrical signals within a shorter time. As the number of ports and the switching rate grow (in scalable switches), synchronization of the electronic circuits and processors becomes more difficult. The parallelism of optical interconnects suggests an alternative technology that may be capable of supporting the enhanced performance needed for contention resolution in an OCRD.

For example, an electrical implementation of a CRD [12] that is based on a polling algorithm to perform contention resolution for 128 inputs at 1.2 Gb/s rates (a 2.5-Tb/s switch), requires 1-ns bit-level synchronization. An optical implementation of this algorithm, however, is capable of achieving higher throughput with an order of magnitude improvement in speed–access number product and requires synchronization at the cell cycle time, 22 ns.

A simple passive optical interconnection network is at the core of this OCRD [25], based on laser array and detector array technologies. The columns of the laser and matching detector array represent the output port number; i.e., there are as many columns as output ports. Similarly, the rows of the array represent the input ports. The lasers are driven by decoders that interpret transmission requests from a specific input port or input buffer. Light from each laser is distributed to a specific set of detectors, based on a deterministic algorithm. Each detector drives a high-speed electronic circuit to enable packet transmission from the input port that wins contention resolution. An enabled input then transmits a packet to the desired output port through the switch. Thus, a packet at an input port requesting transmission to a particular output causes a single laser be turned on, the one in the row representing that input and the column representing the requested output.

There are two basic functions needed in the implementation of this particular OCRD: broadcasting light from any illuminated laser in one column to all the receivers in that matching column, and distributing light from any illuminated laser to a number of adjacent detectors in that column following a simple algorithm. The first is easy to implement using anamorphic optics, imaging light from a point source (any laser in the array) in one dimension to focus light into a column and collimating in the orthogonal direction as shown in Figure 6a. The second function can be achieved by using optical masks or multiwavelength laser arrays and the appropriate bandpass filters in front of the detectors to perform the connections shown in Figure 6b. Both of these can be implemented using passive optical interconnects, where the

PHOTONIC INTERCONNECT APPLICATIONS

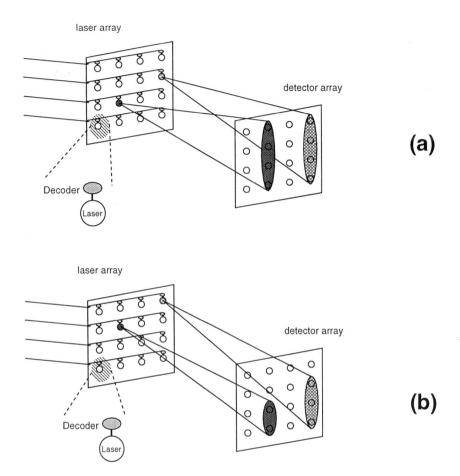

Figure 6 (a) Broadcast interconnection in optical contention resolution device; interconnection of each laser in the array to all detectors in the matching column of the detector array. (b) Selective distributed interconnection in optical contention resolution device; interconnection of each laser in the array to a subset of detectors in the matching column of the detector array.

optics required is simple. The use of elliptical optical imaging systems enables operation in one column of the array independent from that in any other column. The overall interconnect pattern is dependent on which lasers are illuminated within each cycle time, controlled by the decoders that handle transmission requests from the switch input buffers.

Proof-of-principle experiments to verify the operation have been demonstrated [25]. The main advantage of the optical approach over the electrical

CRD is that address decoding does not need to be synchronized at the bit level (1-ns electronic processing for 128 inputs at 1.2 Gb/s) but rather at a cell cycle time (22 ns) because of the independent operation per column. Disadvantages include the large fan-in/fan-out that may create unacceptable insertion loss, limited by detector or receiver sensitivity. Note however, that we are describing an interconnect that spans short distances.

The proof-of-principle experiments were performed for a small number of inputs representing the operation of one column. The demonstrated functionality (the second of the two functions referred to above) was performed in less than 10 ns for signals operating at 600 MHz. Scalability of this technique is dependent on the ability to generate large enough 2-D laser arrays that can be addressed in the manner shown in Figure 6. For a 128-input system a 128×128 laser array would be needed with row addressing and 7-bit decoders that can process requests from the switch inputs.

III. LARGE-SCALE COMPUTER NETWORKS AND SYSTEMS

Many of the justifications for using optical interconnections in telecommunications switching systems are similar to those of high-speed computing. As is true of the broadband switching scenario, the path toward teraflop computing requires a combination of the optimization of both the hardware and the software elements in these systems. With respect to hardware, realization of teraflop computers depends on improvements in the speed and capacity of the processing elements, the location, access time, and size of the memory elements, and the optimization of the interconnection network used to partition the system. At the same time, requirements of flow control and fault tolerance, data organization, and efficient algorithms for such networks present new challenges in the system software [26]. There are several aspects of optical interconnections that make them attractive as an option for overcoming some of the hardware and software challenges in these systems.

One of the trends in architectures for high-speed computing systems has been to introduce parallelism wherever possible. Although the specific allocation of resources is system dependent, most supercomputer architectures take advantage of some degree of parallelism [27]. Sometimes this parallelism is operation oriented (fine grained) as in processor arrays, and in others it is task oriented (coarse grained) as in pipelined systems. In either case, the advantages in using a massively parallel architecture do not come without trade-offs. For example, the increased cabling and delay that result from this architectural choice may compromise the increased computational speed derived from the parallelism.

PHOTONIC INTERCONNECT APPLICATIONS

Capabilities such as wavelength-division multiplexing and the ability to overlay (physically) optical interconnection paths with minimal interference make optics attractive for achieving massively parallel architectures more efficiently than can be done using electrical interconnections. These capabilities address both hardware (interconnection network) and software (data organization, fault tolerance) aspects of the teraflop computing challenge. Additional properties of optical interconnections that make them worth considering in parallel computing systems have been outlined by many authors. Among these attributes are the capability for three-dimensional interconnection [5] and the ability to interconnect without physical contact [28]. These attributes break down many of the barriers that limit the performance of electrical interconnections in these applications.

Direct analogies can be made between the issues discussed relating to the telecommunications switching environment and the high-speed computing environment. In broadband switching, the parallelism stems from the need to perform the same function (call setup and management) simultaneously for many users. This creates bottlenecks in the interconnection networks when resources such as memory are shared. It also creates the need for many copies of the same hardware to perform the subscriber-dedicated functions. In the parallel computer, the users may be relatively few, but the information being processed lends itself to massive parallelism. Sometimes this leads to architectures that consist of thousands of processing elements that perform the same functions on different pieces of the data. Thus, there are a large number of identical hardware elements. The combination of the computational results from the individual processing elements also requires shared resources such as memory. The constraints on the interconnection networks in these two cases are therefore similar.

One aspect of the parallel computing system that distinguishes it from a broadband switching system is the requirement for concurrency of information at each of the processing elements. In a situation where the computational results at some processing elements are dependent on the results at other elements, the timing between elements is critical. Thus, clock distribution plays a major role in achieving required system performance. In broadband switching, the parallel elements in the system are somewhat independent of each other, and so clock distribution applied to these elements is not as critical (although clock distribution does play a role in other aspects of the systems, e.g., in contention resolution as described earlier). The ability to distribute clocks with subnanosecond rise times to many system elements is likely to be a challenge to the physical design of teraflop computing systems. Some authors have begun to address methods for achieving clock distribution for high-speed computing applications using optics [29].

A. Merging of Computing and Telecommunications

Current research on the feasibility of gigabit/second computer networks is driven by parallel advances in computer and communication technologies and the needs of emerging gigabit applications [30]. Examples of applications for gigabit networks include calculations that require large file transfers, remote user access to large databases, and multimedia communications. Despite the continuing improvement in the performance of computers, there are still problems that cannot be solved in a reasonable time, even with a supercomputer. This limitation is one factor that leads to the idea of distributed computing, in which functional tasks within a single problem are distributed among two or more computers that may be separated by large distances. Computations can be separated into two categories: those that require gigabit/second bandwidth for the end-user application and those that require several lower-bandwidth channels to be multiplexed onto the same transmission link. Both of these approaches require a gigabit/second network infrastructure for efficient computation.

The existence of a global high-speed computing network will have an impact on many facets of the society. An example of a computational problem that might be facilitated by a high-capacity information network is the analysis of weather patterns and environmental indicators. In many of these problems, the production and transport of video images that represent the information are potentially the most valuable outputs from such a network. There is no doubt that the medical community (both researchers and clinicians) could benefit from the ability to access databases of laboratory results and to view images created from stored information. Another area that is likely to benefit from the results of a nationwide computing infrastructure would be education. A network interconnecting universities, libraries, and secondary and primary schools would potentially provide everyone access to a huge knowledge base and allow equal quality of education in schools regardless of the geographic location or the economic health of the community in which they are located.

The utility of a widespread computing infrastructure relies heavily on improved computational power at the nodes of the network, the ability to transport large amounts of information across the network, and the ability to switch the information efficiently among users of the network. Optical interconnections have the potential to have an impact on all these areas, particularly as the networks come into existence and expand their functionality and capacity. The first application of optical interconnections will be in the transport of information between network nodes over relatively long distances. It is likely that photonics will play a role in extending the computational and switching capacities at the network nodes.

IV. CONCLUSIONS

The emergence of national and global high-speed information networks is likely to be one of the most revolutionary technological developments during the next decade. Widespread access to databases of information may impact many facets of society including the educational system and the economy. These networks will be based on a synergy of advanced computing, switching, transmission, and information storage techniques. There are many problems to resolve in all these arenas to make the high-speed information networking concepts a reality. These problems include hardware and software issues as well as issues that span across the hardware/software boundaries such as effective protocol implementations, security, interoperability, and system availability. In this chapter, we have addressed some of the hardware issues that need to be resolved to meet the performance requirements of switching and computing equipment that can support the information capacity of and processing speeds in these networks. In particular we have emphasized areas where we believe optical interconnections have the potential to provide enhanced system performance.

Optical fiber interconnections have already played a role in enhancing system performance as relatively long distance intersystem interconnections. The introduction of fiber between systems improved the capacity of the transmission links, and as a result, the throughput bottleneck has moved into the local switching and computing environments. It is likely that the trend of introducing optical solutions at progressively shorter interconnection distances will continue as a means of improving system performance. We reviewed several areas where optical interconnections could be applicable, including interframe and interboard connections and backplane interconnections in some systems. Although the chapter centers around an example of a broadband telecommunications switch, we have tried to draw analogies to high-throughput computing applications as well.

Using a broadband switching model, we have outlined five specific areas that could be enhanced by using optical interconnections. These include the optoelectronic interface to an electronic switch, management of the fibers that terminate on the switch, backplane interconnections within the switch, frame-to-frame interconnections, and the use of optical techniques for contention resolution. The role that the optical interconnect plays in each of these applications varies.

At the optoelectronic interface, optical interconnects are used to terminate efficiently the optical fiber span on an interface card that converts the optical signals into electrical signals that are used to extract relevant communications protocols. This requires that the lasers, detectors, and the clock and data recovery techniques at this interface are designed specifically with the application in mind. For overall system interoperability, the emerging standards for optoelectronic interfaces should be considered in the design and implementation of optical interconnect components.

The management of the many optical fibers that will terminate within a relatively small physical area may be facilitated using optical interconnections. In particular, optical interconnections can be used to distribute the optical signals in a space-efficient manner and to improve system availability by providing spare and redundant switching paths.

The roles that optical interconnections can play at the backplane follow the more traditionally accepted advantages of optics. The improved EMI susceptibility over electrical interconnections allows optical interconnections to be used to increase the interconnect density and thus potentially reduce the number of board layers required for backplane interconnection. In addition, multiplexing techniques can be used at the backplane to achieve enhanced throughput and more efficient use of system resources such as power and space. We have briefly discussed an alternative electrical backplane configuration that utilizes a three-dimensional architecture. Even here, it is possible that free-space optical interconnects might be applicable to enable the interconnect to span between boards in the system.

In frame-to-frame applications, optical interconnects provide a means for achieving high-density connectivity among the modules within a switching fabric. The physical size needed to house these modules for a large switch (serving over 50,000 subscribers) requires several equipment bays that are separated by 10–30 feet. Up to 30% of the power dissipated within the fabric module is because of termination drivers needed for connectivity among the modules. In addition, optical interconnects reduce susceptibility to EMI and crosstalk compared with the equivalent electrical interconnect. High-speed interconnects within the module require operation at low error rate, power consumption, and size, all of which potentially can be achieved by optical interconnects.

Finally, optical contention resolution based on passive optical interconnects offers the potential for scaling the size of future self-routing packet switches that can be used in an FM. The basic limitation in scalability of electrical CRDs is attributed to the severe synchronization requirements that exist in switches with high-speed access and large access numbers. The use of laser and detector array technology provides the opportunity to implement OCRDs with higher speed and larger access numbers than electrical CRDs (an order of magnitude improvement in speed–access number product).

In summary, we have used a model of a high-speed telecommunications switch as a tool to illustrate major physical design issues, particularly those that may be resolved through advances in optical interconnect technology.

In each case the interconnect technology is tailored for the needs of the application. We maintain that the scope and scale of these issues is not restricted to advances in telecommunication networks but also extends to advances in computing networks. While existing electrical interconnect technology is a realistic choice for introduction of small-scale high-speed switches and computing networks, alternative technology is needed to scale to larger sizes in evolving information networks.

Optical interconnects may offer viable alternatives that provide high bandwidth connectivity at reduced drive power levels. Both guided wave and free-space optical interconnects may be practical in achieving the high-density interconnection among the equivalent of line, interline, and feeder boards in high-performance telecommunications and computing systems.

REFERENCES

1. M. L. Dertouzos, Communications, computers and networks, *Scientific American,* September 1992, pp. 63–71.
2. M. Maresca and T. J. Fountain (eds.), Massively parallel computers, special issue of the *Proceedings of the IEEE,* April 1991, pp. 395–401.
3. G. Zorpette, Teraflops galore, *IEEE Spectrum, 29*:27, (1992).
4. S. Kolodziej, Optical Group wants to put US on Top by 1994, *Lightwave Magazine,* September 1992, vol. 9, p. 12.
5. J. Neff, Massive optical interconnections for computer applications, *SPIE,* 1389:27 (1990).
6. T.C. Banwell, R.C. Estes, S.F. Habiby, G. A. Hayward, T. K. Helstern, G. R. Lalk, D.D. Mahoney, D. K. Wilson, and K. C. Young, Jr., Physical design issues for very large ATM switching systems, *J. Sel. Areas Comm., 9:*1227 (1991).
7. CCITT, Synchronous digital hierarchy bit rates, CCITT Recommendation G.707, Blue Book, Geneva, Switzerland, 1989.
8. D. Delisle, and L. Pelamourgues, BISDN and how it works, *IEEE Spectrum,* August 1991, vol. 28, p. 39.
9. W.E. Stephens, M. DePrycker, F.A. Tobagi, and T. Yamaguchi (eds.), Broadband ATM switching systems, special issue of the *IEEE Journal on Selected Areas in Communications,* Volume 9, Number 8, October 1991.
10. K.C. Young, Jr., T.C. Banwell, S.S. Cheng, R.C. Estes, S.F. Habiby, G.A. Hayward, T.K. Helstern, G.R. Lalk, D.D. Mahoney, and D.K. Wilson, Physical design issues for very large ATM switching systems, Proceedings of IEEE Communication Society Global Telecommunications Conference, San Diego, CA, 1990, pp. 1590–1593.
11. J. Jewell, J. P. Harbison, A. Scherer, Y. H. Lee, and L. T. Florez, VCSEL: Design, growth, fabrication, and characterization, *IEEE J. Quantum Electron, QE-27:*1332 (1991).
12. A. Cisneros, Large packet switch and contention resolution device, Proc. of

the XIIIth International Switching Symposium (ISS '90), Stockholm, Sweden, 1990, paper P14, Vol. 3, pp. 72–83.
13. T. T. Lee, M. S. Goodman, and E. Arthurs, A broadband optical multicast switch, Proc. of the XIIIth International Switching Symposium (ISS '90), Stockholm, Sweden, 1990, paper P2, Vol. 3, pp. 7–13.
14. A. Cisneros and C.A. Brackett, A large ATM switch based on memory switches and optical star couplers, *J. Sel. Areas Comm., 9*:1348 (1991).
15. G.R. Lalk, L. Gluck, T.C. Banwell, C.A. Johnston, T.J. Robe, K.A. Walsh, and K.C. Young, Jr., Highly integrated ATM/SONET user–network interface, *Electron. Lett., 27*:2174–2176 (1991).
16. G. Lalk, S.F. Habiby, D.H. Hartman, R.R. Krchnavek, D.K. Wilson, and K.C. Young, Jr., Potential roles of optical interconnections within broadband switching systems, SPIE Proceedings, OPTCON '90, Boston, 1990, paper 1389-31.
17. T. Ohta, T. Sigematsu, Y. Kihara, and H. Kawazoe, Low loss single-mode multi-fiber plastic connector, International Wire and Cable Symposium Proceedings, 1989, pp. 456–462.
18. D.H. Hartman, G.R. Lalk, J.W. Howse, R.R. Krchnavek, Radiant cured polymer optical waveguides on printed circuit boards for photonic interconnection use, *Appl. Opt., 28:*40 (1989).
19. M. Feldman, Holographic optical interconnects for multichip modules, IEEE Workshop on Interconnections within High-Speed Digital Systems," Santa Fe, NM, 1992.
20. D. K. Wilson, A new architecture for packaging wideband communication equipment using a 3-D, orthogonal, edge-to-edge topology, IEEE Globecom 1988 Conference Proceedings, p. 430.
21. J. N. Giacopelli, T. T. Lee, and W. E. Stephens, Scalability study of self-routing packet switch fabrics for very large scale broadband ISDN central offices, IEEE Globecom 1990 Conference Record, San Diego, CA, 1990, paper 805.5, pp. 1609–1614.
22. A. C. Von-Lehmen, T. Banwell, R. R. Cordell, C. Chang-Hasnain, J. W. Mann, J. P. Harbison, and L. T. Florez, High-speed operation of hybrid CMOS-VCSE laser array, *Electron. Lett., 27:*1189 (1991).
23. D. Hartman, S. F. Habiby, L.A. Reith, and G.R. Lalk, Power economy using point-to-point optical interconnect links, SPIE Proceedings, OPTCON '90, Boston, 1990, paper 1390-20.
24. T. Banwell, A. C. Von-Lehmen, and R. R. Cordell, VCSE laser transmitters for parallel data Links," submitted to *IEEE J. Quantum Electron, 29:*635–644, (1993).
25. A. Cisneros, S. F. Habiby, and A. E. Willner, Experimental demonstration of photonic contention resolution, Optical Fiber Communications Conference 1993, Technical Digest, San Jose, CA, pp. 94–96.
26. S.P. Kartashev and S.I. Kartashev, *Supercomputing Systems Architectures, Design, and Performance,* Van Nostrand Reinhold, New York, 1990.
27. M. Maresca, and T.J. Fountain, Massively parallel computers, special issue of the *Proceedings of the IEEE,* Vol. 79, No. 4, April 1991.

28. H.J. Caulfield, The unique advantages of optics over electronics for interconnections, *SPIE, 1390*:399 (1990).
29. D. H. Hartman, P. D. Delfyett, and S. Z. Ahmed, Optical clock distribution using a mode-locked semiconductor laser diode system, Conference Proceedings, OFC '91, San Diego, CA, 1991, p. 210.
30. R. Binder, Network testbeds at gigabit/second speeds, Proceedings Optical Fiber Communications Conference, San Jose, CA, 1992, p. 27.

Index

Add/drop multiplexers, 51
Alberta codes, 126
All-optical multivibrator, 191
All-optical switching devices, ultrafast, 163–212
   challenges, 165–166
   potential applications and technological challenges, 206–210
   routing devices, 166–186
      Kerr gates, 166–169
      nonlinear directional couplers, 169–176
      nonlinear fiber loop mirrors, 176–186
   routing versus logical devices, 164–165
   soliton-dragging logic gates, 186–194
   soliton logic gates, 186–206
      cascade of soliton-dragging and -trapping gates, 202–206
      soliton-trapping logic gates with output energy contrast, 194–202

Amplified spontaneous emission (ASE), 43
AND gate, 118
   soliton-trapping, 200–202
Anisotropic Bragg diffraction, 255–256
Applications of photonic interconnects, 401–431
   large-scale broadband switching systems, 404–424
      broadband switch system architecture, 406–408
      optical interconnect opportunities in a broadband switch, 411–424
      physical design issues in the interface and fabric modules, 408–411
   large-scale computer networks and systems, 424–426
   merging of computing and telecommunications, 426
Asynchronous transfer mode (ATM), 52, 116, 404, 405

433

Bandfilling, 34
Bandwidth, 23
Banyan network, 60, 230, 377, 378
Barium titanate, 251, 267, 303
Beam array generators:
   fixed, 267–275
      Dammann gratings, 267–270
      Fresnel microlens arrays, 270–272
      holographic optical elements, 272–275
   programmable, 275–285
      Fresnel microlens arrays, 279–283
      semiconductor laser arrays, 283–285
      spatial light moldulators, 275–279
Binary phase grating (BPG), 228
Binary step lithographic formation of diffractive optical elements, 340, 342–344
Birefringent fibers, 167
Bismuth silicon oxide, 251, 261–262, 295, 313
Bistable spatial light modulator (BSLM), 395, 396
Bit error rate (BER), 41, 419
   transmission system, noise and, 149–152
Board-level interconnects using free-space optics, 366–368
Boolean operation, 164, 165
Bragg diffraction, 251, 253–256
   anisotropic diffraction, 255–256
   isotropic diffraction, 253–255
Brewster telescope, 229
Brillouin optical amplifier, 93, 97
Broadband communication networks, photonic WD switching for, 108
Broadband intergrated services digital networks, 87
Broadband switching, 57, 58
"Broadcase-and-select" principle, 80

Carrier-sense multiple access with collision detection (CSMA/CD), 117, 119
Cascade-type spot array generation, 228
Cellular logic computers (CLCs), optical implementations of, 383
Chip-level interconnects using free-space optics, 366–368
Chip-to-chip interconnects using integrated optical waveguides, 369–374
Circuit boards, optical interconnects for, 366–374
   board-level interconnects using free-space optics, 366–368
Circuit switches, 116
Clock skew, 166
Code-division multiple access (CDMA) protocols, 121, 123–126
Coherent photonic wavelength-division switching system, 102–104
Compound semiconductors, 2
Computer generated holography, 221
Computing and information processing, 380–387
   cellular logical architecture for optical computers, 383
   computing with guided wave devices controlled by spatial light, 384–385
   large-scale computer networks and systems, 424–426
   merging of computing and telecommunications, 426
   optical neural networks, 386–387
   optical parallel array logic system (OPALS), 382–383
Connection density, comparison between skew and, 291–220
Contention resolution device (CRD), 421–424
Coupling coefficient (k), 12

INDEX

Cross-phase modulation, 189
Crosstalk, 13

Dammann gratings, 230, 267–270
Demimultiplexers, 208
Diffractive optical elements:
  construction techniques for, 340–344
  properties of, 324–331
Diffused waveguides in lithium niobate, 5
Digital cross connects (DXCs), 51, 54
Digital switch, 12, 17, 18
  difference between interference switch and, 18
  principle of operation, 13–14
Distributed Bragg reflector (DBR) lasers, 89–90, 93
Distributed feedback (DFB) filters, photonic WD switching experiments using, 98–102
Distributed feedback (DFB) lasers, 89, 90–92, 93
Drive voltage, 18

Efficiency of diffractive optical element, 326–331
Electrical interconnects, optical interconnects versus, 364
Electron beam lithography, 33
Electronic packaging, 356–360
Erbium-doped fiber amplifier (EDFA), 38–40, 185
Evanescent coupling, 8
Excess loss, 18
Extended generalized shuffle (EGS) network, 234, 235–240

Fabrication of semiconductor PICs, 31–33
Fabry-Perot filters, 93–96
  bypass filters, 196, 198, 199, 200

Fabry-Perot semiconductor laser, 83–84, 89
FET-SEEDs (field effect transistor-SEEDs), 226
Fiber interconnects, 61
Fixed beam array generators, 267–275
  Dammann gratings, 267–270
  Fresnel microlens arrays, 270–272
  holographic optical elements, 272–275
Flame hydrolysis deposition (FHD), fabrication process of silica-based waveguides by, 65–66
Flip-chip bonding, 216
Fourier-plane spot array generators, 228
Frame-to-frame optical interconnects, 419–421
Franz-Keldysh (FK) effect, 34
Free-space digital optics, 213
  benefits of, 213–223
    large interconnect density medium, 221
    lower communication energy, 214–215
    lower on-chip power dissipation per pin-out, 215–216
    lower skew, 219–220
    new connection-intensive architecture, 221–223
  hardware for, 223–232
    beam combination, 231–232
    optical hardware module, 232
    optical interconnects, 230–231
    SEED technology, 223–226
    spot array generation, 227–230
Free-space holographic grating interconnects, 249–321
  background on grating diffraction, 251–256
    Bragg diffraction, 253–256
    Raman-Nath diffraction, 252–253
    thin versus thick diffraction, 251–252

[Free-space holographic grating interconnects]
  background on the photorefractive effect, 257–267
    basic principles, 257–261
    photorefractive grating diffraction, 261–264
    photorefractive grating self-diffraction, 264–265
    self-pumped phase conjugation, 265–267
  fixed beam array generators, 267–275
    Dammann gratings, 267–270
    Fresnel microlens arrays, 270–272
    holographic optical elements, 272–275
  limitations of photorefractive-based interconnects, 304–315
    coherent beam fan-in, 312–315
    grating erasure, 304–306
    interconnect capacity, 306–311
    operation speed and power requirements, 311
  programmable beam array generators, 275–285
    Fresnel microlens arrays, 279–283
    semiconductor laser arrays, 283–285
    spatial light modulators, 275–279
  reconfigurable holographic grating interconnects, 285–303
    general photorefractive interconnect system, 285–299
    optical interconnects using correlation, 299–302
    optical interconnects using self-pumped phase conjugation, 303
    optical interconnects using two-wave mixing, 302–303
Free-space interconnection network, 61, 213, 233–243

[Free-space interconnection network]
  conventional free-space optical interconnects, 331–344
  constructive techniques for substrate-mode diffractive elements, 340–344
  interconnect design considerations, 332–337
  substrate-mode grating components, 337–340
Frequency-division multiple access (FDMA) protocols, 121–123
Frequency transparent, 23
Fresnel microlens arrays:
  fixed beam array generators and, 270–272
  programmable beam array generators and, 279–283
Fresnel-plane spot array generation, 228
Fundamental soliton, 193

GaAs-AlGaAs multiple quantum well (MQW) modulator, 135
Gallium arsenide, 251
Gigabit/second computer networks, 426
Global high-speed computing network, 426
Grating diffraction, 251–256
  Bragg diffraction, 253–256
  photorefractive, 261–264
  photorefractive grating self-diffraction, 264–265
  Raman-Nath diffraction, 252–253
  thin versus thick diffraction, 251–252
Group-velocity walk-off, 182
Guided wave devices controlled by spatial light, computing with, 384–385
Guided wave optics, 1–76
  guided wave optic devices, 1–48
    lithium niobate ($LiNbO_3$), 5–27
    semiconductor devices, 27–48

INDEX

[Guided wave optics]
  optical interconnect, 61–67
    optics on silicon, 64–67
  photonic switching in the context of guided waves, 48–61
  photonic switching systems, 53–61

Hardware for free-space digital optics, 223–232
  beam combination, 231–232
  optical hardware module, 232
  optical interconnects, 230–231
  SEED technology, 223–226
  spot array generation, 227–230
High-electron-mobility transistor (HEMT), 389
High-speed logic based systems, 165
High-speed optical fiber links, 387–391
Holography (see also Free-space holographic grating interconnects):
  computer generated, 221
  for forming diffractive optical elements, 340–342
  holographic optical elements (HOES), 272–275
  multiplexed substrate-mode holographic elements, 344–350
Hybrid tunable lasers, 89

Image-plane spot array generation, 228
Incoherent optical processing, 125–126
Indium tin oxide (ITO) electrode, 278, 281
InGaAsP/InP, 82
Instantaneous bandwidth, 23
Integrated modulator/transmitter, 134–139
Integrated optical waveguides, chip-to-chip interconnects using, 369–374
Integrated optics, 1–2

Interference switches, difference between digital switch and, 18
Ion implantation, 5
Iron-doped $LiNbO_3$ crystal, 309
Isotropic Bragg diffraction, 253–255

Japan, optical interconnects in (see Optical interconnects in Japan)

Kerr gates, 166–169

Large-scale broadband switching systems, 404–424
  broadband switch system architecture, 406–408
  optical interconnect opportunities in a broadband switch, 411–424
    backplane applications, 416–419
    broadband subscriber interface card, 411–412
    contention resolution, 421–424
    fiber management, 412–416
    frame-to-frame optical interconnects, 419–421
    physical design issues in the interface and fabric modules, 408–411
Large-scale computer networks and systems, 424–426
  merging of computing and telecommunications, 426
Laser amplifiers, 2
Laser diodes, surface emitting, 392
Laser-to-waveguide coupling efficiency, 64
Lightwave interconnections employing spatial addressing (LISA), 375, 376
Linear polarization-holding fibers, 3, 4
Liquid crystal light valves (LCLVs), 366

Liquid crystal spatial light modulators, 394–396
Liquid phase epitaxy (LPE), 32
Lithium niobate (LiNbO$_3$) devices, 2, 5–27, 251
   device principles, 6–14
   fabrication of, 14–15
   "LiNbO$_3$-like" PICs, 33
   physical mechanisms available to achieve modulation of the refractive index, 28–29
   polarization problem, 15–19
   switch arrays, 20–27
Lithograhic formation of diffractive optical elements, 340, 342–344
Logic gates with frequency filter, 196–202
Logic systems, optical parallel array, 382–383
LSI chips, optical interconencts for, 366–374
   chip-level interconnects using free-space optics, 366–368
   chip-to-chip interconnects using integrated optical waveguides, 369–374

Mach-Zehnder interferometer, 7, 66–67, 93, 176, 181, 186
Mach-Zehnder tunable filter, 96
Magnetooptic data storage systems, 358–360
Metal-organic-chemical vapor deposition (MOCVD), 32, 391–392
Metal-organic-vapor phase epitaxy (MOVPE), 391
Metal-semiconductor-metal (MSM) photodiodes, 389
Microchannel spatial light modulators (MSLMs), 394
Model of broadband switch architecture, 406, 407
Molecular beam epitaxy (MBE), 32
Monolithic tunable lasers, 89–93
Multichip modules (MCM), 213, 214

Multiple quantum wells (MQW), 2
   MQW modulators, 118, 135
   MQW pin diodes, 223, 224
Multiplexed substrate-mode holographic elements, 344–350
Multiplexers, 208

Neural networks, optical, 386–387
Nonlinear directional couplers (NLDCs), 169–176
   two-core fiber directional coupler, 172–176
Nonlinear fiber loop mirrors, 176–186
   cross-polarized control and signals in loop mirror, 183–186
   loop mirror as a three-dimensional device, 181–182
   solitons in optical loop mirors, 177–181
Nonlinear optical loop mirror (NOLM), 176–177
   cross-polarized control and signals in, 183–186
   solitons in, 177–181
   as a three-dimensional device, 181–182

One-dimensional light source and detection rays, 391–392
Optical bus networks, 357–358
Optical contention resolution devices (OCRDs), 422
Optical cross connections (OXCs), 52, 54, 57
Optical elements and interconnects, substrate-mode, 323–362
   applications of substrate-mode elements, 356–360
   conventional free-space optical interconnects, 331–344
   constructive techniques for substrate-mode diffractive elements, 340–344

INDEX

[Optical elements and interconnects]
  interconnect design considerations, 332–337
  substrate-mode grating components, 337–340
  multiplexed substrate-mode holographic elements, 344–350
  polarization-selective elements, 350–356
  properties of diffractive optical elements, 324–331
Optical filters, tunable-wavelength, 93–97
  characteristics of, 94
Optical hardware module (OHM), 232
Optical interconnects, 57, 61–67, 230–231
  optics on silicon, 64–67
    devices, 66–67
    fabrication, 65–66
    mechanical properties, 64–65
    waveguiding, 64
  using correlation, 299–302
  using self-pumped phase conjugation, 303
  using two-wave mixing, 302–303
Optical interconnects in Japan, 363–400
  categories of optical interconnects, 364–366
  devices for optical interconnects, 391–396
    liquid crystal spatial light modulators, 394–396
    one-dimensional light source and detector arrays, 391–392
    two-dimensional integrated laser diode arrays, 392
    two-dimensional optoelectronic functional devices, 392–394
  high-speed optical fiber links, 387–391
  optical interconnects for circuit boards and LSI chips, 366–374
    board-level and chip-level interconnects using free-space optics, 366–368

[Optical interconnects in Japan]
    chip-to-chip interconnects using integrated optical waveguides, 369–374
  optical interconnects for optical computing and information processing, 380–387
    cellular logical architecture for optical computers, 383
    computing with guided wave devices controlled by spatial light, 384–385
    optical neural networks, 386–387
    optical parallel array logic system (OPALS), 382–383
  optical interconnects for a reconfigurable network, 374–380
  optical versus electrical interconnects, 364
Optical processed TDMA, high-throughput uANS with, 127–164
Optical micro-area networks (uANs) (*see* Time-division optical microarea networks)
Optical (unipolar) pseudoorthogonal codes, 126
Optoelectronic integrated circuits (OEICs), 4, 365

Packet-switched networks, 116
Packet switches, 116
Packet transfer mode (PTM), 52
Passive star network, 104–106
Passive waveguides, 2
Photonic hybrid switching networks, 84–87
Photonic intergrated circuits (PICs), 2
  based on interband resonant effects on free carrier effects, 33–45
  $LiNbO_3$-like, 33
  semiconductor, fabrication of, 31–33

Photonic switching systems, 5–6, 53–61
Photorefractive effect, 257–267
 basic principles, 257–261
 photorefractive grating diffraction, 261–264
 photorefractive grating self-diffraction, 264–265
 self-pumped phase conjugation, 265–267
Photorefractive interconnect system, 285–299
 double-exposure erasure, 291–296
 intensity erasure, 289–290
 limitations of, 304–315
  coherent beam fan-in, 312–315
  grating erasure, 304–306
  interconnect capacity, 306–311
  operation speed and power requirements, 311
  time-average erasure, 296–299
Pin-outs per chip, 215–219
Plain old telephone service (POTS), 404, 405
Planar guided wave interconnect, 61
Plasma-enhanced chemical vapor deposition (PECVD), 65
Pockels readout optical modulators (PROMs), 394–395
Polarization control, 3, 4
Polarization diversity, 3, 4
Polarization-holding fiber, 3
Polarization-independent directional couplers, 17
Polarization scrambling, 3, 4
Polarization sensitive switch arrays, 2
Polarization sensitivity, 19
Prime sequence codes, 126
Programmable beam array generators, 275–285
 Fresnel microlens arrays, 279–283
 semiconductor laser arrays, 283–285
 spatial light modulators, 275–279
Proton exchange, 5, 15

Quantum confined Stark effect (QCSE), 223
 index change, 37
Quantum efficiency, 214
Quantum interference devices (QIDs), 5
Quasiprime codes, 126

Raman effects, 203
Raman-Nath diffraction, 251, 252–253
Random access protocol, 117
Reactive ion beam etching (RIBE), 32
Reactive ion etching (RIE), 32
 fabrication process of silica-based waveguides by, 66
Reconfigurable holographic grating interconnects, 285–303
 general photorefractive interconnect system, 285–299
 optical interconnects using correlation, 299–302
 optical interconnects using self-pumped phase conjugation, 303
 optical interconnects using two-wave mixing, 302–303
Reconfigurable interconnect systems, 356
 optical interconnects for, 374–380
Reference pulse method, wavelength synchronization using, 83
Reflective-type holographic interconnect system, 331
Resistor SEED (R-SEED), 223, 224
Rigorous coupled-wave theory, 350–356
Ring oscillator, 191
Risley prisms, 229
Routing devices, 166–186
 Kerr gate, 166–169
 logical devices versus, 164–165
 nonlinear directional couplers, 169–176
 two-core fiber directional coupler, 172–176

INDEX

[Routing devices]
   nonlinear fiber loop mirrors, 176–186
      cross-polarized control and signals in loop mirror, 183–186
      loop mirror as a three-terminal device, 181–182
      solitons in optical loop mirrors, 177–181

Self-diffraction, photoreactive grating, 264–265
Self-electrooptic effect devices (SEED) technology, 213, 223–227
   FET-SEEDs, 226
   free-space photonic switching systems based on, 231–232
   resistor SEED, 224
   symmetric SEED, 224–226
Self-pumped phase conjugation, 265–267
   optical interconnects using, 303
Self-routing packet switching, 208
Semiconductor devices, 27–48
   coupling efficiency considerations, 47–48
   fabrication of semiconductor PICs, 31–33
   "$LiNbO_3$-like" PICs, 33
   physical mechanisms available to achieve modulation of the refractive index, 28–29
   PICs based on interbank resonant effects and on free carrier effects, 33–45
Semiconductor laser amplifier (SCLA) gates, 30
Semiconductor laser arrays, 283–285
Semiconductor traveling-wave amplifiers, 156
Shared-medium multiple access protocols, 119–127
   CDMA, 123–126

[Shared-medium multiple access protocols]
   FDMA, 121–123
   TDMA, 126–127
Shuffle networks, 213
Signal-to-noise ratio (SNR), 151
Silicon, optics on, 64–67
   devices, 66–67
   fabrication, 65–66
   mechanical properties, 64–65
   waveguiding, 64
Single-mode fiber network, 2
Skew, 219–220
   clock skew, 166
Smart pixels, 213, 222
Soliton logic gates, 185–206
   cascade of soliton-dragging and -trapping gates, 202–206
   soliton-dragging logic gates, 186–194
      all-optical multivibrator or ring oscillator, 191
      time-domain chirp switch architecture, 191–194
   soliton-trapping logical gates with output energy contrast, 194–202
      logic gates with frequency filter, 196–202
      spectral confirmation of siliton trapping, 195–196
Soliton-self-frequency shift (SSFS), 203
Solitons in optical loop mirrors, 177–181
Soliton-trapping AND gate (STAG), 200–202
   cascade of soliton-dragging gate with, 202–206
Space-division (SD) photonic switching networks, 77
Space switches, 5
Space-switching fabric, 20
Spatial light, computing with guided wave devices controlled by, 384–385

Spatial light modulators (SLMs), 275–279, 372–373
  bistable, 395, 396
  liquid crystal, 394–396
Spectral confirmation of soliton trapping, 195–196
Spot array generation, 213, 227–230
  methods of, 228
Star-coupler based WDM network experiments, 104–107
Strictly polarization-independent devices, 3–4
Substrate-mode optical elements and interconnects, 323–362
  application of substrate-mode elements, 356–360
  conventional free-space optical interconnects, 331–344
    constructive techniques for substrate-mode diffractive elements, 340–344
    interconnect design considerations, 332–337
    substrate-mode grating components, 337–340
  multiplexed substrate-mode holographic elements, 344–350
  polarization-selective elements, 350–356
  properties of diffractive optical elements, 324–331
Surface emitting laser arrays, 216, 365–366
Surface emitting laser diodes (SELD), 392
Switching speed, 19
Symmetric SEED (S-SEED), 216, 223, 224–226
Synchronized WD communication network, 83–84
Synchronous optical network (SONET), 404–405
Synchronous transfer mode (STM), 52
  switching systems, 374

Telecommunications, merging computing and, 426
Thin versus thick diffraction, 251–252
Time-division multiple access (TMDA) protocols, 121, 126–127
Time-division multiplexed (TMD) network, 20
Time-division optical microarea networks, 115–162
  uAN access controls, 118–127
    classical access control methods, 118–119
    shared-medium protocols, 119–127
  high-throughput uANs with optically processed TMDA, 127–144
    integrated modulator/transmitter, 134–139
    optical correlation, 139–144
    optical TDMA uAN architecture, 127–129
    variable-integer-delay line demonstration, 129–134
  optical TDMA uAN demonstration at 5Gb/s, 144–153
    experimental setup and results, 144–149
    noise and BER analysis, 149–152
    power budget analysis, 152–153
Time-division switching system (TDS), 52, 72
Time-domain chirp switch (TDCS) architecture, 191–194
Time-resolved correlation data, 142, 143
Time-shift keying, 194
Time slot interchange (TSI) type applications, 51, 208
Time-synchronization circuits, 210
Timing jitter, 166
Titanium indiffusion process, 14–15
Transparency, 19

INDEX

Tunable distributed Bragg reflector lasers, 2
Tunable twin guide (TTG) lasers, 89, 90, 92–93
Tunable wavelength filters, 78
 characteristics of, 94
 optical filters, 93–97
Tunable wavelength local oscillator, 95
Two-core fiber directional coupler, 172–176
Two-dimensionally integrated laser diode arrays, 392
Two-dimensional optoelectronic functional devices, 392–394
Two-wave mixing, optical interconnects using, 302–303

Ultrafast all-optical switching devices, 163–212
 challenges for all-optical switching devices, 165–166
 potential applications and technological challenges, 206–210
 routing devices, 166–186
  Kerr gates, 166–169
  nonlinear directional couplers, 169–176
  nonlinear fiber loop mirrors, 176–186
 routing versus logical devices, 164–165
 soliton logic gates, 186–206
  cascade of soliton-dragging and -trapping gates, 202–206
  soliton-dragging logic gates, 186–194
  soliton-trapping logic dates with output energy contrast, 194–202

Variable-integer-delay line implementation for the optical TDMA coder, 129–134

Vertical to surface transmission electro-photonic (VSTEP) device, 379–380, 392–934

Waveguiding, 64
Wavelength converters, 78, 97–98
Wavelength-division multiplexing (WDM) technology, 50, 66–113
 photonic functional devices for WD switching systems, 88–98
  tunable-wavelength optical filters, 93–97
  wavelength converters, 97–98
  wavelength tunable lasers, 88–93
 photonic WD switching experiments, 98–107
  coherent photonic WD switching system, 102–104
  star-coupler based system experiments, 104–107
  using tunable DFB filters, 98–102
 photonic WD switching network, 78–88
  basic network structure, 78–82
  feasible number of WD channels in photonic WD switching networks, 82–83
  photonic hybrid switching networks, 84–87
  photonic pocket switch using WD switching technologies, 87–88
  wavelength network synchronization, 83–84
Wavelength tunable lasers, 88–93
 hybrid tunable lasers, 89
 monolithic tunable lasers, 89–93

Z-cut lithium niobate, 17